Soybean
Genetics, Molecular Biology and Biotechnology

Soybean
Genetics, Molecular Biology and Biotechnology

Edited by

D.P.S. Verma

Plant Biotechnology Center
Ohio State University
USA

and

R.C. Shoemaker

Department of Agronomy
Iowa State University
USA

CAB INTERNATIONAL

CAB INTERNATIONAL
Wallingford
Oxon OX10 8DE
UK

Tel: +44(0)1491 832111
Fax: +44(0)1491 833508
E-mail: cabi@cabi.org
Telex: 847964 (COMAGG G)

A catalogue record for this book is available from the British
Library.

ISBN 0 85198 984 5

Text set in 10/12 Photina by Techset Composition Ltd, Salisbury
Printed and bound in the UK at Biddles Ltd, Guildford

Contents

Contributors   vii

Preface   ix

1   **Germplasm Diversity within Soybean**   1
R.G. Palmer, T. Hymowitz and R.L. Nelson

2   **Molecular Genetic Mapping of Soybean**   37
R.C. Shoemaker, K.M. Polzin, L.L. Lorenzen and J.E. Specht

3   **Cytoplasmic Genetics in the Legumes (Fabaceae), with Special Reference to Soybean**   57
S. Mackenzie

4   **Plant Transposable Elements: Potential Applications for Gene Tagging in Soybean**   69
L.O. Vodkin

5   **Limitations and Potentials of Genetic Manipulation of Soybean**   91
J.E. Specht and G.L. Graef

6   ***In vitro* Selection and Culture-induced Variation in Soybean**   107
J.M. Widholm

7   **Soybean Seed Composition**   127
N.C. Nielsen

8   Genetic Modification of Soybean Oil Quality   165
    N.S. Yadav

**9   Molecular Genetic Analysis of Soybean Nodulation
      Mutants**   189
    P.M. Gresshoff

**10   Improvement of Soybean for Nitrogen Fixation:
      Molecular Genetics of Nodulation**   219
    A.J. Delauney and D.P.S.Verma

11   Soybean Transformation: Technologies and Progress   249
    J.J. Finer, T.-S. Cheng and D.P.S. Verma

Index   263

Contributors

T.-S. Cheng, National Tainan Teachers' College, Tainan, Taiwan, Republic of China.

A.J. Delauney, Department of Biology, University of West Indies, Cave Hill Campus, Barbados.

J.J. Finer, Horticulture and Crop Sciences Department, Ohio Agricultural Research and Development Center, Ohio State University, Wooster, Ohio 44691, USA.

G.L. Graef, Department of Agronomy, University of Nebraska, Lincoln, Nebraska 68583, USA.

P.M. Gresshoff, Plant Molecular Genetics and Center for Legume Research, University of Tennessee, Knoxville, Tennessee 37901, USA.

T. Hymowitz, Department of Agronomy, University of Illinois, Urbana, Illinois 61801, USA.

L.L. Lorenzen, Department of Agronomy, Iowa State University, Ames, Iowa 50011, USA.

S. Mackenzie, Department of Agronomy, Purdue University, West Lafayette, Indiana 47907, USA.

R.L. Nelson, United States Department of Agriculture, Agricultural Research Service and Department of Agronomy, University of Illinois, Urbana, Illinois 61801, USA.

N.C. Nielsen, United States Department of Agriculture, Agricultural Research Service, Plant Production and Pathology Research Unit, Department of Agronomy, Purdue University, West Lafayette, Indiana 47907, USA.

R.G. Palmer, United States Department of Agriculture, Agricultural Research Service FCR and Departments of Agronomy and Zoology/Genetics, Iowa State University, Ames, Iowa 50011, USA.

K.M. Polzin, Department of Agronomy, Iowa State University, Ames, Iowa 50011, USA.

R.C. Shoemaker, United States Department of Agriculture, Agricultural Research Service FCR and Departments of Agronomy and Zoology/ Genetics, Iowa State University, Ames, Iowa 50011, USA.

J.E. Specht, Department of Agronomy, University of Nebraska, Lincoln, Nebraska 68583, USA.

D.P.S. Verma, Plant Biotechnology Center, Ohio State University, Columbus, Ohio 43210, USA.

L.O. Vodkin, Plant and Animal Biotechnology Laboratory, Department of Agronomy, University of Illinois, 1201 West Gregory, Urbana, Illinois 61801, USA.

J.M. Widholm, Plant and Animal Biotechnology Laboratory, Department of Agronomy, University of Illinois, 1201 West Gregory, Urbana, Illinois 61801, USA.

N.S. Yadav, Agricultural Products, DuPont Co., PO Box 80402, Wilmington, Delaware 19880-0402, USA.

Preface

Legumes in general and soybeans in particular provide high levels of protein for nutrition and the use of soybean is increasing worldwide. Soybean is also a good source of vegetable oil. The soybean crop has been improved significantly both in yield and seed composition over the centuries. As more sophisticated genetic methodologies have evolved, breeding techniques and selection methods have produced useful gene combinations. This approach, however, has tended to miss or overlook the very important trait by which soybean produces protein-rich grain, i.e. symbiotic nitrogen fixation. While this crop is able to fix its own nitrogen and hence grow in nitrogen-poor soils, most of the genetic improvement of this crop, primarily for yield, has been carried out in nitrogen-rich soils. In order to meet the growing demand for a high-protein diet in developing countries, genetic improvement of this trait is essential. The rapid progress that is now being made in the area of molecular biology and plant transformation should allow improvement not only of symbiotic nitrogen fixation, but also of other desirable traits of this important grain legume. Thus it is now possible to alter the protein and oil composition of soybean as well as to produce other foreign proteins in this plant. The combination of desirable genes from diverse sources and the elimination of undesirable genes, by both classical breeding and molecular approaches, would further enhance the quality of soybean. The latter is expected to save time in breeding programmes. To achieve this goal, a basic understanding of the molecular biology of soybean is essential.

This volume focuses on recent progress made towards our understanding of the molecular genetics of soybean and provides a broad review of the subject, from genome diversity to transformation and integration of desired genes using current technologies. Many gaps remain in our knowledge, and many technical difficulties are apparent in handling the soybean genome. However, it is hoped that as technology improves, further successful manipulation of this valuable food crop will be possible.

Desh Pal S. Verma and Randy Shoemaker

Germplasm Diversity within Soybean* 1

R.G. Palmer[1], T. Hymowitz[2] and R.L. Nelson[3]
[1] USDA-ARS FCR and Departments of Agronomy and
Zoology/Genetics, Iowa State University, Ames, Iowa 50011,
USA: [2] Department of Agronomy, University of Illinois,
Urbana, Illinois 61801, USA: [3] USDA-ARS and Department of
Agronomy, University of Illinois, Urbana, Illinois 61801, USA.

Introduction

The genus *Glycine* Willd. is composed of two subgenera: *Glycine* and *Soja* (Moench) F.J. Herm. (Table 1.1). The wild perennial soybeans belong to the subgenus *Glycine* and have a wide array of genomes. The cultivated soybean [*Glycine max* (L.) Merr.] and its wild annual progenitor [*G. soja* (Sieb. and Zucc.)] belong to the subgenus *Soja*, contain $2n = 40$ chromosomes, are cross-compatible, usually produce vigorous fertile F_1 hybrids, and carry similar genomes.

There are more than 100,000 *G. max* accessions, probably less than 10,000 *G. soja* accessions, and approximately 3500 accessions of perennial *Glycine* species in germplasm collections throughout the world. Inasmuch as the only worldwide survey of soybean collections is a decade old (Juvik *et al.*, 1985), the exact numbers are unknown. Major *Glycine* collections exist in Australia, Brazil, China, Germany, India, Indonesia, Japan, Russia, South Korea, and the United States. Many other smaller but important collections exist throughout Asia and Europe. It is not known how many of the accessions are duplicated among collections, but the percentage of duplication in each species group is certain to be inversely proportional to the total number of accessions. It is likely that less than half of the *G. max* accessions are unique. Active collection programmes for each of the three groups indicate that the number of unique accessions will continue to grow.

The USDA Soybean Germplasm Collection is one of the largest collections and the largest outside Asia. It consists of five major subcollections:

* This is a joint contribution of USDA-ARS FCR, and Journal Paper No. J-15857 of the Iowa Agriculture and Home Economics Experiment Station, Ames, Iowa; Project No. 2985.

Table 1.1. List of species in the genus *Glycine* Willd., three-letter code, 2n, standard (PI), genome symbols and distribution.

Species	Code	2n	Standard	Genome	Distribution
Subgenus *Glycine*					
1. *G. albicans* Tind. and Craven	ALB	40	–	–	Australia
2. *G. arenaria* Tind.	ARE	40	505204	–	Australia
3. *G. argyrea* Tind.	ARG	40	505151	A_2A_2	Australia
4. *G. canescens* F.J. Herm.	CAN	40	440932	AA	Australia
5. *G. clandestina* Wendl.	CLA	40	440948	A_1A_1	Australia
6. *G. curvata* Tind.	CUR	40	505166	C_1C_1	Australia
7. *G. cyrtoloba* Tind.	CYR	40	440963	CC	Australia
8. *G. falcata* Benth.	FAL	40	505179	FF	Australia
9. *G. hirticaulis* Tind. and Craven	HIR	40	–	–	Australia
		80	–	–	Australia
10. *G. lactovirens* Tind. and Craven	LAC	40	–	–	Australia
11. *G. latifolia* (Benth.) Newell and Hymowitz	LAT	40	378709	B_1B_1	Australia
12. *G. latrobeana* (Meissn.) Benth.	LTR	40	483196	A_3A_3	Australia
13. *G. microphylla* (Benth.) Tind.	MIC	40	440956	BB	Australia
14. *G. pindanica* Tind. and Craven	PIN	40	–	–	Australia
15. *G. tabacina* (Labill.) Benth.	TAB	40	373990	B_2B_2	Australia
		80	–	Complex[1]	Australia, West Central and South Pacific Islands
16. *G. tomentella* Hayata	TOM	38	440998	EE	Australia
		40	–	DD[2]	Australia, Papua New Guinea
		78	–	Complex[3]	Australia, Papua New Guinea
		80	–	Complex[4]	Australia, Papua New Guinea, Philippines, Taiwan
Subgenus *Soja* (Moench) F.J. Herm.					
17. *G. soja* Sieb. and Zucc.	SOJ	40		GG	China, Russia, Taiwan, Japan, Korea (wild soybean)
18. *G. max* (L.) Merr.	MAX	40		GG	Cultigen (soybean)

[1] Allopolyploids (A and B genome) and segmental allopolyploids (B genome); [2] At least three genomic groups; [3] Allopolyploids (D and E, A and E, or any other unknown combination); [4] Allopolyploids (A and D genomes or any other unknown combination).

introduced *G. max*, introduced *G. soja*, perennial *Glycine* species, Genetic Collection, and Domestic Cultivar Collection. All of the collections are kept at the University of Illinois, Urbana-Champaign campus.

Germplasm – Subgenus *Soja*

The Domestic Cultivar Collection is divided into three sections: old, modern, and private. The 202 cultivars named or released before 1945 were introduced strains or were developed in the United States or Canada; however, their ancestry is uncertain. There are currently 347 modern cultivars released after 1945 by public institutions in both the United States and Canada. There are 28 privately released cultivars in the Collection. The intent of establishing this subcollection was to preserve selected, privately developed cultivars that have unique pedigrees, have been widely used in production, or have been used in published research. Many of the current entries do not meet those criteria but were part of the Domestic Cultivar Collection before 1970.

The Genetic Collection consists of the Type Collection, the Isoline Collection, and Crop Science Society of America registered germplasm releases. The current purpose of the Type Collection is to preserve mutations and variants not available in any other germplasm accession. This collection has not always been defined to include only unique variation. Some of the older entries do not meet this present criterion but have been maintained for their historical value. There are 145 lines in this collection, and approximately 75% of them have been analysed genetically. The most common phenotypic classes are represented by chlorophyll mutations with 45 entries, followed by mutations and variants affecting fertility (29), seed pigments (13), and leaflet form (11). The Isoline Collection contains nearly 500 near-isogenic lines developed from 11 recurrent parents and involves more than 80 genes. The present collection was developed almost exclusively by R.L. Bernard, former Curator of the USDA Soybean Germplasm Collection and Research Geneticist with USDA-ARS. The most common recurrent parents are cvs Clark (268), Harosoy (125), and Williams (63). This subcollection has been widely distributed and continues to be an important resource with many research applications. Recently, the Collection began to preserve all germplasm releases registered in *Crop Science*. There are now approximately 70 such lines or populations in the Collection.

Both of the introduced annual species are maintained as pure lines. Each accession in the Collection is descended from a single seed of the original seed lot. The only genetic variation that can exist within an accession is the result of heterozygosity in that original seed. Because both *G. max* and *G. soja* are almost completely self-pollinated, accessions within the Collection generally can be assumed to be homozygous and homogeneous. Having genetically uniform accessions in the Collection greatly facilitates evaluation. It is also critical in the maintenance of the Collection because contaminants can be more easily detected and removed.

All the accessions in the USDA Soybean Germplasm Collection are assigned to a maturity group which provides a general indication of the area of adaptation. There are 13 designations: 000, 00, 0, and roman numerals I to X. These maturity groups are assigned based on maturity data collected at one of three locations: Urbana, Illinois (40°07'N), Stoneville, Mississippi (33°25'N) or Isabela, Puerto Rico (18°30'N). The time to maturity is affected by both the length of the photoperiod and temperature. Because all these sites are less than 250 metres above sea level, these classifications indicate the general latitude of adaptation for spring planting of full-season cultivars. Maturity group 000 is adapted to the highest latitudes, generally greater than 49°N. Maturity group X is adapted to very low latitudes of less than 15°N.

The natural range of G. soja extends through most of China except for portions of the west and the far south, the Korean Peninsula, Japan and the far southeastern region of Russia. Glycine soja is reported to have been collected as far south as 24°30'N in Guangdong province (Chen et al., 1984). No G. soja has been reported from the coastal areas of south China, Hainan Island (Li and Satoh, 1990) or southern Taiwan. The most southerly collection site for accessions in the USDA Collection is 24°44'N in Taiwan. The USDA Collection has accessions from the N.I. Vavilov Institute of Plant Industry that were collected in Amur, Russia, from latitudes greater than 50°N (Bernard et al., 1989). Glycine soja has been reported in far northern China at nearly 53°N (Fushan and Chang, 1984).

The USDA Collection has G. soja accessions from 13 provinces in China (Table 1.2), but nearly 70% of these accessions originated in the three northeastern provinces (Heilongjiang, Jilin and Liaoning). There are 41 prefectures and the Kanto-Tozan region of central Honshu represented from Japan (Table 1.3), with the most accessions from northern and central Honshu and Kyushu. The northern part of South Korea is better represented in the Collection than the southern part (Table 1.4). The Collection has no accessions from Cholla Nam and only nine from Cheju. Actual collection sites for most of the accessions from Russia are not available, but the range of G. soja is thought to be limited to the areas east of Heilongjiang and Jilin, China. Most of the accessions in the G. soja Collection come from South Korea and the least from China, but the greatest range of maturity groups comes from China (Table 1.5). The limited range of maturity groups from Russia and South Korea is expected but lack of early maturity groups in Japan is somewhat unusual. Accessions from Hokkaido, Japan, were collected above 42°N but, when grown at Urbana, Illinois, were classified as maturity groups III and IV, much later than would have been predicted based solely on latitude.

Glycine max is a domesticated species and does not survive without human intervention. Glycine max lines have been moved in commerce for centuries. Interpreting origin data for G. max germplasm is more difficult than for G.

Table 1.2. Origin and number of accessions of *Glycine soja* in the USDA Soybean Germplasm Collection from China.

Province	Number of accessions
Beijing	4
Hebei	1
Heilongjiang	37
Henan	3
Jiangsu	13
Jilin	54
Liaoning	31
Ningxia	4
Shaanxi	4
Shandong	6
Shanghai	4
Shanxi	8
Zhejiang	1
Total	170

soja germplasm and should be done cautiously. It may be a reasonable assumption that accessions sharing a common, modern, geographical origin would be more similar than those with disparate origins, but that may not always be the case. The origin of the ancestral germplasm of a given region will provide the definition for what genetic diversity is possible, but both natural and human selection will help determine what actually exists. Ongoing research with newly available Chinese germplasm may provide information about the relatedness of primitive cultivars and the importance of geographical origin in predicting diversity.

The *G. max* accessions in the USDA Soybean Germplasm Collection come from more than 60 countries but those countries in which the wild soybean is native are considered the primary sources of *G. max* germplasm. Nearly 85% of the *G. max* germplasm comes from China, Japan, Korea, and Russia (Table 1.6). Approximately 30% of the germplasm listed from China has been received from third-party countries, and until recently more than 85% of all of the Chinese germplasm came from the three northeastern provinces (Heilongjiang, Jilin, and Liaoning). Recent exchanges with China have added more than 1300 accessions from central and southern China. The large number of accessions from Korea come almost exclusively from South Korea.

The soybean also has a history of ancient cultivation in Asia from Afghanistan to Vietnam and as far south as Indonesia. These countries are important sources of soybean germplasm, accounting for 7% of the Collection. The

Table 1.3. Origin and number of accessions of *Glycine soja* in the USDA Soybean Germplasm Collection from Japan.

Prefecture or region	Number of accessions
Aichi	35
Akita	36
Aomori	10
Chiba	4
Ehime	3
Fukoka	1
Fukui	1
Fukuoka	2
Fukushima	5
Gifu	1
Gumma	1
Hiroshima	1
Hokkaido	9
Hyogo	48
Ibaraki	9
Ishikawa	1
Iwate	18
Kagawa	1
Kagoshima	8
Kanagawa	1
Kanto-Tozan	4
Kochi	2
Kumamoto	37
Kyoto	1
Miyazaki	1
Nagano	14
Nagasaki	2
Nara	4
Niigata	4
Okayama	4
Saga	1
Saitama	3
Shimame	2
Shizuoka	3
Tochigi	7
Tokushima	1
Tokyo	3
Tottori	1
Toyama	1
Yamagata	1
Yamaguchi	2
Yamanashi	2
Total	295

Table 1.4. Origin and number of accessions of *Glycine soja* in the USDA Soybean Germplasm Collection from South Korea.

Province	Number of accessions
Cheju	9
Cholla Puk	25
Chungchong Puk	41
Chungchong Nam	30
Kangwon	58
Kyonggi	122
Kyongsang Nam	24
Kyongsang Puk	44
Total	353

Table 1.5. Origin and number of accessions of *Glycine soja* in the USDA Soybean Germplasm Collection by country and maturity group.

Maturity group	China	Japan	Russia	South Korea	Taiwan	Total
000	15	0	68	0	0	83
00	11	0	55	0	0	66
0	24	0	17	0	0	41
I	15	0	46	0	0	61
II	35	0	54	2	0	91
III	26	2	8	0	0	36
IV	19	23	0	41	0	83
V	15	63	0	282	0	360
VI	9	128	0	27	2	166
VII	1	74	0	1	1	77
VIII	0	2	0	0	0	2
IX	0	3	0	0	0	3
X	0	0	0	0	3	3
Total	170	295	248	353	6	1072

Table 1.6. Origin and number of accessions of *Glycine max* in the USDA Soybean Germplasm Collection.

Country	Number of accessions
China	3692
Japan	2858
Korea	3485
Russia	905
All other Asia	864
Europe	825
Africa	170
Americas	174
South Pacific	30
Unknown	56
Total	13059

remaining portion of the Collection has origins outside Asia (Table 1.6). Although these accessions are from secondary sources, useful genetic diversity has been found in this material.

The relatively narrow band of latitude in which a soybean line is adapted, defined as maturity groups in the United States and Canada, provides a natural genetic isolation mechanism for soybean germplasm. Soybean germplasm moved from low latitudes to high latitudes may not even flower and movement of germplasm in the opposite direction, without artificial manipulation of the photoperiod, will produce small plants with few seeds. These differences in adaptation are important factors to consider when assessing genetic diversity. The number of accessions in the USDA Soybean Germplasm Collection adapted to the extremes in latitudes is a very small percentage of the collection. Maturity groups 000 to 0 are less than 12% of the Collection, and maturity groups VIII, IX, and X are only 7% of the Collection (Table 1.7).

The *G. max* accessions in the USDA Soybean Germplasm Collection are routinely characterized for approximately 20 seed and plant descriptive traits, 10 agronomic traits, and 7 seed composition traits. Data for accessions in maturity groups earlier than IV are published in USDA Technical Bulletins (Nelson *et al.*, 1987, 1988; Juvik *et al.*, 1989b; Coble *et al.*, 1991; Bernard *et al.*, in press). Similar evaluation data are available for 200 *G. soja* accessions acquired by the Collection before 1985 in maturity groups 000 to IV (Juvik *et al.*, 1989a). Data have been collected for *G. max* accessions in maturity groups VI, IX, and X and will be published soon. Accessions in maturity groups V, VII, and VIII will be evaluated in the near future. In addition, the germplasm has been evaluated for resistance to many of the major insect and disease problems in the United States. These data are available on the Germplasm Re-

Table 1.7. Number of accessions of *Glycine max* in the USDA Soybean Germplasm Collection by maturity group.

Maturity group	Number of accessions
000	126
00	419
0	976
I	1252
II	1514
III	1475
IV	3082
V	2001
VI	794
VII	442
VIII	338
IX	349
X	291
Total	13059

sources Information Network (GRIN) operated by the USDA–Agricultural Research Service. Scientists in Canada, Mexico, and the United States can directly access this database by using telecommunications. Any scientist connected to Internet can access the database or can download a personal computer version of the database for specific species. As of September 1994 this can be done by connecting to the Agricultural Genome Gopher Service (gopher probe.nalusda.gov.) and selecting the following menu items: Plant Genome Information, Species Specific Gophers, Soybase Gopher, Germplasm Information and then *Glycine max, Glycine soja,* or *Glycine* other. The personal computer version of the database also can be ordered on diskettes. For more information about GRIN or Soybase, contact the Database Management Unit, USDA-ARS, BARC-West, or the National Agriculture Library, Beltsville, Maryland 20705, USA.

The number of accessions in a germplasm collection is not necessarily related to the genetic diversity contained in that collection but often is used as a first approximation. For several reasons, it is difficult to estimate the genetic variability that exists within a germplasm collection. It is possible to measure phenotypic variability for many traits, but there are problems with these data. There are rarely enough resources for sufficient replication to get reliable information. Many plant and seed traits are highly influenced by the environment, even those generally regarded as qualitative traits. Extensive testing that demonstrates the same phenotype does not necessarily indicate identical genotypes. Carefully constructed core collections can be a way of facilitating the evaluation and utilization of the diversity within a collection. Such a core may contain as many as 70% of the alleles of the Collection with only 10% of the accessions (Brown, 1989a). The core provides a more manageable unit of accessions for detailed study and a starting place for specific germplasm evaluations. A core collection has been established for the perennial *Glycine* in Australia (Brown, 1989a) and is currently being developed for *G. max* accessions in the USDA Soybean Germplasm Collection.

Data from the USDA Soybean Germplasm Collection indicate a large range of values for most traits and wide adaptation to environments. Oil percentages exist from 7% in *G. soja* to nearly 28% in *G. max*, and protein percentages range from less than 35% to more than 50%. Seed weight in *G. max* can be as large as 45 cg seed^{-1} and as small as 1 cg seed^{-1} in *G. soja*. *Glycine max* accessions have been collected at elevations as high as 2800 metres in Nepal and at latitudes over 58°N in Sweden (Bernard *et al.*, 1989). There are many traits for which great diversity can be cited and many variable traits that have not yet been quantified. Two salient examples of desired variation that have not yet been found in the Collection are certain changes in the fatty acid composition of oil and the amino acid composition of protein. New technologies provide the means to genetically engineer plants but, equally important, provide the science necessary to identify and exploit existing diversity. The constraint of available resources will not allow germplasm collections to grow without limits. Developing economical and useful measures of genetic diversity will be critical to maintaining germplasm collections that must continue to meet the expanded needs of the research community with diminishing resources.

Germplasm – Subgenus *Glycine*

The subgenus *Glycine* (Table 1.1) is composed of 16 wild perennial species (Newell and Hymowitz, 1980; Tindale, 1984, 1986a,b; Singh and Hymowitz, 1985a; Tindale and Craven, 1988, 1993). Wild perennials are diverse morphologically, cytologically, and genomically (Hymowitz and Singh, 1987; Singh *et al.*, 1988, 1989, 1992a,b).

Investigations on wild perennial *Glycine* species in the past have been restricted by the lack of experimental material. Before 1977, about 50 wild perennial accessions were maintained as part of the USDA Soybean Germplasm Collection. Since 1977, extensive plant exploration trips have been conducted by US and/or Australian scientists to Papua New Guinea, the Philippines, Taiwan, the Pescadores Islands (Taiwan), Ryukyu Islands (Japan), Marianas, Tonga, Fiji, Vanuatu, New Caledonia, Indonesia, and within Australia. At present, the US collection at the University of Illinois, Urbana-Champaign campus, has about 900 accessions and the Australian collection at Canberra has about 2500 accessions. Dr Anthony H.D. Brown and Dr Theodore Hymowitz are the curators of the comprehensive Australian and United States wild *Glycine* germplasm collections, respectively. Core collections are available at both sites (Brown, 1989b). All taxonomic activity within the subgenus *Glycine* is coordinated by Dr Mary Tindale, National Herbarium of New South Wales, Sydney, Australia.

The US wild perennial *Glycine* collection database also can be accessed via Internet. Connect to the Agricultural Genome Gopher and then follow directions as provided in the previous section.

Utilization of wild species for genetic improvement of their cultivated counterparts is steadily increasing in various crops (Singh, 1993). Exploitation of the wild progenitors is a reasonable approach because a cultigen (e.g. the soybean) and its wild progenitor (*G. soja*) are genetically members of the same species, and gene transfer between them is a relatively easy task. Use of other wild species, such as those belonging to the secondary or tertiary gene pools of the cultigen (Harlan and de Wet, 1971), is much more difficult because various types of isolating mechanisms that prevent gene flow between different biological units must be overcome. On the other hand, because of genetic remoteness and unique selection pressures on these wild species in comparison with the cultigen, there is a good possibility that the wild species possess variation in economically valuable characteristics that may be missing in the cultivated germplasm.

Investigations have shown that several wild perennial *Glycine* accessions carry resistance to brown spot caused by *Septoria glycines* Hermmi (Lim and Hymowitz, 1987). In addition, the wild perennial *Glycine* species carry resistance to soybean rust, *Phytophthora* root rot, yellow mosaic virus, and powdery mildew (Singh *et al.*, 1974; Mignucci, 1975; Burdon and Marshall, 1981; Harrison *et al.*, 1989; Hartman *et al.*, 1992; Schoen *et al.*, 1992). Accessions have been identified that are salt tolerant (Hymowitz *et al.*, 1987; Pantalone and Kenworthy, 1989; Shannon *et al.*, 1989), are tolerant to certain herbicides (Loux *et al.*, 1987; Hart *et al.*, 1988; White *et al.*, 1990), and can be regenerated from protoplast, leaf cotyledonary, petiole, and hypocotyl tissue (Grant, 1984; Newell and Luu, 1985; Hammatt *et al.*, 1986, 1989; Hymowitz *et al.*, 1986).

Extensive cytogenetic studies have been conducted during the past 16 years with the objective of establishing the genomic relationships among the wild perennial *Glycine* species (Singh *et al.*, 1988, 1992a). Genome symbols were assigned to 11 wild perennial diploid species on the basis of crossability, hybrid viability, meiotic chromosome association in intra- and inter-specific F_1 hybrids, and seed protein profiles (Table 1.1; Fig. 1.1).

Crossability rate is an excellent measure for assessing the degree of affinity between parental species. The only exception so far to this rule involves the two species, *G. cyrtoloba* and *G. curvata*. Both species are alike morphologically (Tindale, 1984, 1986a,b), and carry similar chloroplast DNA (Doyle *et al.*, 1990a,b) and seed protein profiles. In spite of such a close similarity be-

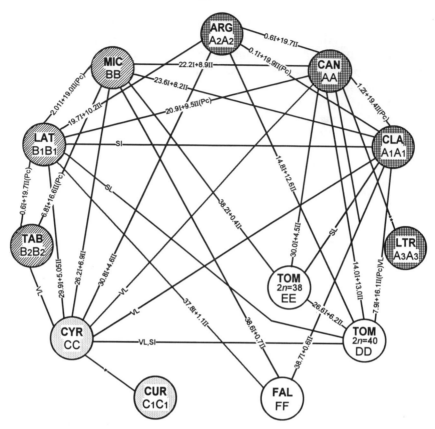

Fig. 1.1. Summary of genomic relationships based on cytogenetics and seed protein profiles among 11 of the 16 wild perennial species of the subgenus *Glycine*. For three letter codes see Table 1.1; PC, paracentric inversion; SI, seed inviability; SL, seedling lethality; VL, vegetative lethality; *, **, genomic relationships based on seed protein profiles. All the species in the periphery carry 2n = 40 chromosomes.

tween *G. cyrtoloba* and *G. curvata*, Singh *et al.* (1992a) were unable to hybridize these species even though a large number of flowers were pollinated. This result suggests that the crossability barrier between *G. cyrtoloba* and *G. curvata* probably is physiological and genic rather than chromosomal.

Intraspecific hybrids within diploid wild perennial *Glycine* showed normal meiosis and were fertile (Fig. 1.1). However, *G. tomentella* (2n = 38, 40) was an exception. Three distinct genomic groups – D3, D4, and D5 – were found in 40-chromosome *G. tomentella*, and one genomic group was observed in aneudiploid (2n = 38) *G. tomentella*. Hybrids within a group were fertile, and between the groups, were either sterile or hybrid seed did not germinate (Kollipara *et al.*, 1993).

The separation of morphologically distinct species having the same genome letter symbol with differing number subscripts is due primarily to a paracentric inversion (Fig. 1.1). Pod abortion was common in F_1 hybrids between diploid species with different genome designations. Immature seeds 19 to 21 days after pollination were germinated through culture (Singh and Hymowitz, 1985a,b; Singh *et al.*, 1988), and hybrids were weak, slow growing, and completely sterile.

Investigations on geographical distribution, cytotypes, cytogenetics, isozyme banding patterns, seed protein profiles, chloroplast DNA variation and restriction fragment length polymorphism (RFLP) revealed that 80-chromosome *G. tabacina* and 78- and 80-chromosome *G. tomentella* evolved through allopolyploidization, most likely through multiple independent events, and are species complexes (Singh *et al.*, 1987b, 1989, 1992b; Doyle *et al.*, 1990a; Kollipara *et al.*, 1994). Two distinct groups were recorded in 80-chromosome *G. tabacina*. Despite morphological similarity, three (T_1, T_5, T_6) and four (T_2, T_3, T_4, T_7) genomic groups were observed among aneutetraploid ($2n = 78$) and tetraploid ($2n = 80$) *G. tomentella* accessions, respectively.

Wild perennial *Glycine* species have not been exploited in soybean breeding programmes because of the extremely low crossability and the need to employ immature seed-rescue techniques to obtain F_1 hybrids. Several sterile intersubgeneric F_1 hybrids were reported in the literature (Broué *et al.*, 1982; Newell and Hymowitz, 1982; Brown *et al.*, 1985; Singh and Hymowitz, 1985c; Newell *et al.*, 1987; Singh *et al.*, 1987a). The results suggest that wild perennial *Glycine* species constitute the tertiary gene pool of soybean (Harlan and de Wet, 1971). Artificially synthesized amphiploids ($2n = 118$) of *G. max* ($2n = 40$) × *G. tomentella* ($2n = 78$, genome DDEE) produced a few one- or two-seeded pods (Hymowitz and Singh, 1984; Newell *et al.*, 1987; Shoemaker *et al.*, 1990).

Singh *et al.* (1990) obtained backcross−1 (BC_1)-derived progeny with $2n = 76$ (expected $2n = 79$) from the amphiploid ($2n = 118$ × soybean cv. Clark 63). All the BC_2 plants were vigorous but sterile, although pods with seeds were recovered after backcrossing with Clark 63. The BC_2 plants were obtained through immature seed-rescue procedures (Singh *et al.*, 1987a).

In the BC_2 to BC_4 plants, the elimination of chromosomes from *G. tomentella* was gradual, which is common during the synthesis of alien addition lines (Singh *et al.*, 1993). The range of chromosome number was $2n = 41$ to $2n = 52$ among BC_3 plants and $2n = 40$ to $2n = 64$ among BC_4 plants. Phenotypes of BC_3 and BC_4 plants were dissimilar but generally closely resembled the soybean cv. Clark 63. All $2n = 40$ plants were fertile. Alien addition line development will make it feasible to test for gene introgression from wild perennial *Glycine* species into soybean.

Germplasm – Characterization of Diversity

Cultivar development has played an important role in establishing the soybean as a major crop throughout the world. The successful commercialization of soybean cultivars has been attained by the integration of three factors: (i) an understanding of photoperiodic response among soybeans (Hartwig, 1972, 1973); (ii) the application of controlled hybridization and subsequent pureline selection that began in the late 1930s (Johnson and Bernard, 1963); and (iii) utilization of plant introductions, directly or through selections, as parents in controlled hybridizations (Hymowitz and Bernard, 1991).

About 137 cultivars originated as direct plant introductions or were selections from plant introductions (Specht and Williams, 1984). The first cultivar released that originated as a selection made among progenies of controlled hybridizations was the cv. Pagoda, released in 1939 (cited by Bernard *et al.*, 1988).

Soybean breeding and loss of diversity

Increases in soybean productivity can be attributed to genetic improvements in cultivars interacting with changing agronomic practices. Delannay *et al.* (1983) reported on the relative genetic contributions among ancestral soybean lines to 158 cultivars of North American origin. Ten plant introductions contributed more than 80% of the northern gene pool, and seven plant introductions contributed more than 80% of the southern gene pool. Many of the predominant introductions originated from the same geographic area; however, the genetic improvement for yield was accompanied by a loss in genetic diversity.

Specht and Williams (1984) reported that only 16 plant introductions were the maternal or cytoplasmic ancestor of the 136 cultivars released since 1939. Essentially five ancestors served as the ultimate cytoplasm source for the bulk of the 136 cultivars (Cheng and Hadley, 1986).

Luedders (1977) summarized the progress in soybean breeding that had been made from 1917 to 1971 by evaluating 18 cultivars in maturity groups I to IV for yield, lodging, and plant height. The 1962 to 1971 cultivar releases yielded 16% and 45% more than the 1943 to 1953 and the pre-1938 releases. A linear response of cultivar yield on cultivar release year indicated a 16.1 kg ha^{-1} annual gain in genetic yield improvement. The cultivars also showed better standability and plant height increased very slightly.

Wilcox *et al.* (1979) evaluated ten soybean cultivars from maturity groups II and III released over a 50-year period. Currently grown cultivars yielded about 25% more than the original plant introductions in both maturity groups. The annual genetic yield gain was 11.7 kg ha^{-1}. In a study of 18 cultivars from maturity groups VI to VIII released during the period 1914 to 1973, Boerma (1979) reported an annual genetic yield gain of 13.7 kg ha^{-1}.

In a study of 240 cultivars released between 1924 and 1980, Specht and Williams (1984) calculated the average annual genetic yield gain of 19.1, 17.4, 13.8, 29.1, 17.2, and 22.5 kg ha^{-1} for maturity groups 00 to IV, respectively.

Salado-Navarro *et al.* (1993) grew 18 cultivars of maturity groups VI to VIII released between 1945 and 1983 in four environments in Florida and four environments in Argentina. The annual genetic yield gain was positively significant in only two of the US environments. In Argentina, no increase in yield was observed in relation to year of cultivar release.

These reports indicated that currently grown cultivars yield appreciably better than the original plant introductions, but genetic improvement for yield generally was accompanied by a loss in genetic diversity. Gizlice *et al.* (1993a, 1994) calculated that approximately 21% and 26% of the genetic base has been eroded in both northern and southern soybean gene pools, respectively. The genetic relationships for many North American cultivars have been calculated (Allen and Bhardwaj, 1987; Lohnes and Bernard, 1991; Carter *et al.*, 1993).

Estimates of genetic similarity or distance between populations of plants are useful in planning crosses. Cox *et al.* (1985) used coefficients of parentage and similarity indices in soybeans. Correlations were higher for soybean cultivars released in the 1970s than for earlier-released cultivars.

Gizlice *et al.* (1993a) studied pedigree analyses of 258 public soybean cultivars released between 1947 and 1988. They noted that cultivars released after 1983 carried about 50% more genes in common than the cultivars released before 1954. Previously, St. Martin (1982) calculated the effective number of lines recombined in each cycle of cultivar development by determining coefficients of parentage of 27 cultivars released between 1976 and 1980. St. Martin (1982) stated that the relatively low effective number of lines recombined (11 to 15) indicated that soybean breeders have made progress at the expense of loss of substantial genetic variation originally available in the germplasm.

To validate the agronomic utility of genetic distance estimates, Gizlice *et al.* (1993b) correlated genetic distances between parents with agronomic variance of their progenies. The positive heterotic responses observed were not predicted by genetic similarity estimates based upon isozyme or restriction fragment length polymorphism data or by the combination of the two. They were predicted by diversity estimates based upon phenology.

The suggestion has been made that the yield level of soybean would be greater if commercially grown soybeans had a broader genetic base (Hartwig, 1972, 1973). Nelson *et al.* (1987) noted that, in evaluations of more than 3000 plant introductions across seven maturity groups, only three yielded more than the appropriate highest-yielding check cultivars. Cregan *et al.* (1994) evaluated more than 2000 plant introductions of maturity groups II to V and selected 113 for further evaluation. When plant introductions were compared with ancestral cultivars, the latter generally produced superior yields. However, in maturity group II tests, the highest yielding genotypes were plant introductions. The mean yield of the 13 plant introductions from Korea was significantly higher than the mean yields of those from Japan and northeastern China. The results of Cregan *et al.* (1994) suggest that the original introductions that formed the genetic base of northern maturity group II soybean germplasm were not necessarily the most productive genotypes.

Diversity for 'yield' traits within G. max

Expanding genetic diversity beyond ancestral lines means introducing germplasm through controlled hybridizations, selection, evaluation, and eventual release of elite lines or cultivars. In general, plant introductions used as parents may bring in a few favourable genes, but they also contribute a considerable number of unfavourable genes. Hymowitz and Bernard (1991) summarized information on 35 near-isogenic public cultivars that were developed by substituting one or two genes into certain cultivars by backcrossing. It is not surprising that, for all these cultivars except one, the trait introduced was pest resistance. Additional backcross-derived near-isogenic public cultivars that have unique traits include 'Kunitz' (Williams 82 isoline), which lacks the Kunitz trypsin inhibitor (Bernard *et al.*, 1991), and five 'Century' near-isogenic lines lacking one or more of the genes encoding seed lipoxygenase (Davies and Nielsen, 1987).

Soybean breeders view yield as the main objective of their plant breeding programmes. They are interested in using germplasm as a source for genes affecting quantitative traits. Their strategy has been to identify recombinant phenotypes in segregating progenies that possess most of the favourable genes contributed by the elite parent plus the few favourable, perhaps unique, genes contributed by the plant introduction. With continued evaluation, higher-yielding genotypes might be identified that would be superior to the adapted parent. The efforts to increase productivity by the introduction of 'exotic' germplasm have been numerous; however, the resulting selections have seldom attained or exceeded the yield levels of the currently grown cultivars.

Two-way (adapted × exotic) and three-way [(adapted × exotic) × adapted] crosses were made and evaluated for seed protein and oil and seed yield. Generally, the three-way crosses produced more of the superior lines than the two-way crosses (Thorne and Fehr, 1970a,b).

Hartwig (1972) crossed 12 agronomically acceptable plant introductions to cv. Hill. Bulk populations were evaluated in the F_3, F_4, and F_5 generations, but seed yields averaged 20% less than the yield of Hill.

Five populations ranging from 0% plant introduction to 100% were synthesized to determine what percentage of plant introduction germplasm should be used in a soybean improvement programme when the plant introductions were chosen for good agronomic performance (Schoener and Fehr, 1979). Their data suggested that about 50% plant introduction germplasm would be suitable for populations intended for long-term recurrent selection. Kenworthy and Brim (1979) showed that, after three cycles of recurrent selection, populations of greater diversity and productivity could be obtained.

Two sets of diverse two-way, three-way, and four-way crosses were evaluated for agronomic characteristics over a two-year period (Khalaf *et al*, 1984). The objective was to determine which cross generated the most variability. Results showed that the three-parent crosses were the most variable and had higher frequencies of agronomically desirable lines than the other combinations.

Vello *et al*. (1984) used 40 plant introductions and 40 domestic cultivars and elite breeding lines, intermated systematically during four generations to form five populations containing from 0% to 100% plant introduction germplasm. These researchers concluded that the use of plant introductions for short-term improvement of yield is not likely to be as productive as selection in populations developed from existing cultivars and elite breeding lines.

St. Martin and Aslam (1986) evaluated eight plant introductions and eight cultivars and elite breeding lines as parents to produce 48 bulk populations. They concluded that the changes in seed yield and seed oil percentage were consistent with additive inheritance.

Utilizing plant introductions in a cultivar development programme might be more successful if selection systems were used to identify plant introductions that have good combining abilities with adapted genotypes. Reese *et al*. (1988) selected nine plant introductions on the basis of phenotype and 14 plant introductions on the basis of yield. Each of these 23 plant introductions were crossed to four adapted lines. The results indicated that selecting plant introductions for yield was the more efficient method to identify desirable plant introductions for cultivar development programmes.

Sweeney and St. Martin (1989) selected plant introductions on the basis of maturity and yield per se to represent four different origin groups. They used two separate testcross experiments to determine whether germplasm strains differing in geographic origin also differed in their potential to increase yield and genetic diversity in soybean improvement programmes. The results suggested that, where germplasm lines have been selected for performance, additional selection based upon geographic origin may help in predicting the magnitude of genetic variability. Sweeney and St. Martin noticed that Korean lines had, on average, greater variances for yield than lines from other geographic origins.

Continued advances in cultivar development will depend upon the genetic improvement for yield beyond the level possible with the domestic gene pool of elite breeding lines and cultivars.

Diversity for value-added traits within G. max

Special value-added traits utilizing G. *max* germplasm have been bred into high-yielding lines, and selections have been made for traits for human consumption (Brar and Carter, 1993; Carter and Shanmugasundaram, 1993). Examples are found for tofu and other food products, cv. HP201 (Fehr *et al.*, 1990b); soy sprouts and the fermented product, natto, cv. SS202 (Fehr *et al.*, 1990a); and vegetable soybeans and the fermented product, miso, cv. LS301 (Fehr *et al.*, 1990c).

Diversity for pest resistance/tolerance within G. max

Plant introductions are used routinely as differentials to identify races of cyst nematode (Schmitt and Shannon, 1992). Moderately resistant to resistant plant introductions to races of soybean cyst nematode have been identified in large germplasm screenings (Anand *et al.*, 1988; Young, 1990; Rao-Arelli *et al.*, 1992). The first cultivar released with resistance to certain races of soybean cyst nematode was Pickett (Brim and Ross, 1966).

Cultivar development with insect resistance is complicated by the presence of large genotype × environment interactions (McKenna *et al.*, 1987). The first two insect-resistant cultivars, both resistant to the Mexican bean beetle, were released in 1990; Crockett (Bowers, 1990), and Lamar (Hartwig *et al.*, 1990). Mexican bean beetle resistance in PI 171451 and PI 229358 (both *G. max*) had been identified 20 years previously.

Table 1.8. Representative examples of plant introductions and their descendants used in pest reaction studies in soybean.

Trait	Reference	Year
Bacterial pustule	Hartwig and Lehman	1951
	Palmer *et al.*	1992
Brown stem rot	Nelson *et al.*	1989
Cyst nematode	Caldwell *et al.*	1960
	Matson and Williams	1965
	Anand and Gallo	1984
Downy mildew	Lim	1989
Peanut mottle virus	Boerma and Kuhn	1976
	Roane *et al.*	1983
Phomopsis seed decay	Zimmerman and Minor	1993
Phytophthora rot	Bernard *et al.*	1957
	Hartwig *et al.*	1968
	Buzzell and Anderson	1981
Soybean mosaic virus	Kiihl and Hartwig	1979
	Bowers and Goodman	1982
	Roane *et al.*	1983
Soybean rust	McLean and Byth	1980
	Hartwig and Bromfield	1983
Yellow mosaic virus	Singh and Malick	1978

Rowan *et al.* (1993) reported differences among plant introductions for the defoliating insects velvetbean caterpillar, soybean looper, beet armyworm, and corn earworm. Powell and Lambert (1993) used near-isogenic susceptible and resistant lines for pubescence type to screen for reaction to corn earworms. Egg predation by the beneficial predator bigeyed bug was not adversely affected by plant pubescence type.

A summary of representative examples of plant introductions and their descendants used in pest reaction studies in soybean is given in Table 1.8.

Diversity for physiological traits within G. max

Soybean researchers have tried to identify associations between physiological traits and seed yield. Several groups have surveyed the germplasm for length of the seed-filling period. Considerable variation in the seed-filling period of plant introductions of maturity groups III to V was reported by Reicosky *et al.* (1982). Egli *et al.* (1984) noted a significant genotype × year interaction for seed-filling period. Nelson (1986) reported that the currently defined R-stages of reproductive growth provided reasonable estimates of the duration of seed fill. A positive relationship between seed-filling period and yield was reported by Smith and Nelson (1986) using experimental lines and by Nelson (1987) among germplasm accessions. Pfeiffer *et al.* (1991) reported that selection of plant introductions on the basis of a seed-filling period longer than the adapted cultivar was not superior to the selection of plant introductions based upon yield. Salado-Navarro *et al.* (1993) stated that seed yield with 18 cultivars from maturity groups VI to VIII was not consistently associated, within and across environments, with any of the measured seed growth traits.

Specific leaf weight has been positively correlated with leaf photosynthesis in several species and has therefore been suggested as a potentially useful selection criterion. Buttery and Buzzell (1972) reported that, in soybean, specific leaf weight was a useful characteristic for indirect selection for yield. Lugg and Sinclair (1979) evaluated 373 soybean lines and found a linear relationship between the mean specific leaf weight at each sampling date and the radiation received in the week before sampling at growth stage R-2. Nelson and Schweitzer (1988) surveyed 231 maturity group I accessions and found variation in specific leaf weight.

Devine (1987) tested 537 *G. soja* plant introductions and 691 *G. max* plant introductions from maturity groups 00 to X with several bradyrhizobial strains. Devine concluded that there has not been a substantial introgression of genes controlling nodulation from Korean *G. soja* into Korean *G. max*. The frequency of ineffective nodulation was higher in *G. soja* than in *G. max*. Thus, the use of *G. soja* in breeding programmes could increase the probability of undesirable ineffective nodulation. Improvement of host-plant symbiotic nitrogen fixation relationships will require additional research.

Leghaemoglobins from a genetically diverse selection of 69 cultivars and plant introductions and 18 *G. soja* plant introductions were measured (Fuchsman and Palmer, 1985). All genotypes consisted of the same set of major leghaemoglobins as determined by analytical isoelectric focusing.

Plant introductions were screened to identify lines that were affected by the susceptibility to specific *Agrobacterium* strains and by the ability to regenerate plants from transformed cells (Delzer *et al.*, 1990). Two plant introductions were identified.

Variation in sensitivity to photoperiod in soybean, a short-day plant, was investigated by Criswell and Hume (1972), Polson (1972), and Nissly *et al.* (1981). Generally, early-maturing accessions were less sensitive to photoperiod than were later-maturing accessions. These researchers identified germplasm that varied little in flowering date regardless of photoperiod. This germplasm should be valuable for breeding programmes because of their potentially wider adaptation to different latitudes.

A summary of representative examples of plant introductions and their descendants used in qualitative genetic studies in soybean is given in Table 1.9.

Soybean breeding and use of G. soja

The wild annual soybean *G. soja* has been used as a parent in crosses with *G. max*. Gai *et al.* (1982) reported that in *G. max* × *G. soja* crosses, three backcrosses to *G. max* resulted in between 16.2 to 18.3% of the lines meeting the six agronomic criteria that they had chosen. Ertl and Fehr (1985) and Carpenter and Fehr (1986), using the same *G. max* × *G. soja* crosses of Gai *et al.*, determined that three backcrosses to *G. max* were necessary to obtain a reasonable number of lines similar to the recurrent parent for yield and lodging resistance.

LeRoy *et al.* (1991a,b) used *G. soja* plant introductions to transfer low seed weight into *G. max*. More efficient progress was made if selection for low seed weight was practiced between backcross generations. Selection for small seed size was effective both in Iowa and in Puerto Rico.

Advanced generation soybean breeding lines have been developed that have the 'impermeable' seed coat trait from *G. soja* (Hartwig and Potts, 1987). 'Impermeable' seed coats have been shown to maintain seed quality in the field during warm, rainy periods.

Two *G. soja* and *G. max* crosses were evaluated by using isozymes to determine if selection based upon genotypes at marker loci would increase the efficiency of gene transfer and plant improvement (Graef *et al.*, 1989). Marker associations were observed for specific enzyme genotypes and every quantitative trait studied in both crosses, but the specific isozyme genotype–quantitative trait relationship was population-specific. Suàrez *et al.* (1991) used a subset of the lines developed by Graef *et al.* (1989) to evaluate the phenotypic effects on yield and seed composition traits associated with isozyme marker

Table 1.9. Representative examples of plant introductions and their descendants used in qualitative genetic studies in soybean.

Trait	Reference	Year
Aluminum tolerance	Campbell and Carter	1990
Chlorophyll deficiency	Woodworth	1921
	Terao	1918
Dwarfness	Kilen and Hartwig	1975
	Boerma and Jones	1978
Flavonol glycosides	Buttery and Buzzell	1975
Flower colour	Hartwig and Hinson	1962
	Groose and Palmer	1991
Growth of stem, petiole, inflorescence	Bernard	1972
	Albertsen *et al.*	1983
Herbicide reaction	Kilen and He	1992
Iron-deficiency chlorosis	Weiss	1943
	Fehr and Cianzio	1980
Leaf form	Fehr	1972
	Bernard and Weiss	1973
	Rode and Bernard	1975
Linolenic acid	Rennie *et al.*	1988
	Rennie and Tanner	1989
Pod colour	Bernard	1967
Pubescence type	Ting	1946
	Bernard and Singh	1969
Rhizobium response	Caldwell	1966
	Vest	1970
	Vest and Caldwell	1972
	Devine and Breithaupt	1981
	Devine	1985
Root fluorescence in ultraviolet light	Delannay and Palmer	1982
Sterility	Johns and Palmer	1982
Time of flowering and maturity	Bernard	1971
	Buzzell and Voldeng	1980
	Voldeng and Saindon	1991
Ultraviolet-B radiation	Reed *et al.*	1992

loci. Statistically significant associations were found between particular isozyme genotypes and every trait analysed. All associations were population-
specific, except for an association between phosphoglucomutase (EC 2.7.5.1)
and maturity.

Voldeng (1981, 1985, 1989) has used G. *soja* as a source of the small-seed
phenotype. The cultivars Nattawa, Canatto, and Nattosan are natto-type cultivars developed for the Japanese food export market. The small seed size cv.
Vance, which possibly has G. *soja* as a parent, and the cv. Camp, a selection
from Vance, are natto-type cultivars (G.R. Buss, Virginia, 1994, personal communication). Carter *et al.* (1995) used Vance as a parent and has released the
small seed size cv. Pearl.

Diversity for cytogenetic and molecular traits within G. max *and* G. soja

Cytogenetic investigations of interspecific hybrids of G. *max* and G. *soja* have
shown that both species have similar genomes. The species do differ by paracentric inversions (Ahmad *et al.*, 1977). Inversions have been found in G. *max*
accessions in low frequency but at a higher frequency in G. *soja*, particularly
in accessions from Korea and Japan (Palmer, 1985). *Glycine max* and G. *soja*
accessions from China and Russia differ by a reciprocal chromosome interchange (Palmer and Heer, 1984; Palmer, 1985; Palmer *et al.*, 1987). An idiogram
of the pachytene chromosomes of soybean has been constructed (Singh and
Hymowitz, 1988).

Plant introductions and their descendants have been sources of qualitatively inherited traits (Palmer and Kilen (1987). The traditional genetic linkage map has many mutant traits that originated in plant introductions
(Palmer and Hedges, 1993). Table 1.10 gives a summary of representative examples of protein and isozyme markers.

Molecular characterization of soybean germplasm has been surveyed
predominantly with isozymes and proteins, chloroplast DNA and mitochondrial DNA. Perry and McIntosh (1991) and Perry *et al.* (1991) evaluated more
than 2000 G. *max* plant introductions from 78 countries for 17 morphological
traits and five enzymes. The Korean and Chinese accessions were the most
diverse morphologically, but these accessions were not separable on the basis
of isozyme data. Griffin and Palmer (1995) screened more than 1200 G. *max*
and G. *soja* plant introductions for eight enzymes. The number of alleles per
locus and average gene diversity was greater in the G. *soja* sample than in the
G. *max* sample.

Ribosomal gene repeats varied among G. *max* germplasm and among G.
soja accessions (Doyle, 1988). Greater variation in ribosomal gene repeats occurred among the perennial wild species than among accessions from subgenus *Soja* (Doyle and Beachy, 1985).

The first report of soybean mitochondrial DNA (mtDNA) polymorphism
was reported by Sisson *et al.* (1978). They found four groups of mtDNA among

Table 1.10. Representative examples of plant introductions and their descendants used in protein and isozyme studies in soybean.

Trait	Reference	Year
α-amylase	Kiang	1981
β-amylase	Orf and Hymowitz	1976
Bowman-Birk inhibitor	Domagalski *et al.*	1992
Kunitz inhibitor	Singh *et al.*	1969
	Orf and Hymowitz	1977, 1979
Lipoxygenase	Hildebrand and Hymowitz	1981, 1982
	Kitamura *et al.*	1983
	Davies and Nielsen	1987
Phytic acid	Raboy and Dickinson	1993
Seed storage proteins	Kitamura *et al.*	1984
	Davies *et al.*	1985
	Cho *et al.*	1989a,b
	Nielsen *et al.*	1989

the ten accessions examined. Grabau *et al.* (1989) observed that 19 cultivars maternally derived from cv. Lincoln had the cv. Mandarin cytoplasm, which differed from 38 cultivars derived from other cytoplasms. A 10.5 kb *Hind*III fragment was absent from the cv. Lincoln class of mtDNA but was present in the mtDNA of the other cytoplasms tested. Additional mtDNA research has divided the *G. max* germplasm into four cytoplasmic groups (Grabau *et al.* (1992).

Chloroplast DNA analyses in the cultivated and the wild annual soybean identified six RFLP patterns or groups (Shoemaker *et al.*, 1986; Close *et al.*, 1989). Maternal inheritance of chloroplast DNA in the genus *Glycine* subgenus *Soja* was demonstrated by Hatfield *et al.* (1985) and Corriveau and Coleman (1988).

Restriction fragment length polymorphism (RFLP) analyses of 58 lines indicated that molecular diversity was least among cultivated soybeans and greatest between accessions of *G. max* and *G. soja* (Keim *et al.*, 1989). Southern germplasm has been characterized by using RFLPs (Skorupska *et al.*, 1993). RFLPs were associated with cultivar pedigree and showed relationship to ancestral genotypes.

Molecular genetic linkage maps have been constructed that are based upon *G. max* × *G. soja* crosses (Shoemaker *et al.*, 1992; Rafalski and Tingey, 1993; Shoemaker and Olson, 1993).

Conclusions

The genus *Glycine* is divided into two subgenera, *Soja* and *Glycine*. In addition to the 18 species in the germplasm collection, the subcollections of cultivars, isolines, and the genetic collection are used by scientists of many diverse disciplines.

Per se evaluation data of the cultivated soybean accessions are documented and available, and the performance of accessions used as parents in crosses has been recorded. Various strategies have been employed to assess the agronomic potential of the germplasm, but no one method has been consistently and efficiently used by plant breeders.

The diverse array of variability available in the annual soybean, *G. soja*, and the wild perennial *Glycine* species has been recognized. The characterization and subsequent transfer of desirable traits from these species will help to broaden the genetic and agronomic diversity of modern cultivars.

References

Ahmad, Q.N., Britten, E.J. and Byth, D.E. (1977) Inversion bridges and meiotic behavior in species hybrids of soybean. *Journal of Heredity* 68, 360–364.

Albertsen, M.C., Curry, T.M., Palmer, R.G. and LaMotte, C.E. (1983) Genetics and comparative growth morphology of fasciation in soybeans (*Glycine max* (L.) Merr.). *Botanical Gazette* 144, 263–275.

Allen, F.L. and Bhardwaj, H.L. (1987) Genetic relationships and selected pedigree diagrams of North American soybean cultivars. *University of Tennessee-Knoxville Agricultural Experiment Station Bulletin* No. 652.

Anand, S.C. and Gallo, K.M. (1984) Identification of additional germplasm with resistance to race 3 of the soybean cyst nematode. *Plant Disease* 68, 593–595.

Anand, S.C., Gallo, K.M., Baker, I.A. and Hartwig, E.E. (1988) Soybean plant introductions with resistance to races 4 or 5 of soybean cyst nematode. *Crop Science* 28, 563–564.

Bernard, R.L. (1967) The inheritance of pod color in soybeans. *Journal of Heredity* 58, 165–168.

Bernard, R.L. (1971) Two major genes for time of flowering and maturity in soybeans. *Crop Science* 11, 242–244.

Bernard, R.L. (1972) Two genes affecting stem termination in soybeans. *Crop Science* 12, 235–239.

Bernard, R.L. and Singh, B.B. (1969) Inheritance of pubescence type in soybeans: glabrous, curly, dense, sparse, and puberulent. *Crop Science* 9, 192–197.

Bernard, R.L. and Weiss, M.G. (1973) Qualitative genetics. In: Caldwell, B.E. (ed.) *Soybeans: Improvement, Production, and Uses*. American Society of Agronomy Monograph 16, Madison, Wisconsin, pp. 117–154.

Bernard, R.L., Smith, P.E., Kaufmann, M.J. and Schmitthenner, A.F. (1957) Inheritance of resistance to phytophthora root and stem rot in the soybean. *Agronomy Journal* 49, 391.

Bernard, R.L., Juvick, G.A., Hartwig, E.E. and Edwards, Jr., C.J. (1988) Origins and pedigrees of public soybean varieties in the United States and Canada. *USDA Technical Bulletin* No. 1746.

Bernard, R.L., Juvik, G.A. and Nelson, R.L. (1989) *USDA Soybean Germplasm Collection Inventory* Vol. 2. International Agricultural Publications. INTSOY Series No. 31.

Bernard, R.L., Hymowitz, T. and Cremeens, C.R. (1991) Registration of 'Kunitz' soybean. *Crop Science* 31, 232.

Bernard, R.L., Cremeens, C.R., Cooper, R.L., Collins, F.I., Krober, O.A., Athow, K.L., Laviolette, F.A., Coble, C.J. and Nelson, R.L. (in press). Evaluation of the USDA Soybean Germplasm Collection: maturity groups 000 to IV (FC 01.547 to PI 266.807). *USDA Technical Bulletin.*

Boerma, H.R. (1979) Comparison of past and recently developed soybean cultivars in maturity groups VI, VII, and VIII. *Crop Science* 19, 611–613.

Boerma, H.R. and Jones, B.G. (1978) Inheritance of a second gene for brachytic stem in soybeans. *Crop Science* 18, 344–346.

Boerma, H.R. and Kuhn, C.W. (1976) Inheritance of resistance to peanut mottle virus in soybeans. *Crop Science* 16, 533–534.

Bowers, G.R., Jr. (1990) Registration of 'Crockett' soybean. *Crop Science* 30, 427.

Bowers, G.R., Jr. and Goodman, R.M. (1982) New sources of resistance to seed transmission of soybean mosaic virus in soybeans. *Crop Science* 22, 155–156.

Brar, G.S. and Carter, Jr., T.E. (1993) Soybean *Glycine max* (L.) Merrill. In: Kalloo, G. and Bergh, B.O. (eds), *Genetic Improvement of Vegetable Crops*. Pergamon Press, Oxford, UK, pp. 427–463.

Brim, C.A. and Ross, J.P. (1966) Registration of Pickett soybeans. *Crop Science* 6, 305.

Broué, P., Douglas, J., Grace, J.P. and Marshall, D.R. (1982) Interspecific hybridization of soybeans and perennial *Glycine* species indigenous to Australia via embryo culture. *Euphytica* 31, 715–724.

Brown, A.H.D. (1989a) Core collections: A practical approach to genetic resources management. *Genome* 31, 818–824.

Brown, A.H.D. (1989b) The case for core collections. In: Brown, A.H.D., Frankel, O.H., Marshall, D.R., and Williams, J.T. (eds), *The Use of Plant Genetic Resources*. Cambridge University Press, Cambridge, UK, pp. 136–156.

Brown, A.H.D., Grant, J.E., Burdon, J.J., Grace, J.P. and Pullen, R. (1985) Collection and utilization of wild perennial *Glycine*. In: Shibles, R.S. (ed.) *Proceedings of the World Soybean Research Conference III*, Westview Press, Boulder, Colorado. pp. 345–352.

Burdon, J.J. and Marshall, D.R. (1981) Evaluation of Australian native plant species of *Glycine* for resistance to soybean rust. *Plant Disease* 65, 44–45.

Buttery, B.R. and Buzzell, R.I. (1972) Some differences between soybean cultivars observed by growth analysis. *Canadian Journal of Plant Science* 52, 13–20.

Buttery, B.R. and Buzzell, R.I. (1975) Soybean flavonol glycosides: Identification and biochemical genetics. *Canadian Journal of Botany* 53, 219–224.

Buzzell, R.I. and Anderson, T.R. (1981) Another major gene for resistance to *Phytophthora megasperma* var. *sojae* in soybeans. *Soybean Genetics Newsletter* 8, 30–33.

Buzzell, R.I. and Voldeng, H.D. (1980) Inheritance of insensitivity to long daylength. *Soybean Genetics Newsletter* 7, 26–29.

Caldwell, B.E. (1966) Inheritance of a strain-specific ineffective nodulation in soybeans. *Crop Science* 6, 427–428.

Caldwell, B.E., Brim, C.A. and Ross, J.P. (1960) Inheritance of resistance of soybeans to cyst nematode, *Heterodera glycines*. *Agronomy Journal* 52, 635–636.

Campbell, K.A.G. and Carter, T.E., Jr. (1990) Aluminum tolerance in soybean, I. Genotypic correlation and repeatability of solution culture and greenhouse screening methods. *Crop Science* 30, 1049–1054.

Carpenter, J.A. and Fehr, W.R. (1986) Genetic variability for desirable agronomic traits in populations containing *Glycine soja* germplasm. *Crop Science* 26, 681–686.

Carter, T.E., Jr. and Shanmugasundaram, S. (1993) Vegetable soybean (*Glycine*). In: Williams, J.T. (ed.) *Pulses and Vegetables*. Chapman and Hall, London, pp. 219–239.

Carter, T.E., Jr., Gizlice, Z. and Burton, J.W. (1993) Coefficient-of-parentage and genetic-similarity estimates for 258 North American soybean cultivars released by public agencies during 1945–88. *USDA Technical Bulletin* No. 1314.

Carter, T.E., Jr., Huie, E.B., Burton, J.W., Farmer, F.S. and Gizlice, Z. (1995) Registration of 'Pearl' soybean. *Crop Science* 35 (in press).

Chen, R.K., Xu, S.C., Chen, Z.F. and Li, H.Q. (1984) Investigations on wild soyabeans in Fujian province. *Fujian Agricultural Science and Technology* 2, 2–5.

Cheng, S.H. and Hadley, H.H. (1986) Cytoplasm sources of soybean cultivars grown in United States and Canada. *Soybean Genetics Newsletter* 13, 166–175.

Cho, T.-J., Davies, C.S., Fischer, R.L. and Nielsen, N.C. (1989a) Inheritance and organization of glycinin genes in soybean. *Plant Cell* 1, 329–337.

Cho, T.-J., Davies, C.S., Fischer, R.L., Turner, N.E., Goldberg, R.B. and Nielsen, N.C. (1989b) Molecular characterization of an aberrant allele for the *Gy3* glycinin gene, A chromosomal rearrangement. *Plant Cell* 1, 338–350.

Close, P.S., Shoemaker, R.C. and Keim, P. (1989) Distribution of restriction site polymorphism within the chloroplast genome of the genus *Glycine*, subgenus *Soja*. *Theoretical and Applied Genetics* 77, 768–776.

Coble, C.J., Sprau, G.L., Nelson, R.L., Orf, J.L., Thomas, D.I. and Cavins, J.F. (1991) Evaluation of the USDA Soybean Germplasm Collection: maturity groups 000 to IV (PI 490.765 to PI 507.573). *USDA Technical Bulletin* No. 1802.

Corriveau, J.L. and Coleman, A.W. (1988) Rapid screening method to detect potential biparental inheritance of plastid DNA and results for over 200 angiosperm species. *American Journal of Botany* 75, 1443–1458.

Cox, T.S., Kiang, Y.T., Gorman, M.B. and Rodgers, D.M. (1985) Relationship between coefficient of parentage and genetic similarity indices in the soybean. *Crop Science* 25, 529–532.

Cregan, P.B., Yocum, J.O., Justin, G.R., Buss, G.R., Kenworthy, W.J., Wiskand, E.L. and Camper, H.M., Jr. (1994) Relative performance of soybean cultivars, ancestral cultivars and plant introductions. *Soybean Genetics Newsletter* 21, 168–183.

Criswell, J.G. and Hume, D.J. (1972) Variation in sensitivity to photoperiod among early maturing soybean strains. *Crop Science* 12, 657–660.

Davies, C.S. and Nielsen, N.C. (1987) Registration of soybean germplasm that lacks lipoxygenase isozymes. *Crop Science* 27, 370–371.

Davies, C.S., Coates, J. and Nielsen, N.C. (1985) Inheritance and biochemical analysis of four electrophoretic variants of β-conglycinin from soybean. *Theoretical and Applied Genetics* 71, 351–358.

Delannay, X. and Palmer, R.G. (1982) Four genes controlling root fluorescence in soybean. *Crop Science* 22, 278–281.

Delannay, X., Rogers, D.M. and Palmer, R.G. (1983) Relative genetic contributions among ancestral lines to North American soybean cultivars. *Crop Science* 23, 944–949.

Delzer, B.W., Somer, D.A. and Orf, J.H. (1990) *Agrobacterium tumefaciens* susceptibility and plant regeneration of 10 soybean genotypes in maturity groups 00 to II. *Crop Science* 20, 320–322.

Devine, T.E. (1985) Nodulation of soybean [*Glycine max* (L.) Merr.] plant introduction lines with the fast-growing rhizobial strain USDA 205. *Crop Science* 25, 354–356.

Devine, T.E. (1987) A comparison of rhizobial strain compatibilities of *Glycine max* and its progenitor species *Glycine soja. Crop Science* 27, 635–639.

Devine, T.E. and Breithaupt, B.H. (1981) Frequencies of nodulation response alleles *Rj2* and *Rj4* in soybean plant introduction and breeding lines. *USDA Technical Bulletin* 1268.

Domagalski, J.M., Kollipara, K.P., Bates, A.H., Brandon, D.L., Friedman, M. and Hymowitz, T. (1992) Nulls for the major soybean Bowman-Birk protease inhibitor in the genus *Glycine. Crop Science* 32, 1502–1505.

Doyle, J.J. (1988) 5S ribosomal gene variation in the soybean and its progenitor. *Theoretical and Applied Genetics* 75, 621–624.

Doyle, J.J. and Beachy, R.N. (1985) Ribosomal gene variation in soybean (*Glycine*) and its relatives. *Theoretical and Applied Genetics* 70, 369–376.

Doyle, J.J., Doyle, J.L. and Brown, A.H.D. (1990a) A chloroplast-DNA phylogeny of the wild perennial relatives of soybean (*Glycine* subgenus *Glycine*), congruence with morphological and crossing groups. *Evolution* 44, 371–389.

Doyle, J.J., Doyle, J.L. and Brown, A.H.D. (1990b) Chloroplast DNA phylogenetic affinities of newly described species in *Glycine* (Leguminosae, Phaseoleae). *Systematic Botany* 15, 466–471.

Egli, D.B., Orf, J.H. and Pfeiffer, T.W. (1984) Genotypic variation for duration of seedfill in soybean. *Crop Science* 24, 587–592.

Ertl, D.S. and Fehr, W.R. (1985) Agronomic performance of soybean genotypes from *Glycine max × Glycine soja* crosses. *Crop Science* 25, 589–592.

Fehr, W.R. (1972) Genetic control of leaflet number in soybeans. *Crop Science* 12, 221–224.

Fehr, W.R. and Cianzio, S.R. (1980) Registration of AP9(SI)C2 soybean germplasm. *Crop Science* 20, 677.

Fehr, W.R., Cianzio, S.R. and Welke, G.A. (1990a) Registration of 'SS202' soybean. *Crop Science* 30, 1361.

Fehr, W.R., Cianzio, S.R. and Welke, G.A. (1990b) Registration of 'HP201' soybean. *Crop Science* 30, 1361.

Fehr, W.R., Cianzio, S.R. and Welke, G.A. (1990c) Registration of 'LS301' soybean. *Crop Science* 30, 1363–1364.

Fuchsman, W.H. and Palmer, R.G. (1985) Conservation of leghemoglobin heterogeneity and structures in cultivated and wild soybean. *Canadian Journal of Botany* 63, 1951–1956.

Fushan, L. and Chang, R. (1984) The distribution and types of wild soybean in China. In: Wong, S., Boethel, D.J., Nelson, R.L., Nelson, W.L. and Wolf, W.J. (eds), *Proceedings of the Second US–China Soybean Symposium*. Office of International Cooperation and Development, USDA, Washington, DC, pp. 262–265.

Gai Junyi, Fehr, W.R. and Palmer, R.G. (1982) Genetic performance of some agronomic characters in four generations of a backcrossing programme involving *Glycine max* and *Glycine soja*. *Acta Genetica Sinica* 9, 44–56.

Gizlice, Z., Carter, T.E., Jr. and Burton, J.W. (1993a) Genetic diversity in North American soybean, I. Multivariate analysis of founding stock and relation to coefficient of parentage. *Crop Science* 33, 614–620.

Gizlice, Z., Carter, T.E., Jr. and Burton, J.W. (1993b) Genetic diversity in North American soybean, II. Prediction of heterosis in F2 populations of southern founding stock using genetic similarity measures. *Crop Science* 33, 620–626.

Gizlice, Z., Carter, T.E., Jr. and Burton, J.W. (1994) Genetic base for North American public soybean cultivars released between 1947 and 1988. *Crop Science* 34, 1143–1151.

Grabau, E.A., Davis, W.H. and Gengenbach, B.G. (1989) Restriction fragment length polymorphism in a subclass of the 'Mandarin' soybean cytoplasm. *Crop Science* 29, 1554–1559.

Grabau, E.A., Davis, W.H., Phelps, N.D. and Gengebach, B.G. (1992) Classification of soybean cultivars based on mitochondrial DNA restriction fragment length polymorphisms. *Crop Science* 32, 271–274.

Graef, G.L., Fehr, W.R. and Cianzio, S.R. (1989) Relation of isozyme genotype to quantitative characters in soybean. *Crop Science* 29, 683–688.

Grant, J.E. (1984) Plant regeneration from cotyledonary tissue of *Glycine canescens*, a perennial wild relative of soybean. *Plant Cell, Tissue and Organ Culture* 3, 169–173.

Griffin, J.D. and Palmer, R.G. (1995) Variability of 13 isozyme loci in the USDA soybean germplasm collections. *Crop Science* 35, 897–904.

Groose, R.W. and Palmer, R.G. (1991) Gene action for anthocyanin pigmentation in soybean. *Journal of Heredity* 82, 498–501.

Hammatt, N., Davey, M.R. and Nelson, R.S. (1986) Plant regeneration from seedling cotyledons, leaves and petioles of *Glycine clandestina*. *Physiologia Plantarum* 68, 125–128.

Hammatt, N., Jones, B. and Davey, M.R. (1989) Plant regeneration from seedling explants and cotyledon protoplasts of *Glycine argyrea* Tind. *In Vitro* 25, 669–672.

Harlan, J.R. and de Wet, J.M.J. (1971) Toward a rational classification of cultivated plants. *Taxon* 20, 509–517.

Harrison, W.J., Kenworthy, W.J., Thomas, C.A. and Thomison, P.R. (1989) Infection of *Glycine* species with *Phytophthora*. *Agronomy Abstracts*, 84.

Hart, S.E., Kenworthy, W.J. and Glenn, D.S. (1988) Mechanisms of 2,4-D resistance in perennial *Glycine* species. *Agronomy Abstracts*, 82–83.

Hartman, G.L., Wang, T.C. and Hymowitz, T. (1992) Sources of resistance to soybean rust in perennial *Glycine* species. *Plant Disease* 76, 396–399.

Hartwig, E.E. (1972) Utilization of soybean germplasm strains in a soybean improvement programme. *Crop Science* 12, 856–859.

Hartwig, E.E. (1973) Varietal development. In: Caldwell, B.E. (ed.) *Soybeans, Improvement, Production, and Uses*. American Society of Agronomy Monograph 16, Madison, Wisconsin, pp. 187–210.

Hartwig, E.E. and Bromfield, K.R. (1983) Relationships among three genes conferring specific resistance to rust in soybeans. *Crop Science* 23, 237–239.

Hartwig, E.E. and Hinson, K. (1962) Inheritance of flower color of soybeans. *Crop Science* 2, 152–153.

Hartwig, E.E. and Lehman, S.G. (1951) Inheritance of resistance to the bacterial pustule disease in soybeans. *Agronomy Journal* 43, 226–229.

Hartwig, E.E. and Potts, H.C. (1987) Development and evaluation of impermeable seed coats for preserving soybean seed quality. *Crop Science* 27, 506–508.

Hartwig, E.E., Keeling, B.L. and Edwards, C.J., Jr. (1968) Inheritance of reaction to phytophthora rot in the soybean. *Crop Science* 8, 634–635.

Hartwig, E.E., Lambert, L. and Kilen, T.C. (1990) Registration of 'Lamar' soybean. *Crop Science* 30, 231.

Hatfield, P.M., Shoemaker, R.C. and Palmer, R.G. (1985) Maternal inheritance of chloroplast DNA within the genus *Glycine*, subgenus *Soja*. *Journal of Heredity* 76, 373–374.

Hildebrand, D.F. and Hymowitz, T. (1981) Two soybean genotypes lacking lipoxygenase-1. *Journal of the American Oil Chemists' Society* 58, 583–586.

Hildebrand, D.F. and Hymowitz, T. (1982) Inheritance of lipoxygenase-1 activity in soybean seeds. *Crop Science* 22, 851–853.

Hymowitz, T. and Bernard, R.L. (1991) Origin of the soybean and germplasm introduction and development in North America. In: Shands, H.L. and Wiesner, L.E. (eds) *Use of Plant Introductions in Cultivar Development. Part 1.* Special Publication 17. Crop Science Society of America, Madison, Wisconsin, pp. 147–164.

Hymowitz, T. and Singh, R.J. (1984) A soybean × *Glycine tomentella* hybrid: Progress and problems. *Soybean Genetics Newsletter* 11, 90.

Hymowitz, T. and Singh, R.J. (1987) Taxonomy and speciation. In: Wilcox, J.R. (ed.) *Soybeans, Improvement, Production and Uses*, 2nd edn. American Society of Agronomy Monograph 16, Madison, Wisconsin, pp. 23–48.

Hymowitz, T., Chalmers, N.L., Constanza, S.H. and Saam, M.M. (1986) Plant regeneration from leaf explants of *Glycine clandestina* Wendl. *Plant Cell Reports* 3, 192–194.

Hymowitz, T., Woolley, J.T. and Peters, D.B. (1987) Preliminary investigations on the salt tolerance of wild perennial *Glycine* species. *Soybean Genetics Newsletter* 14, 271–272.

Johns, C.W. and Palmer, R.G. (1982) Floral development of a flower-structure mutant in soybeans, *Glycine max* (L.) Merr. (Leguminosae). *American Journal of Botany* 69, 829–842.

Johnson, H.W. and Bernard, R.L. (1963) Soybean genetics and breeding. In: Norman, A.G. (ed.) *The Soybean.* Academic Press, New York, pp. 1–73.

Juvik, G.A., Bernard, R.L. and Kauffman, H.E. (1985) *Directory of Germplasm Collections.* 1. II. Food Legumes (Soyabean). International Board for Plant Genetic Resources. Rome, Italy.

Juvik, G.A., Bernard, R.L., Chang, R. and Cavins, J.F. (1989a) Evaluation of the USDA Wild Soybean Germplasm Collection: maturity groups 000 to IV (PI 65.549 to PI 483.464). *USDA Technical Bulletin* No. 1761.

Juvik, G.A., Bernard, R.L., Orf, J.H., Cavins, J.F. and Thomas, D.I. (1989b) Evaluation of the USDA Soybean Germplasm Collection: maturity groups 000 to IV (PI 446.893 to PI 486.355). *USDA Technical Bulletin* No. 1760.

Keim, P., Shoemaker, R.C. and Palmer, R.G. (1989) Restriction fragment length polymorphism diversity in soybean. *Theoretical and Applied Genetics* 77, 786–792.

Kenworthy, W.J. and Brim, C.A. (1979) Recurrent selection in soybeans, I. Seed yield. *Crop Science* 19, 315–318.

Khalaf, A.G.M., Brossman, G.D. and Wilcox, J.R. (1984) Use of diverse populations in soybean breeding. *Crop Science* 24, 358–360.

Kiang, Y.T. (1981) Inheritance and variation of amylase in cultivated and wild soybeans and their wild relatives. *Journal of Heredity* 72, 382–386.

Kiihl, R.A.S. and Hartwig, E.E. (1979) Inheritance of reaction to soybean mosaic virus in soybeans. *Crop Science* 19, 372–375.

Kilen, T.C. and Hartwig, E.E. (1975) Short internode character in soybeans and its inheritance. *Crop Science* 15, 878.

Kilen, T.C. and He, G. (1992) Identification and inheritance of metribuzin tolerance in wild soybean. *Crop Science* 32, 684–685.

Kitamura, K., Davies, C.S., Kaizuma, N. and Nielsen, N.C. (1983) Genetic analysis of a null-allele for lipoxygenase-3 in soybean seeds. *Crop Science* 23, 924–927.

Kitamura, K., Davies, C.S. and Nielsen, N.C. (1984) Inheritance of alleles for *Cgyl* and *Gy4* storage protein genes in soybean. *Theoretical and Applied Genetics* 68, 253–257.

Kollipara, K.P., Singh, R.J. and Hymowitz, T. (1993) Genomic diversity in aneudiploid (2n = 38) and diploid (2n = 40) *Glycine tomentella* Hayata revealed by cytogenetic and biochemical methods. *Genome* 36, 391–396.

Kollipara, K.P., Singh, R.J. and Hymowitz, T. (1994) Genomic diversity and multiple origins of tetraploid (2n = 78,80), *Glycine tomentella* Hayata. *Genome* 37, 448–459.

LeRoy, A.R., Cianzio, S.R. and Fehr., W.R. (1991a) Direct and indirect selection for small seed of soybean in temperate and tropical environments. *Crop Science* 31, 697–699.

LeRoy, A.R., Fehr, W.R. and Cianzio, S.R. (1991b) Introgression of genes for small seed size from *Glycine soja* into *G. max. Crop Science* 31, 693–697.

Li, W.B. and Satoh, C. (1990) The distribution and classification of Chinese wild soyabeans. *Agriculture and Horticulture* 65, 803–808.

Lim, S.M. (1989) Inheritance of resistance to *Peronospora manshurica* races 2 and 33 in soybean. *Phytopathology* 79, 877–879.

Lim, S.M. and Hymowitz, T. (1987) Reactions of perennial wild species of the genus *Glycine* subgenus *Glycine* to *Septoria glycines. Plant Disease* 71, 891–893.

Lohnes, D.G. and Bernard, R.L. (1991) Ancestry of US/Canadian commercial cultivars developed by public institutions. *Soybean Genetics Newsletter* 18, 243–255.

Loux, M.M., Leibl, R.A. and Hymowitz, T. (1987) Examination of the wild perennial *Glycine* collection for tolerance to glyphosate. *Soybean Genetics Newsletter* 14, 268–271.

Luedders, V.D. (1977) Genetic improvement in yield of soybeans. *Crop Science* 17, 268–271.

Lugg, D.G. and Sinclair, T.R. (1979) A survey of soybean cultivars for variability in specific leaf weight. *Crop Science* 19, 887–892.

Matson, A.L. and Williams, L.F. (1965) Evidence of a fourth gene for resistance to the soybean cyst nematode. *Crop Science* 5, 477.

McKenna, T., Lambert, L., Ouzts, J. and Kilen, T.C. (1987) Field cage evaluation of wild soybean, *Glycine soja* (Sieb. & Zucc.) for resistance to three lepidopterous defoliators. *Proceedings of the Mississippi Entomology Association Insect Control Conference* 5, 53–55.

McLean, R.J. and Byth, D.E. (1980) Inheritance of resistance to rust *Phakopsora pachyrhizi* in soybeans. *Australian Journal of Agricultural Research* 31, 951–956.

Mignucci, J.S. (1975) Powdery mildew of soybeans. M.S. Thesis. University of Illinois, Urbana, Illinois.

Nelson, R.L. (1986) Defining the seed-filling period in soybeans to predict yield. *Crop Science* 26, 132–135.

Nelson, R.L. (1987) The relationship between seed-filling period and seed yield in selected soybean germplasm accessions. *Field Crops Research* 15, 245–250.

Nelson, R.L. and Schweitzer, L.E. (1988) Evaluating soybean germplasm for specific leaf weight. *Crop Science* 28, 647–649.

Nelson, R.L., Amdor, P.J., Orf, J.H., Lambert, J.W., Cavins, J.F., Kleiman, R., Laviolette, F.A. and Athow, K.A. (1987) Evaluation of the USDA Soybean Germplasm Collection: maturity groups 000 to IV (PI 273.483 to PI 427.107). *USDA Technical Bulletin* No. 1718.

Nelson, R.L., Amdor, P.J., Orf, J.H. and Cavins, J.F. (1988) Evaluation of the USDA Soybean Germplasm Collection: maturity groups 000 to IV (PI 427.136 to PI 445.845). *USDA Technical Bulletin* No. 1726.

Nelson, R.L., Nickell, C.D., Orf, J.H., Tachibana, H., Gritton, E.T., Grau, C.R. and Kennedy, B.W. (1989) Evaluating soybean germplasm for brown stem rot resistance. *Plant Disease* 73, 110–114.

Newell, C.A. and Hymowitz, T. (1980) A taxonomic revision of the genus *Glycine* subgenus *Glycine* (Leguminosae). *Brittonia* 32, 63–69.

Newell, C.A. and Hymowitz, T. (1982) Successful wide hybridization between the soybean and a wild perennial relative, *G. tomentella* Hayata. *Crop Science* 22, 1062–1065.

Newell, C.A. and Luu, H.T. (1985) Protoplast culture and plant regeneration in *Glycine canescens* F.J. Herm. *Plant Cell Tissue Organ Culture* 4, 145–149.

Newell, C.A., Delannay, X. and Edge, M.E. (1987) Interspecific hybrids between the soybean and wild perennial relatives. *Journal of Heredity* 78, 301–306.

Nielsen, N.C., Dickinson, C.D., Cho, T.-J., Thanh, V.H., Scallon, B.J., Fischer, R.L., Sims. T.L., Drews, G.N. and Goldberg, R.B. (1989) Characterization of the glycinin gene family in soybean. *Plant Cell* 1, 313–328.

Nissly, C.R., Bernard, R.L. and Hittle, C.N. (1981) Variation in photoperiod sensitivity for time of flowering and maturity among soybean strains of maturity group III. *Crop Science* 21, 833–836.

Orf, J.H. and Hymowitz, T. (1976) The gene symbols *Spl-a* and *Spl-b* assigned to Larsen and Caldwell's seed protein bands A and B. *Soybean Genetics Newsletter* 3, 27–28.

Orf, J.H. and Hymowitz, T. (1977) Inheritance of a second trypsin inhibitor variant in seed protein of soybeans. *Crop Science* 17, 811–813.

Orf, J.H. and Hymowitz, T. (1979) Inheritance of the absence of the Kunitz trypsin inhibitor in seed protein of soybeans. *Crop Science* 19, 107–109.

Palmer, R.G. (1985) Soybean Cytogenetics. In: Shibles, R.S. (ed.) *Proceedings of World Soybean Research Conference III.* Westview Press, Boulder, Colorado, pp. 337–344.

Palmer, R.G. and Heer, H.E. (1984) Agronomic characteristics and genetics of a chromosome interchange in soybean. *Euphytica* 33, 651–663.

Palmer, R.G. and Hedges, B.R. (1993) Linkage map of soybean [(*Glycine max* (L.) Merr.].

In: O'Brien, S.J. (ed.) *Genetic Maps.* Cold Spring Harbor Laboratory Publisher, Cold Spring Harbor, New York, pp. 6.139–6.148.

Palmer, R.G. and Kilen, T.C. (1987) Qualitative genetics and cytogenetics. In: Wilcox, J.R. (ed.) *Soybeans, Improvement, Production and Uses*, 2nd edn. American Society of Agronomy Monograph 16, Madison, Wisconsin, pp. 135–209.

Palmer, R.G., Newhouse, K.E., Graybosch, R.A. and Delannay, X. (1987) Chromosome structure of the wild soybean. Accessions from China and the Soviet Union of *Glycine soja* Sieb. and Zucc. *Journal of Heredity* 78, 243–247.

Palmer, R.G., Lim, S.M. and Hedges, B.R. (1992) Testing for linkage between the *Rxp* locus and nine isozyme loci in soybean. *Crop Science* 32, 681–683.

Pantalone, V.R., III and Kenworthy, W.J. (1989) Salt tolerance in *Glycine max* and perennial *Glycine. Soybean Genetics Newsletter* 16, 145–146.

Perry, M.C. and McIntosh, M.S. (1991) Geographic patterns of variation in the USDA Soybean Germplasm Collection, I. Morphological traits. *Crop Science* 31, 1350–1355.

Perry, M.C., McIntosh, M.S. and Stoner, A.K. (1991) Geographic patterns of variation in the USDA Soybean Germplasm Collection, II. Allozyme frequencies. *Crop Science* 31, 1356–1360.

Pfeiffer, T.W., Suryati, D. and Egli, D.B. (1991) Soybean plant introductions selected for seed filling period or yield: Performance as parents. *Crop Science* 31, 1418–1421.

Polson, D.E. (1972) Day-neutrality in soybeans. *Crop Science* 12, 773–776.

Powell, J.E. and Lambert, L. (1993) Soybean genotype effects on big-eyed bug feeding on corn earworm in the laboratory. *Crop Science* 33, 556–559.

Raboy, V. and Dickinson, D.B. (1993) Phytic acid levels in seed of *Glycine max* and *G. soja* as influenced by phosophorus status. *Crop Science* 33, 1300–1305.

Rafalski, A. and Tingey, S. (1993) Map of soybean (*Glycine max*). In: O'Brien, S.J. (ed.) *Genetic Maps.* Cold Spring Harbor Laboratory Publisher, Cold Spring Harbor, New York. pp. 6.149–6.156.

Rao-Arelli, A.P., Anand, S.C. and Wrather, J.A. (1992) Soybean resistance to soybean cyst nematode race 3 is conditioned by an additional dominant gene. *Crop Science* 32, 862–864.

Reed, H.E., Teranaura, A.H. and Kenworthy, W.J. (1992) Ancestral US soybean cultivars characterized for tolerance to ultraviolet-B radiation. *Crop Science* 32, 1214–1219.

Reese, Jr., P.F., Kenworthy, W.J., Cregan, P.B. and Yocum, J.O. (1988) Comparison of selection systems for the identification of exotic soybean lines for use in germplasm development. *Crop Science* 28, 237–241.

Reicosky, D.A., Orf, J.H. and Poneleit, C. (1982) Soybean germplasm evaluation for length of the seed filling period. *Crop Science* 22, 319–322.

Rennie, B.D. and Tanner, J.W. (1989) Mapping a second fatty acid locus to soybean linkage group 17. *Crop Science* 29, 1081–1083.

Rennie, B.D., Zilka, J., Cramer, M.M. and Beversdorf, W.D. (1988) Genetic analysis of low linolenic acid levels in the soybean line PI 361.088B. *Crop Science* 28, 655–657.

Roane, C.W., Tolin, S.A. and Buss, G.R. (1983) Inheritance of reaction to two viruses in the soybean cross 'York' and 'Lee' 68'. *Journal of Heredity* 74, 289–291.

Rode, M.W. and Bernard, R.L. (1975) Inheritance of bullate leaf. *Soybean Genetics Newsletter* 2, 44–46.

Rowan, G.B., Boerma, H.R., All, J.N. and Todd, J.W. (1993) Soybean maturity effect on expression of resistance to lepidopterous insects. *Crop Science* 33, 433–436.

Salado-Navarro, L.R., Sinclair, T.R. and Hinson, K. (1993) Changes in yield and seed growth traits in soybean cultivars released in the Southern USA from 1945 to 1983. *Crop Science* 33, 1204–1209.

Schmitt, D.P. and Shannon, G. (1992) Differentiating soybean responses to *Heterodera glycines* races. *Crop Science* 32, 275–277.

Schoen, D.J., Burdon, J.J. and Brown, A.H.D. (1992) Resistance of *Glycine tomentella* to soybean leaf rust *Phakopsora pachyrhizi* in relation to ploidy level and geographic distribution. *Theoretical and Applied Genetics* 83, 827–832.

Schoener, C.S. and Fehr, W.R. (1979) Utilization of plant introductions in soybean breeding populations. *Crop Science* 19, 185–188.

Shannon, M.C., Shouse, P.J., Zidan, A.T. and Hymowitz, T. (1989) Variability in salt tolerance among wild relatives of soybean. *Agronomy Abstracts*, 99.

Shoemaker, R.C. and Olson, T.C. (1993) Molecular linkage map of soybean [(*Glycine max* (L.) Merr.]. In: O'Brien, S.J. (ed.) *Genetic maps*. Cold Spring Harbor Laboratory Publisher, Cold Spring Harbor, New York, pp. 6.131–6.138.

Shoemaker, R.C., Hatfield, P.M., Palmer, R.G. and Atherly, A.G. (1986) Chloroplast DNA variation in the genus *Glycine* subgenus *Soja*. *Journal of Heredity* 77, 26–30.

Shoemaker, R.C., Heath, M.S., Skorupska, H., Delannay, X, Edge, M. and Newell, C.A. (1990) Fertile progeny of a hybridization between soybean [*Glycine max* (L.) Merr.] and *G. tomentella* Hayata. *Theoretical and Applied Genetics* 80, 17–23.

Shoemaker, R.C., Guffy, R.D., Lorenzen, L.L. and Specht, J.E. (1992) Molecular genetic mapping of soybean: map utilization. *Crop Science* 32, 1091–1098.

Singh, B.B. and Malick, A.S. (1978) Inheritance of resistance to yellow mosaic in soybean. *Indian Journal of Genetics and Plant Breeding* 38, 258–261.

Singh, B.B., Gupta, S.C. and Singh, B.D. (1974) Sources of field resistance to rust and yellow mosaic diseases of soybean. *Indian Journal of Genetics and Plant Breeding* 34, 400–404.

Singh, L., Wilson, C.M. and Hadley, H.H. (1969) Genetic differences in soybean trypsin inhibitors separated by disc electrophoresis. *Crop Science* 9, 489–491.

Singh, R.J. (1993) *Plant Cytogenetics*. CRC Press, Boca Raton, Florida.

Singh, R.J. and Hymowitz, T. (1985a) The genomic relationships among six wild perennial species of the genus *Glycine* subgenus *Glycine* Willd. *Theoretical and Applied Genetics* 71, 221–230.

Singh, R.J. and Hymowitz, T. (1985b) Intra- and interspecific hybridization in the genus *Glycine* subgenus *Glycine* Willd., chromosome pairing and genomic relationships. *Zeitschrift für Pflanzenzüchtung* 95, 289–310.

Singh, R.J. and Hymowitz, T. (1985c) An intersubgeneric hybrid between *Glycine tomentella* Hayata and the soybean, *G. max* (L.) Merr. *Euphytica* 34, 187–192.

Singh, R.J. and Hymowitz, T. (1988) The genomic relationship between *Glycine max* (L.) Merr. and *G. soja* Seib. and Zucc. as revealed by pachytene chromosome analysis. *Theoretical and Applied Genetics* 76, 705–711.

Singh, R.J., Kollipara, K.P. and Hymowitz, T. (1987a) Intersubgeneric hybridization of soybeans with a wild perennial species, *Glycine clandestina* Wendl. *Theoretical and Applied Genetics* 74, 391–396.

Singh, R.J., Kollipara, K.P. and Hymowitz, T. (1987b) Polyploid complexes of *Glycine ta-*

bacina (Labill.) Benth. and *G. tomentella* Hayata revealed by cytogenetic analysis. *Genome* 29, 490–497.

Singh, R.J., Kollipara, K.P. and Hymowitz, T. (1988) Further data on the genomic relationships among wild perennial species (2n = 40) of the genus *Glycine* Willd. *Genome* 30, 166–176.

Singh, R.J., Kollipara, K.P. and Hymowitz, T. (1989) Ancestors of 80- and 78-chromosome *Glycine tomentella* Hayata intersubgeneric hybrids. *Genome* 32, 796–801.

Singh, R.J., Kollipara, K.P. and Hymowitz, T. (1990) Backcross-derived progeny from soybean and *Glycine tomentella* Hayata intersubgeneric hybrids. *Crop Science* 30, 871–874.

Singh, R.J., Kollipara, K.P. and Hymowitz, T. (1992a) Genomic relationships among diploid wild perennial species of the genus *Glycine* Willd. subgenus *Glycine* revealed by cytogenetics and seed protein electrophoresis. *Theoretical and Applied Genetics* 85, 276–282.

Singh, R.J., Kollipara, K.P., Ahmad, F. and Hymowitz, T. (1992b) Putative diploid ancestors of 80-chromosome *Glycine tabacina* (Labill.) Benth. *Genome* 35, 140–146.

Singh, R.J., Kollipara, K.P., and Hymowitz, T. (1993) Backcross (BC2-BC4)-derived fertile plants from *Glycine max* (L.) Merr. and *G. tomentella* intersubgeneric hybrids. *Crop Science* 33, 1002–1007.

Sisson, V.A., Brim, C.A. and Levings, C.S., III (1978) Characterization of cytoplasmic diversity in soybean by restriction endonuclease analysis. *Crop Science* 18, 991–996.

Skorupska, H.T., Shoemaker, R.C., Warner, A., Shipe, E.R. and Bridges, W.C. (1993) Restriction fragment length polymorphism in soybean germplasm of the southern USA. *Crop Science* 33, 1169–1176.

Smith, J.R. and Nelson, R.L. (1986) Relationship between seed-filling period and yield among soybean breeding lines. *Crop Science* 26, 469–472.

Specht, J.E. and Williams, J.H. (1984) Contribution of genetic technology to soybean productivity – retrospect and prospect. In: Fehr, W.R. (ed.) *Genetic Contributions to Yield Gains of Five Major Crop Plants*. Special Publication 7. Crop Science Society of America and American Society of Agronomy, Madison, Wisconsin, pp. 48–74.

St. Martin, S.K. (1982) Effective population size for the soybean improvement programme in maturity groups 00 to IV. *Crop Science* 22, 151–152.

St. Martin, S.K. and Aslam, M. (1986) Performance of progeny of adapted and plant introduction soybean lines. *Crop Science* 26, 753–756.

Suárez, J.C., Graef, G.L., Fehr, W.R. and Cianzio, S.R. (1991) Association of isozyme genotypes with agronomic and seed composition traits in soybean. *Euphytica* 52, 137–146.

Sweeney, P.M. and St. Martin, S.K. (1989) Testcross evaluation of exotic soybean germplasm of different origins. *Crop Science* 29, 289–293.

Terao, H. (1918) Maternal inheritance in the soybean. *American Naturalist* 52, 51–56.

Thorne, J.C. and Fehr, W.R. (1970a) Incorporation of high-protein, exotic germplasm into soybean populations by 2- and 3-way crosses. *Crop Science* 10, 652–655.

Thorne, J.C. and Fehr, W.R. (1970b) Exotic germplasm for yield improvement in 2-way and 3-way soybean crosses. *Crop Science* 10, 677–678.

Tindale, M.D. (1984) Two new eastern Australian species of *Glycine* Willd. (Fabaceae). *Brunonia* 7, 207–213.

Tindale, M.D. (1986a) A new North Queensland species of *Glycine* Willd. (Fabaceae). *Brunonia* 9, 99–103.

Tindale, M.D. (1986b) Taxonomic notes on three Australian and Norfolk Island species of *Glycine* Willd. (Fabaceae, Phaseolae) including the choice of a neotype for *G. clandestina* Wendl. *Brunonia* 9, 179–191.

Tindale, M.D. and Craven, L.A. (1988) Three new species of *Glycine* (Fabaceae, Phaseolae) from north-western Australia, with notes on the amphicarpy in the genus. *Australian Systematic Botany* 1, 399–410.

Tindale, M.D. and Craven L.A. (1993) *Glycine pindanica* (Fabaceae, Phaseolae), a new species from West Kimberley, Western Australia. *Australian Systematic Botany* 6, 371–376.

Ting, C.L. (1946) Genetic studies on the wild and cultivated soybeans. *Journal of the American Society of Agronomy* 38, 381–393.

Vello, N.A., Fehr, W.R. and Bahrenfus, J.B. (1984) Genetic variability and agronomic performance of soybean populations developed from plant introductions. *Crop Science* 24, 511–514.

Vest, G. (1970) *Rj3*-a gene conditioning ineffective nodulation in soybean. *Crop Science* 10, 34–35.

Vest, G. and Caldwell, B.E. (1972) *Rj4*-a gene conditioning ineffective nodulation in soybean. *Crop Science* 12, 692–693.

Voldeng, H.D. (1981) *Glycine max* (L.) Merr. cv. Nattawa, License No. 2114. Agriculture Canada, Ottawa.

Voldeng, H.D. (1985) *Glycine max* (L.) Merr. cv. Canatto, License No. 2500. Agriculture Canada, Ottawa

Voldeng, H.D. (1989) *Glycine max* (L.) Merr. cv. Nattosan, License No. 3116. Agriculture Canada, Ottawa

Voldeng, H.D. and Saindon, G. (1991) Registration of seven long-day length insensitive soybean genetic stocks. *Crop Science* 31, 1399.

Weiss, M.G. (1943) Inheritance and physiology of efficiency in iron utilization in soybeans. *Genetics* 28, 253–268.

White, R.H., Leibl, R.A. and Hymowitz, T. (1990) Examination of 2,4-D tolerance in perennial *Glycine* species. *Pesticide Biochemistry and Physiology* 38, 153–161.

Wilcox, J.R., Schapaugh, W.T., Jr., Bernard, R.L., Cooper, R.L., Fehr, W.R. and Neihaus, M.H. (1979) Genetic improvement of soybeans in the midwest. *Crop Science* 19, 803–805.

Woodworth, C.M. (1921) Inheritance of cotyledon, seed-coat, hilum, and pubescence colors in soy-beans. *Genetics* 6, 487–553.

Young, L.D. (1990) Soybean germplasm evaluated for resistance to races 3,5, and 14 of soybean cyst nematode. *Crop Science* 30, 735–736.

Zimmerman, M.S. and Minor, H.C. (1993) Inheritance of *Phomopsis* seed decay resistance in soybean PI 417.479. *Crop Science* 33, 96–100.

Molecular Genetic Mapping of Soybean $\boxed{2}$

R.C. Shoemaker[1], K.M. Polzin[2], L.L. Lorenzen[2]
and J.E. Specht[3]

[1] USDA-ARS FCR and Departments of Agronomy and
Zoology/Genetics, Iowa State University, Ames, Iowa 50011, USA:
[2] Department of Agronomy, Iowa State University, Ames, Iowa
50011, USA: [3] Department of Agronomy, University of Nebraska,
Lincoln, Nebraska 68583, USA.

Introduction

The soybean first emerged as a domesticated plant around the 11th century B.C. (Hymowitz, 1970), although references to the soybean appear in books written 4500 years ago (Smith and Huyser, 1987). As such, it is one of man's oldest cultivated crops. Originally planted in this country as a forage crop, it became important also as an oilseed crop by the early 1900s (Smith and Huyser, 1987). Today it is one of the world's leading sources of seed oil and seed protein (Burton, 1985).

Soybean genetic map development proceeded slowly relative to genetic maps of crops of similar economic importance. This has been due largely to inherent difficulties in performing successful sexual crosses, a lack of cytogenetic markers and the unavailability of proper genetic stocks. Today, just 63 morphological, pigmentation or isoenzyme markers have been assigned to the classical genetic map (Palmer and Hedges, 1993). In contrast, the application of molecular genetic mapping techniques has allowed for the development of detailed molecular genetic maps for soybean in a relatively short period. Several of these maps have been published and they include hundreds of RFLP, RAPD and classical markers (Lark et al., 1993; Rafalski and Tingey, 1993; Shoemaker and Olson, 1993)

Genome Organization

The amount of DNA in an unreplicated haploid genome differs by more than 600-fold among various angiosperm species (Bennett et al., 1982). Though certainly not the largest or most complex of the plant genomes, the genome size of the soybean is many times larger than that of *Arabidopsis*, and many times more complex.

The soybean genome contains an estimated 1.29 × 109 bp (Gurley *et al.*, 1979) to 1.81 × 109 bp (Goldberg, 1978) for 1 N DNA content. The genome is approximately 40–60% repetitive sequences (Goldberg, 1978; Gurley *et al.*, 1979). The majority (65–70%) of single-copy sequences are estimated to have a short period interspersion with single-copy sequences of 1.1–1.4 kb alternating with repetitive sequence elements of 0.3–0.4 kb (Gurley *et al.*, 1979). Analysis of pachytene chromosomes has shown that over 35% of the soybean genome is made up of heterochromatin; the short arms of six of the 20 bivalents appear to be completely heterochromatic (Singh and Hymowitz, 1988).

A detailed analysis of the distribution of repetitive sequences along regions of the soybean genome was conducted using lambda clones carrying sequences homologous to a late embryogenic abundant gene, *sle1*, and to a genomic clone, A071. These results suggest that the distribution of repetitive sequences along the genome can vary significantly by region (Fig. 2.1).

Each clone was subcloned and restriction mapped to identify consecutive fragments ranging in size from 0.4–2 kb covering the entire clone. These fragments were used to construct dot blots and were used as probes. Regions containing highly repetitive sequences were identified by screening dot blots with labelled soybean genomic DNA under both high and low stringency conditions. Fragments which were not considered highly repetitive sequence by the above method were used as probes against five digests of soybean genomic DNA and the number of hybridizing bands scored. Multiple digests were used to avoid overestimation of copy number due to an enzyme cutting within a probe fragment sequence and thus generating more than one hybridizing fragment per locus. Fragments were classified as follows: single-copy, one band; low-copy, 2–10 bands; repetitive, >10 bands.

The distribution of repetitive and low-copy sequence surrounding the sle and A071 loci under high stringency conditions is shown (Fig. 2.1). Both *sle1* and *sle2* contained primarily low or single-copy sequence downstream of the sle gene. The A071 loci exhibited more difference in the distribution of repetitive and low-copy sequence near the A071 loci. Two loci, A071-1 and A071-5, were located in regions consisting almost entirely of low and single-copy sequence while A071-2, A071-4, and A071-10 all contained large blocks of highly repetitive sequence. Clones A071-2, A071-4 and A071-10 did appear similar in their central regions. However, these sequences do not crosshybridize suggesting that this is only coincidence or that these regions have undergone significant divergence since duplication. Therefore, while the regions flanking the sle loci appear similar in structure with regard to the distribution of repetitive sequence, the A071 loci do not. These findings highlight two important facts concerning soybean genome organization: (i) stretches of repetetive sequences do not seem to be too large to easily traverse or span, and (ii)

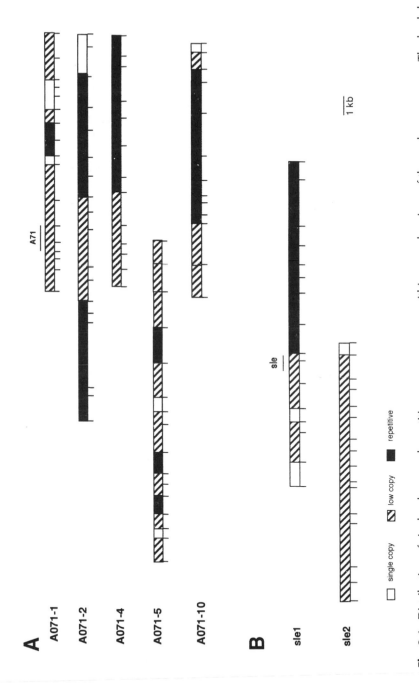

Fig. 2.1. Distribution of single-, low-, and repetitive-copy sequences within several regions of the soybean genome. The lambda clones represented in the figure were selected using the probes sle and A071 (see text).

single- and low-copy sequences seem to be well-dispersed throughout the genome. Similar results have since been observed with additional clones from additional regions.

Genome Duplication

For most characters the soybean behaves genetically like a diploid. We therefore generally regard 2n = 40 as the diploid chromosome number.

Most genera of the Phaseoleae have a genome complement of 2n = 22. Lackey (1980) suggested that *Glycine* was probably derived from a diploid ancestor (n = 11) which underwent an aneuploid loss to n = 10 and subsequent polyploidation to yield the present 2n = 2x = 40 genome size (Palmer and Kilen, 1987).

Genetic evidence of gene duplication also suggests that soybean is a polyploid (Buttery and Buzzell, 1976; Palmer and Kilen, 1987). Hymowitz and Singh (1987) have suggested that the soybean be regarded as a stable tetraploid with diploidized genomes. Soybean possesses many examples of qualitative traits controlled by two loci (Zobel, 1983; Palmer and Kilen, 1987).

There is a strong tendency for polyploids to evolve into a diploid state through sequence diversification and chromosome rearrangement (Leipold and Schmidtke, 1982). This type of genome rearrangement was observed during the analysis and comparison of the leghaemoglobin genes of *Phaseolus vulgaris* and *Glycine max* (Lee and Verma, 1984). Also, studies of rRNA gene sequences in soybean have indicated the rRNA gene sites have been eliminated during diploidation (Skorupska *et al.*, 1989).

Many examples of duplicated qualitative genes are known for soybean (Buttery and Buzzell, 1976; Palmer and Kilen, 1987). During the construction of the molecular genetic map many examples of duplicated loci have also been observed. In fact, it is likely that the vast majority of the soybean genome exists in multiple copies.

Major changes in the amount of nuclear DNA and in the structure of chromosomes occur during evolution (Stebin, 1966; Ohno, 1970). Two mechanisms for increasing the amount of nuclear DNA are regional duplication of chromosomal segments and global duplication of the entire genome (i.e. polyploidization) (Ohno, 1970). Both processes can be detected by the presence of duplicate loci (loci which are present more than once in the genome). Regional duplication is evidenced by duplication of one or a few loci and is the probable mechanism for the evolution of clustered multigene families (Lundin, 1993). Following regional duplication both the original locus and its copy are located on the same chromosome. However, in a few cases subsequent translocations or excision/insertion events may result in the two loci being present on different chromosomes (Childs *et al.*, 1981). In contrast, polyploidization duplicates the entire genome resulting in many duplicate loci whose members are almost always present on different chromosomes (Lundin, 1993). Also, unlike regional duplication, polyploidization often results in major genome restructuring as subsequent genetic rearrangements restore the genome to a diploid state (diploidization) (Ohno, 1970). Both regional duplication and polyploidization result in paralogous regions, regions within a genome that contain homologous loci. Paralogous regions derived from polyploidization are termed homoeologous regions.

It is important to remember that failure to detect duplicate loci does not preclude their existence. Most probes hybridize to multiple fragments, but duplicate loci can only be detected if a polymorphism is present between parental genotypes for both loci. In working with many independent populations we have found that some cross combinations yield a much higher percent of 'new' locus detection than other populations using previously mapped probes. Because soybean probes detect multiple fragments in a Southern hybridization experiment, and because each fragment could represent a different locus on a chromosome, it is important to map each polymorphism. Without this information, the map location of the polymorphic fragments must remain ambiguous and the framework map developed in the original mapping population will be of limited value. Those duplications we have been able to map have provided us with information on the organization and structure of the soybean genome and have made each probe more applicable in a wide range of populations (Fig. 2.2).

An analysis of duplicated (paralogous) genomic regions can provide a wealth of information about the sequence divergence and rearrangements associated with events which occur during the genomic evolution of the tetraploid (diploidization). Additionally, knowledge of organization of paralogous regions allow us to potentially transfer information from map-rich regions to map-poor regions within a genome, i.e. to predict linkages. Finally, the analysis of duplicated regions within a genome may aid in our analysis of complex traits and gene families.

Molecular Genetic Map

To make full use of the potential of a genetic map it is necessary to integrate conventional markers into molecular maps. A few classical markers have been integrated into the soybean molecular genetic map, mainly because those markers fortuitously segregated in the original mapping populations, and were thus simultaneously mapped with the molecular markers (Keim *et al.*, 1990; Diers *et al.*, 1992a; Lark *et al.*, 1993; Shoemaker and Olson, 1993). In some cases, a molecular marker analysis of an F_2 population that had been specifically created to segregate for a classical marker of interest resulted in the discovery of linkage between that classical marker and a previously mapped molecular marker (Landau-Ellis *et al.*, 1991; Diers and Shoemaker,

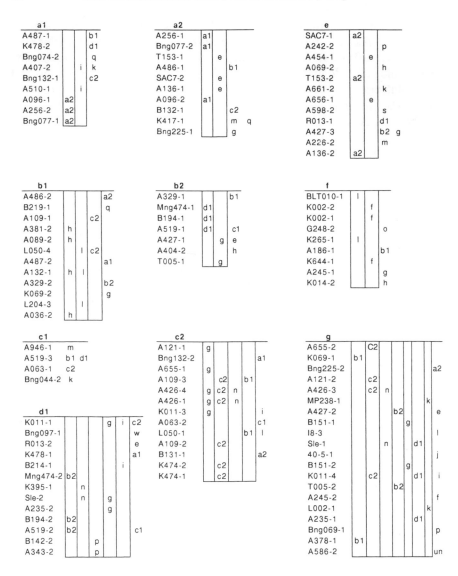

1992; Weisemann *et al.*, 1992). In other cases, an analysis of near-isogenic lines followed by analyses of a segregating population resulted in the association of a molecular marker with genes (Muehlbauer *et al.* 1991; Diers *et al.* 1992c). By inference, this established that the classical marker was a member of the specific molecular linkage group. Unfortunately, the number of molecular markers segregating in these special F_2 populations was usually too few to provide the kind of multi-point linkage data that was needed to establish the position of the classical marker relative to all of the molecular markers of the given group.

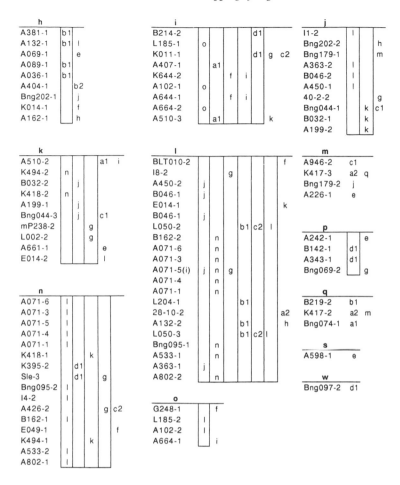

Fig. 2.2. (and opposite) Soybean genetic markers known to be duplicated on other linkage groups. The letter designation at the top of each table indicates the linkage group where the markers reside. The markers are listed in the order in which they are found on the linkage group. To the right of each marker is shown the linkage group(s) where it is known to be duplicated.

Since 1992, several specific populations have been generated for the purpose of conducting multi-point linkage analysis and a full integration of classical markers into the molecular genetic map (Nickell *et al.*, 1994; Polzin *et al.*, 1994; Shoemaker and Specht, 1995). The development of the software program 'JoinMap' (Stam, 1993) has allowed for the co-analysis of independent data sets from these and the original mapping population to yield a best-fit 'joined' map.

The original public molecular genetic map was constructed from an F_2 population derived from a cross between *G. max* breeding line A81-356022 and *G. soja* Plant Introduction 468916 provided by W. Fehr, Iowa State University (Shoemaker and Olson, 1993). Subsequently another genetic map was developed from a cross between 'Clark' and 'Harosoy' near-isogenic lines specially constructed (J. Specht, University of Nebraska) so as to differ among 20 morphological, isoenzyme and pigmentation markers. The integration of this map and the original public molecular genetic map, both developed using a LOD score of 3 as a significance level, resulted in a map containing approximately 415 markers and the correlation of about half of the 19 classical linkage groups to molecular linkage groups. (Shoemaker and Specht, 1994). This map is shown in Fig. 2.3. However, when referring to maps it is important to remember that a 'joined' map is only a 'best guess' of linear relationships between markers and of distances between markers. Most genetic maps should be used only as guidelines to positioning, not as absolutes.

QTL Mapping

Saturated genetic maps allow us to evaluate genetic regulation and interactions of genes or chromosomal regions that affect complex agronomic traits. A broad range of studies have been conducted to identify quantitative trait loci (QTL) affecting agronomically important traits in soybean. To date, most of these have been conducted in populations developed for map construction and only fortuitously segregating for the traits to be genetically dissected into QTLs. As a consequence, even though the maps contain many markers, with the exception of Concibido *et al.* (1994) these results may represent only preliminary examples of QTL mapping for these traits (Table 2.1).

The probability of an RFLP detected by a cDNA probe being a direct measure of a QTL effect for any given agronomically important trait has been estimated to be approximately 1 in 200 (Beckmann and Soller, 1985). Thus, any QTLs detected by random genomic clones, if they are indeed real QTLs, probably represent indirect measures of the QTL effect, which in turn are affected by the degree of linkage disequilibrium between the QTL and the markers, and the environment × genotype interaction (Gillespie and Turelli, 1989). Therefore, the direct application of molecular markers in constructing desirable QTL allelic combinations, across a wide range of genotypes and environments, is probably quite limited. However, mapping QTLs and construction of NILs (near-isogenic lines) containing alternative forms of the QTL (i.e. positive vs negative effects) does provide researchers with a means to understand the nature of the QTL, i.e. the biological function of a specific QTL in the absence of confounding genotypic effects.

The mechanisms by which populations maintain variation in quantitative traits is still a mystery (Gillespie and Turelli, 1989). Through detailed analyses of QTL using multiple genotypic combinations and multiple environments, we may eventually begin to understand these mechanisms.

Table 2.1. Quantitative trait loci mapping studies in soybean.

Trait	Reference
Hard-seededness	Keim *et al.* (1990)
Seed protein (%)	Diers *et al.* (1992a)
	Mansur *et al.* (1993)
	Lark *et al.* (in press)
Seed oil (%)	Diers *et al.* (1992a)
	Mansur *et al.* (1993)
	Lark *et al.* (in press)
Fatty acid (%)	Diers and Shoemaker (1992)
Seed weight	Mansur *et al.* (1993)
Reproductive traits	Keim *et al.* (1990)
	Mansur *et al.* (1993)
Morphological traits	Keim *et al.* (1990)
	Mansur *et al.* (1993)
Nutrient efficiency (Fe)	Diers *et al.* (1992b)
Cyst nematode resistance	Concibido *et al.* (1994)

By understanding variation in quantitative traits we may be better able to direct breeding programmes to more efficiently achieve breeding objectives.

Map-based Genotype Analyses

The development of detailed molecular genetic maps for soybean provides for a wide range of map-based applications (Shoemaker *et al.*, 1992; Shoemaker and Olson, 1993). The application of markers comprising an RFLP map to genotypes used in breeding programmes, is essential in establishing the value of a map. Because mapping populations are often selected because they demonstrate a high degree of genotypic polymorphism, it is likely that many markers comprising an RFLP map will have limited usefulness in day to day breeding programmes. It is therefore important to identify those markers which can constitute a core set, useful in evaluating soybean breeding lines.

The genetic foundation of modern soybean cultivars is somewhat limited. Only 12 ancestors account for more than 88% of the germplasm of the northern gene pool (Delannay *et al.*, 1983; Specht and Williams, 1984). This limited number of contributors to current soybean germplasms and relatively short history of cultivar development (Fehr, 1987) make it possible to conduct detailed molecular genetic analyses of soybean pedigrees.

Shoemaker *et al.* (1992) discussed the application of molecular genetic maps for a retrospective analysis of the soybean genome among breeding lines in an attempt to correlate genomic regions to breeder manipulations of agronomic traits. Four cultivars were evaluated using markers from only two

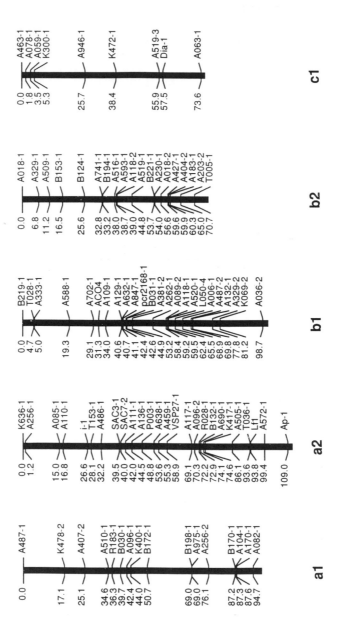

Fig. 2.3. The molecular genetic map of soybean.

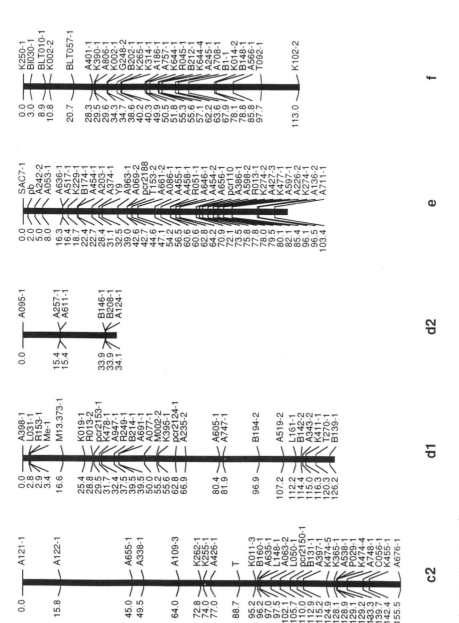

Fig. 2.3. *(continued).*

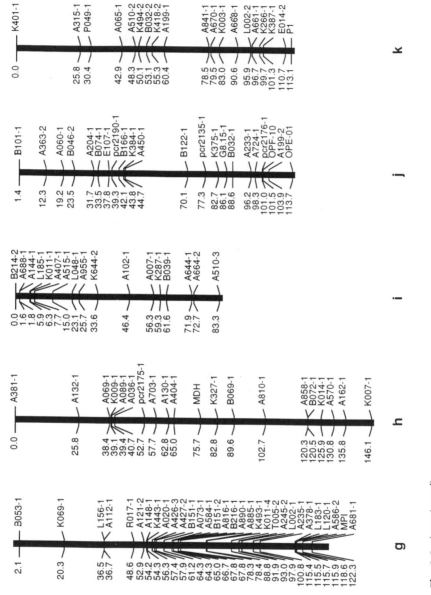

Fig. 2.3. (continued).

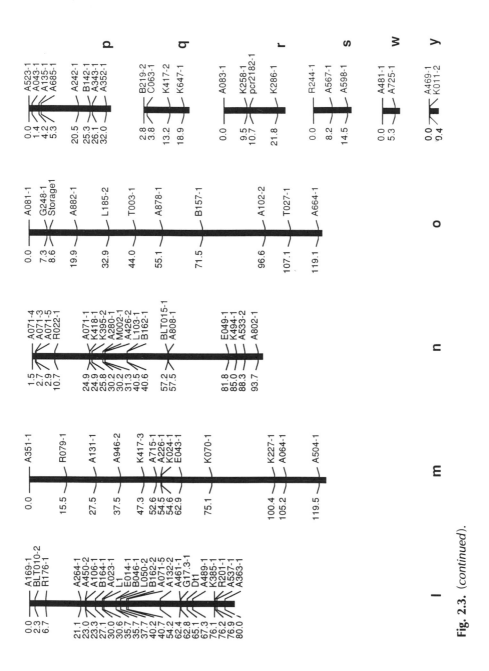

Fig. 2.3. (*continued*).

linkage groups to demonstrate that molecular pedigree assessment had potential for identifying desirable linkage blocks. These analyses were only preliminary and provided no definitive picture of genome manipulation during cultivar development.

Lorenzen *et al.* (1995) conducted a detailed analysis of the genotypes of 64 cultivars released between 1939 and the early 1990s. These represented 'milestone' cultivars, based on commercial success and acres of production, and other breeding lines involved in their pedigrees or derived from them. Of the 216 loci evaluated among these cultivars, only 44% showed any polymorphism using the probe/enzyme combination used in construction of the *max* × *soja* genetic map. Therefore, only those probes detecting polymorphisms are likely to be useful in evaluating most soybean breeding lines.

A detailed genotype analysis coupled to a molecular genetic map and pedigree information provides an interesting picture of soybean cultivar development. For example, Fig. 2.4 shows the pedigree of Hark, with 18 abbreviated linkage groups. Loci names and positions have been left off this figure for space consideration. Twenty-nine of Hark's tested loci can be traced to its immediate parents, Hawkeye (K) and Harosoy (H). Hawkeye contains 25 loci which can be traced to AK Harrow (A) and Mandarin Ottawa (O). Characters to the left of the linkage groups in Fig. 2.4 indicate the parental source of each informative locus. It is important to note that all linkage groups in this pedigree were not equally informative. Sometimes, as in linkage group 'H' of Hawkeye, the parents had all of the alleles at these markers in common, so no information could be obtained about the origin of this linkage group. Additionally, many times an entire linkage group had only one informative marker, as in linkage group 'A' of Hark. The origin of the entire linkage group should not be based on this one marker, but limited to the region that is informative. Alternately, some linkage groups had several informative markers. Linkage group 'F' of Hark had six informative markers, allowing a more precise conclusion of the origin of this linkage group.

In Hark, Hawkeye contributed region(s) to the top of five linkage groups, the bottom of three linkage groups, and the middle of two linkage groups. Additionally, Hawkeye contributed the one informative locus available on linkage group 'X'. Harosoy contributed region(s) on the top of three linkage groups, the bottom of five linkage groups, and the middle of four linkage groups. At least one locus is informative on all linkage groups with the exceptions of 'O', 'S', and 'W', which had one uninformative marker located on each.

Hawkeye, the female parent of Hark, is the result of a cross between Mukden and Richland. Mukden contributed to the top of one linkage group, the bottom of five linkage groups, and the centre of four linkage groups, and the centre of three linkage groups. No information is known about the origin of six linkage groups.

Finally, Harosoy is the result of a single backcross of Mandarin Ottawa × AK Harrow. Mandarin Ottawa contributed to the top of three linkage

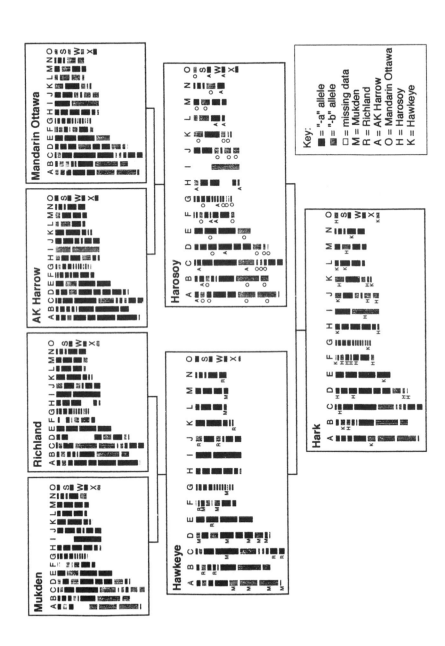

Fig. 2.4. Map-based pedigree of the soybean cultivar Hark. Letters to the left of linkage groups indicate the parental source of the chromosomal segment: AK Harrow (A), Harosoy (H), Hawkeye (K), Mukden (M), Mandarin Ottawa (O), and Richland (R).

groups, the bottom of seven linkage groups, and the middle of seven linkage groups. AK Harrow contributed to the top of four linkage groups, the bottom of three linkage groups, and the centre of four linkage groups. Additionally, AK Harrow contributed to linkage groups 'S' and 'W'.

Additionally, this type of analysis has the potential to provide a wide range of information on soybean cultivar development. It can (i) determine linkage blocks passed on from generation to generation; (ii) provide an estimate of genetic relatedness; (iii) provide an estimate of the amount of genetic recombination taking place during breeding programmes; and (iv) provide an opportunity to conduct pedigree-based mapping, or the detection of cosegregation of genes with markers.

The data collected represents approximately 14,000 data points. This data is publicly available through the soybean genome database project and can be accessed by anonymous file transfer protocol (ftp) from mendel.agron.iastate.edu (IP address 129.186.20.43).

The Soybean Genome Database

The volume of genetic data being accumulated is burgeoning at a staggering rate. It has become impossible for a laboratory or an individual researcher to keep abreast of the latest research findings. Only through the development of electronic storage, querying capabilities, and retrieval can the scientific community take advantage of the massive amounts of information being generated.

Through the USDA-ARS Plant Genome Office a soybean genome database has been developed. The database currently contains available information on germplasm accessions, i.e. information from the Germplasm Resource Information Network and the Plant Variety Protection Office. It also contains information on pathology, genes, genetic linkage studies, probes, RFLP patterns, genetic maps from several sources (classical and molecular), QTLs, and metabolic pathways of nodulation, nitrogen uptake and utilization, and lipid metabolism. The database is complete with images of RFLP patterns and can be accessed through the National Agricultural Library. The models used to structure the database are being used to develop other legume databases.

The Future of Soybean Molecular Genetic Mapping

In spite of the inherent difficulties of working with a complex genome such as the soybean's, the soybean mapping effort has maintained a competetive edge over many equally important crop plants. This has been due mainly to the financial support of the USDA Agricultural Research Service and various state and national commodity groups and the aggressive efforts of public and private researchers throughout the country.

Continuing efforts toward the utilization of the genetic map are still hindered by relatively serious considerations. The foremost considerations are the lack of large amounts of genetic polymorphism between most breeding lines and the confusion caused by probes detecting polymorphisms among duplicated loci. The first problem can be overcome through the application of technologies designed to detect polymorphisms with a much higher degree of sensitivity. These types of technologies have been demonstrated through the use of microsatellites or single sequence repeats (Akkaya *et al.*, 1992; Rongwen *et al.*, 1995), the coupling of endonuclease cleavage and amplification of arbitrary DNA sequences (Caetano-Anolles *et al.*, 1993), and the amplification of random DNA stretches followed by denaturing gradient gel electrophoresis (He *et al.*, 1992). All of these techniques or approaches can greatly increase the chances of detecting polymorphisms, even among closely related genotypes. The problem of confusion caused by probes detecting polymorphisms among duplicate loci can be obviated by the development and accessibility of a database that identifies and characterizes alleles present in germplasm, the patterns to be expected in that germplasm, and the association of those patterns or polymorphisms to specifically identified loci.

There remains no doubt that the soybean molecular genetic map will provide the means by which breeding programmes can be revolutionized and the development of improved beans can be accelerated. However, the true power of the genetic map will not be realized by the molecular biologists working at the bench. It will only be realized as the professional breeders adapt these technologies to their own programmes in ways that will solve the problems that only they can identify.

References

Akkaya, M.S., Bhagwat, A.A., and Cregan, P.B. (1992) Length polymorphisms of simple sequence repeat DNA in soybean. *Genetics* 132, 1131–1139.

Beckmann, J. and Soller, M. (1985) Restriction fragment length polymorphisms in plant genetic improvement. *Oxford Surveys of Plant Molecular and Cellular Biology* 3, 196–250.

Bennett, M.D., Smith, J.B. and Heslop-Harrison, J.S. (1982) Nuclear DNA amounts in angiosperms. *Proceedings of the Royal Society* B216, 179–199.

Burton, J.W. (1985) Breeding soybeans for improved protein quantity and quality. In: Shibles, R. (ed.) *Proceeddings of the 3rd World Soybean Research Conference.* Westview Press, Boulder, Colorado, pp. 361–367.

Buttery, B.R. and Buzzell, R.I. (1976) Flavonol glycoside genes and photosynthesis in soybeans. *Crop Science* 16, 547–550.

Caetano-Anolles, G., Bassam, B.J. and Gresshoff, P.M. (1993) Enhanced detection of polymorphic DNA by multiple arbitrary amplicon profiling of endonuclease-digested DNA: identification of markers tightly linked to the supernodulation locus in soybean. *Molecular and General Genetics* 241, 57–64.

Childs, G., Maxson, R., Cohn, R.H. and Kedes, L. (1981) Orphons: dispersed genetic lements derived from tandem repetitive genes of eucaryotes. *Cell* 23, 651–663.

Concibido, V.C., Denny R.L., Boutin, S.R., Hautea, R., Orf, J. H., and Young, N.D.(1994) DNA marker analysis of loci underlying resistance to soybean cyst nematode (*Heterodera glycines* Ichinohe). *Crop Science* 34, 240–246.

Delannay, X., Rodgers, D.M. and Palmer, R.G. (1983) Relative genetic contributions among ancestral lines to North American soybean cultivars. *Crop Science* 23, 944–949.

Diers, B.W. and Shoemaker, R.C. (1992) Restriction fragment length polymorphism analysis of soybean fatty acid content. *Journal of the American Oil Chemists Society* 69, 1242–1244.

Diers, B.W., Keim, P., Fehr, W.R. and Shoemaker, R.C. (1992a) RFLP analysis of soybean seed protein and oil content. *Theoretical and Applied Genetics* 83, 608–612.

Diers, B.W., Cianzio, S.R. and Shoemaker, R.C. (1992b) Possible identification of quantitative trait loci affecting iron efficiency in soybean. *Journal of Plant Nutrition* 15, 2127–2136.

Diers, B.W., Mansur, L., Imsande, J. and Shoemaker, R.C. (1992c) Mapping of *Phytophthora* resistance loci in soybean with restriction fragment length polymorphism markers. *Crop Science* 32, 377–383.

Fehr, W.R. (1987) Breeding methods for cultivar development. In: Wilcox, J.R. (ed.) *Soybeans: Improvement, Production, and Uses.* 2nd edn. No. 16. American Society of Agronomy, Inc., Crop Science Society of America, Inc., and Soil Science Society of America, Inc., Madison, Wisconsin, pp. 249–293.

Gillespie, J.H. and Turelli, M. (1989) Genotype–environment interactions and the maintenance of polygenic variation. *Genetics* 121, 129–138.

Goldberg, R.B. (1978) DNA sequence organization in the soybean plant. *Biochemical Genetics* 16, 45–68.

Gurley, W.B., Hepburn, A.G. and Key, J.L. (1979) Sequence organization of the soybean genome. *Biochemica et Biophysica Acta* 561, 167–183.

He, S., Ohm, H. and Mackenzie, S. (1992) Detection of DNA sequence polymorphisms among wheat varieties. *Theoretical and Applied Genetics* 84, 573–578.

Hymowitz, T. (1970) On the domestication of the soybean. *Economic Botany* 24, 408–421.

Hymowitz, T. and Singh, R.J. (1987) Taxonomy and speciation. In: Wilcox, J.R. (ed.), *Soybeans: Improvement, Production, and Uses.* 2nd edn. No. 16. American Society of Agronomy, Inc., Crop Science Society of America, Inc., and Soil Science Society of America, Inc., Madison, Wisconsin, pp. 23–48.

Keim, P., Diers, B.W., Olson, T.C. and Shoemaker, R.C. (1990) RFLP mapping in soybean: Association between marker loci and variation in quantitative traits. *Genetics* 126, 735–742.

Lackey, J.A. (1980) Chromosome numbers in the Phaseoleae (Fabaceae:Faboideae) and their relation to taxonomy. *American Journal of Botany* 67, 595–602.

Landau-Ellis, D., Angermuller, S., Shoemaker, R.C. and Gresshoff, P.M. (1991) The genetic locus controlling supernodulation in soybean (*Glycine max* L.) co-segregates tightly with a cloned molecular marker. *Molecular and General Genetics* 228, 221–226.

Lark, K.G., Weisemann, J.M., Matthews, B.F., Palmer, R., Chase, K. and Macalma, T. (1993) A genetic map of soybean (*Glycine max* L.) using an intraspecific cross of two cultivars: 'Minsoy' and 'Noir 1'. *Theoretical and Applied Genetics* 86, 901–906.

Lark, K.G., Orf, J. and Mansur, L.M. (1994). Epistatic expression of quantitative trait loci (QTL) in soybean [*Glycine max* (L) Merr.] Determined by QTL association with RFLP alleles. *Theoretical and Applied Genetics* 88, 486–489.

Lee, J.S. and Verma, D.S. (1984) Structure and chromosomal arrangement of leghemoglobin genes in kidney bean suggest divergence in soybean leghemoglobin gene loci following tetraploidization. *EMBO Journal* 12, 2745–2752.

Leipold, M. and Schmidtke, J. (1982) Gene expression in phylogenetically polyploid organisms. In: Dover, G. and Flavell, R. (eds) *Genome Evolution*. Academic Press, New York, pp. 219–236.

Lorenzen, L.L., Boutin, S., Young, N. and Specht, J.E. (1995) Soybean pedigree analysis using map-based molecular markers. *Crop Science* (in press).

Lundin, L.G. (1993) Evolution of the vertebrate genome as reflected in paralogous chromosomal regions in man and the house mouse. *Genomics* 16, 1–19.

Mansur, L., Lark, K.G., Kross, H. and Oliveira, A. (1993) Interval mapping of quantitative trait loci for reproductive, morphological and seed traits of soybean (*Glycine max* L.). *Theoretical and Applied Genetics* 86, 907–913

Muehlbauer, G.J., Staswick, P.E., Specht, J.E., Graef, G.L., Shoemaker, R.C. and Keim, P. (1991) RFLP mapping using near-isogenic lines in the soybean [*Glycine max* (L.) Merr.]. *Theoretical and Applied Genetics* 81, 189–198.

Nickell, A.D., Wilcox, J.R., Lorenzen, L.L., Cavins, J.F., Guffy, R.D. and Shoemaker, R.C. (1994) The Fap2 locus in soybean maps to linkage group D. *Journal of Heredity* 85, 160–162.

Ohno, S. (1970) *Evolution by Gene Duplication*. Springer Verlag, New York.

Palmer, R.G. and Kilen, T.C. (1987) Qualitative genetics and cytogenetics. In: Wilcox, J.R. (ed.), *Soybeans: Improvement, Production, and Uses*. 2nd edn. No. 16. American Society of Agronomy, Inc., Crop Science Society of America, Inc., and Soil Science Society of America, Inc., Madison, Wisconsin, pp. 135–209.

Palmer, R.G. and Hedges, B.R. (1993) Linkage map of soybean (*Glycine max* L. Merr.). In: O'Brien, S.J. (ed.) *Genetic Maps: Locus Maps of Complex Genomes*. Cold Spring Harbor Laboratory Press, New York, pp. 6.139–6.148.

Polzin, K.M., Lohnes, D.G., Nickell, C.D. and Shoemaker, R.C. (1994) Rps2, Rmd, and Rj2 are integrated into linkage group J of the soybean molecular map. *Journal of Heredity* 85, 300–303.

Rafalski, A. and Tingey, S. (1993) RFLP Map of soybean (*Glycine max*) 2N = 40. In: O'Brien, S.J. (ed.) *Genetic Maps: Locus Maps of Complex Genomes*. Cold Spring Harbor Laboratory Press, New York, pp. 6.149–6.156.

Rongwen, J., Akkaya, M.S., Bhagwat, A.A., Lavi, U. and Cregan, P.B. (1995) The use of microsatellite DNA markers for soybean genotype identification. *Theoretical and Applied Genetics* 90, 43–48.

Shoemaker, R.C. and Olson, T.C. (1993) Molecular linkage map of soybean (*Glycine max* L. Merr.). In: O'Brien, S.J. (ed.) *Genetic Maps: Locus Maps of Complex Genomes*. Cold Spring Harbor Laboratory Press, New York, pp. 6.131–6.138.

Shoemaker, R.C. and Specht, J.E. (1995) Integration of the soybean molecular and classical genetic linkage groups. *Crop Science* 35, 436–446.

Shoemaker, R.C., Guffy, R.D., Lorenzen, L.L. and Specht, J.E. (1992) Molecular genetic mapping of soybean: map utilization. *Crop Science* 32, 1091–1098.

Singh, R.J. and Hymowitz, T. (1988) The genomic relationship between *Glycine max* (L.) Merr. and *G. soja* Sieb. and Zucc. as revealed by pachytene chromosomal analysis. *Theoretical and Applied Genetics* 76, 705–711.

Skorupska, H., Albertsen, M.C., Langholz, K.D. and Palmer, R.G. (1989) Detection of ribosomal RNA genes in soybean, *Glycine max* (L.) Merr., by in situ hybridization. *Genome* 32, 1091–1095.

Smith, K.J. and Huyser, W. (1987) World distribution and significance of soybean. In: Wilcox, J.R. (ed.) *Soybeans: Improvement, Production, and Uses*, 2nd edn. No. 16. American Society of Agronomy, Inc., Crop Science Society of America, Inc., and Soil Science Society of America, Inc., Madison, Wisconsin, pp. 1–22.

Specht, J.E. and Williams, J.H. (1984) Contribution of genetic technology to soybean productivity: retrospect and prospect. In: Fehr, W.F. (ed.) *Genetic Contributions to Yield Gains of Five Major Crop Plants*. American Society of Agronomy, Madison, Wisconsin, pp. 49–74.

Stam, P. (1993) Construction of integrated genetic linkage maps by means of a new computer package: Join Map. *Plant Journal* 3(5), 739–744.

Stebin, G.L. (1966) Chromosomal variation and evolution. *Science* 152, 1463–1469.

Weisemann, J.M., Matthews, B.F. and Devine, T.E. (1992) Molecular markers located proximal to the soybean cyst nematode resistance gene, Rhg4. *Theoretical and Applied Genetics* 85, 136–138.

Zobel, R.W. (1983) Genic duplication: A significant constraint to molecular and cellular genetic manipulation in plants. *Comments on Molecular and Cellular Biophysics* 1, 355–364.

Cytoplasmic Genetics in the Legumes (Fabaceae), with Special Reference to Soybean

3

S. Mackenzie

Department of Agronomy, Purdue University, West Lafayette,
Indiana 47907, USA.

Introduction

The study of cytoplasmic genetics has provided some important clues about plant evolution, nuclear–cytoplasmic genetic cooperation and numerous essential functions supplied by the mitochondrial and chloroplast genomes in higher plants. The implications of cytoplasmic genetic manipulation to crop improvement may be more substantial than was once thought. The availability of cytoplasmic male sterile mutants in most crop species greatly facilitates hybrid seed production. With the depletion of genetic variation in many major crops, the opportunity afforded by intercrossing to enhance variability in self-pollinated species is significant. Furthering our understanding of nuclear–cytoplasmic incompatibilities, and their physiological repercussions in wide interspecific crossing efforts, becomes essential as breeding strategies change.

The legumes, comprising a plant family of tremendous genetic diversity, have not been as well investigated in the area of cytoplasmic genetics as some other plant families. This is unfortunate because members of this family have served to demonstrate some important aspects of cytoplasmic genetics and nuclear–cytoplasmic interaction not observed in others. In this chapter, I will focus on some of the more important aspects of plant evolution, species diversity, nuclear–cytoplasmic interaction and organellar inheritance that have been observed in this unusual family of plants, and will relate these to soybean where appropriate.

The Chloroplast Genome

The chloroplast genome of higher land plants is a circular DNA molecule 120–180 kb in size containing two inverted repeats of 10–76 kb. This pair of inverted repeats, containing the ribosomal RNA genes, is represented in the more than 33 angiosperm families examined (Palmer *et al.*, 1987) as well as gymnosperms and ferns. Only in the legume family has this inverted repeat structure been altered. The legumes can be divided into the tribes that have retained this inverted repeat, Phaseoleae (mung bean, common bean, soybean) and Genisteae (lupin), and those that have undergone deletion of a part or all of one repeat, Vicieae (broad bean, pea), Trifolieae (alfalfa, subclover), and Tephrosieae (wisteria) (Palmer *et al.*, 1987). The deletion of a repeat is, in at least some cases, a relatively precise event. Whereas the entire repeat may be lost, the flanking single copy sequences are retained intact (Wolfe, 1988). Why this unusual genomic alteration has occurred only once in the evolution of land plants, and its effect on the chloroplast evolution of the legumes, is not known.

Chloroplast genome maps have been established for a number of legume species (for example Chu *et al.*, 1981; Mubumbila *et al.*, 1983; Shinozaki *et al.*, 1984; Singh *et al.*, 1985; Michalowski *et al.*, 1987). Surprisingly, the chloroplast sequence divergence within the legume family is higher than in most other angiosperm families tested (Palmer *et al.*, 1983). This sequence diversity has allowed the evaluation of species relationships within the genus *Glycine* (Shoemaker *et al.*, 1986; Close *et al.*, 1989; Doyle *et al.*, 1990) and the fate of soybean cytoplasms during outcrossing (Lee *et al.*, 1993, 1994). In most plant species, the chloroplast genome is inherited maternally. The legumes comprise one of few angiosperm families with members that differ in cytoplasmic inheritance pattern. The majority of legume species evaluated demonstrate strict maternal inheritance of the chloroplast and mitochondrial genomes. In *Medicago sativa*, however, biparental chloroplast inheritance is clearly evident (Smith *et al.*, 1986; Schumann and Hancock, 1989; Masoud *et al.*, 1990). This combining of paternal and maternal chloroplast contributions produces an unusual condition of heteroplasmy that can apparently be maintained indefinitely (Lee *et al.*, 1988; Johnson and Palmer, 1989). It has been suggested that a selective advantage may exist for some chloroplast types over others in a heteroplasmic condition (Fitter and Rose, 1993). Also, the degree of biparental chloroplast transmission can range from predominantly paternal to predominantly maternal depending, in part, on parents used in the cross (Schumann and Hancock, 1989; Masoud *et al.*, 1990). Mitochondrial inheritance in *Medicago* remains strictly maternal, implying independent mechanisms for regulation of chloroplast and mitochondrial transmission in a cytoplasm.

Chloroplast genome replication is an important area of current research in higher plant systems. In soybean, a DNA helicase and DNA polymerase have been purified from chloroplast extracts (Cannon and Heinhorst, 1990; Heinhorst *et al.*, 1990). Two forms of the chloroplast DNA polymerase exist in soybean, demonstrating characteristics of γ DNA polymerases (Heinhorst *et al.*, 1990). Analysis of chloroplast DNA replication in soybean, using two-dimensional gel electrophoresis and *in vitro* deoxynucleotide incorporation studies, have allowed the tentative identification of an origin of replication and have provided support for a model of replication initiation from double D-loop structures separated in the genome by distances of 4–5 kb (Hedrick *et al.*, 1993). Evidence for a model of chloroplast DNA replication initiation at double D-loop structures has also been provided by electron microscopy and deoxynucleotide incorporation studies in pea (Kolodner and Tewari, 1975), where a putative origin of replication has been identified (Meeker *et al.*, 1988), as well as *Chlamydomonas* (Waddell *et al.*, 1984; Wu *et al.*, 1986), *Oenothera* (Chiu and Sears, 1992), and maize (Gold *et al.*, 1987). One of the important features of soybean (*Glycine max*) that lends this particular species to the biochemical and molecular analysis of plastid DNA replication is the capacity of this species to grow photoautotrophically in suspension culture (Horn *et al.*, 1983).

The coding capacity of the chloroplast, determined by sequencing the entire chloroplast genome in tobacco (Shinozaki *et al.*, 1986), does not appear to vary significantly in different higher plant species. Over 95% of the gene products required for plastid function are nuclear-encoded and imported to the organelle. Consequently, crossing strategies can significantly influence photosynthetic properties of a line. An understanding of the functions required for plastid maintenance, maturation and nuclear–chloroplast interaction should permit a more directed approach to the enhancement of photosynthetic capacity.

The Mitochondrial Genome

The mitochondrial genomes of species in the legume family are not well characterized with respect to genomic organization, size diversity, or recombinational activity. A physical map of the mitochondrial genome of common bean (*Phaseolus vulgaris* L.) has been developed (Janska and Mackenzie, 1993) for the study of cytoplasmic male sterility and is described in the following section. Studies of mitochondrial genetics in legume species suggest that this family shares many features in common with the better characterized maize, *Brassica*, *Oenothera*, and petunia genomes.

A number of important mitochondrial genes have been sequenced in legumes, with all demonstrating remarkable conservation with corresponding genes in other dicots and monocot species (for example Spielmann and Stutz, 1983; Grabau, 1985, 1987; Grabau *et al.*, 1988; Herdenberger *et al.*, 1988; Grabau and Gengenbach, 1989; Macfarlane *et al.*, 1990). Gene expression studies indicate multiple transcript start sites for some of these genes (Brown *et al.*, 1991), occurrence of both *cis*- and *trans*-splicing of intron sequences (for example Wintz *et al.*, 1989; Pereira de Souza *et al.*, 1992), a general consensus transcription initiation sequence (Brown *et al.*, 1991), conservation of sequences at some translation initiation sites (Pring *et al.*, 1992) and evidence of intragenic recombination events (Morgens *et al.*, 1984; Grabau *et al.*, 1988; Chanut *et al.*, 1993). All of these mitochondrial features are common among many plant families. What has proven particularly useful among legumes is the information derived from the mitochondrial diversity present within and among species.

Mitochondrial genome diversity studies have been conducted in a few of the most common legume species, providing information about the process of plant domestication as well as nuclear–cytoplasmic evolution in higher plants. In common bean, two centres of origin have been identified, characterized by the Mesoamerican and Andean gene pools (Gepts, 1988). Although highly conserved in mitochondrial genome configuration within a gene pool, diversity exists between the two pools (Khairallah *et al.*, 1990). The pattern of diversity, when compared with wild undomesticated selections from Central and South America, suggests that *Phaseolus vulgaris* L. has undergone two distinct domestication events. The mitochondrial rearrangement/mutation events distinguishing the two gene pools accompanied, or occurred soon after, the domestication events (Khairallah *et al.*, 1992). Similar investigations of mitochondrial diversity have been conducted in soybean to reveal four maternal lineages among the soybean cultivars whose ancestors were predominant in the original North American soybean gene pool (Grabau *et al.*, 1989, 1992).

When broader comparisons of mitochondrial coding capacity are made within the legume family, an interesting pattern is observed. Evidence obtained from mitochondrial studies provides the basis for current hypotheses addressing the evolution of nuclear-mitochondrial interdependence. Whereas in nearly all other plant species investigated to date, subunit 2 of the cytochrome oxidase complex (*coxII*) is encoded within the mitochondrion, some legume species have apparently transferred this function to the nucleus. More importantly, within the legume family can be found a number of presumptive intermediates in this cellular gene transfer process. Consequently, common bean, soybean and pea contain a mitochondrial copy of *coxII* as well as a nuclear copy of the gene (Nugent and Palmer, 1991; Covello and Gray, 1992). In pea, the mitochondrial copy is functional. Mung bean and cowpea have apparently evolved one step further in this process, containing only the nuclear form of *coxII*, with no mitochondrial form of the gene present. DNA sequence analysis of the nuclear form of *coxII* in these species suggests that this gene transfer event has likely occurred via an RNA intermediate (Nugent and Palmer, 1991; Covello and Gray, 1992). The process by which expressional inactivation of the mitochondrial form of *coxII* occurs in those species containing both nuclear and mitochondrial forms is not clear. This question is of particular interest in soybean where certain chimeric mitochondrial gene forms exist, probably derived by past intragenic recombination events (Grabau *et al.*, 1988). One of these chimeric forms consists of the 5' end of *coxII* in association with *atp6* (ATPase subunit 6). This chimeric sequence appears to be transcriptionally active in soybean (Grabau *et al.*, 1988).

In those legume species that have been characterized, the mitochondrial genome is approximately 400–450 kb in size (Levings and Pring, 1979; Bailey-Serres *et al.*, 1987; Janska and Mackenzie, 1993). Based on the presence of recombinationally active repeated sequences in soybean (Chanut *et al.*, 1993) as well as mapping data in common bean (Janska and Mackenzie, 1993), it appears likely that at least some legumes contain a mitochondrial genome that undergoes recombinational events to give rise to subgenomic DNA molecules. One study, in soybean, suggests that these subgenomic DNA molecules are circular (Synenki *et al.*, 1978). Some legume species also contain plasmid DNAs within the mitochondrion. In broad bean (*Vicia faba*), three plasmid DNAs, 1704, 1695 and 1476 bp in length, have been observed within fertile and male-sterile plant lines (Wahleithner and Wolstenholme, 1987, 1988). These plasmids are circular and contain a number of inverted repeated sequences, some apparently important to their replication. Although the function of these three unusual mitochondrial inclusions is not known, evidence exists of transcriptional activity as well as sequence homology to the nuclear genome (Flamand *et al.*, 1992).

Cytoplasmic Male Sterility

Cytoplasmic male sterility (cms) is a maternally inherited phenotype that results, in most cases, from a mitochondrial mutation or abnormality. Mitochondrial dysfunction is evident as a plant's inability to produce and/or shed viable pollen. Two examples of cms have been fairly well described in legumes, both demonstrating unusual features distinct from all other known cms systems to date.

In common bean, the cms phenotype is associated with the presence of a unique 2.4-kb sequence, designated *pvs* (*Phaseolus vulgaris* sterility sequence), within the mitochondrial genome (Mackenzie and Chase, 1990; Johns *et al.*, 1992). This sequence apparently originated from an insertion event immediately 3' to an intact *atpA* gene (alpha subunit of mitochondrial ATPase) (Chase and Ortega, 1992; Johns *et al.*, 1992). The *pvs* sequence encodes at least two cotranscribed open reading frames, *orf98* and *orf239*, predicted to encode peptides of 98 and 239 amino acids, respectively. The *pvs* sequence is transcriptionally active, and ELISA, protein gel blot, and immunocytochemistry analyses using polyclonal antibodies developed against ORF239 indicate that this gene product is expressed only in developing reproductive tissues of the male-sterile bean line (Abad *et al.*, 1995).

The sterility-associated sequence, *pvs*, is present within the mitochondrial genome of a number of lines of *P. vulgaris*, and can be observed in lines of *P. coccineus* and *P. polyanthus* as well (Hervieu *et al.*, 1993). This unusual observation implies that the cms-inducing mutation occurred during the evolution of the genus *Phaseolus* prior to the divergence of these three species

(Hervieu *et al.*, 1994). The maintenance of this unusual insertion in such a large number of lines is attributable to a couple of possible factors. Nuclear genes are present in *Phaseolus* spp. that override the effect of *pvs*; at least four nuclear fertility restorer genes are currently available and under investigation (Mackenzie and Bassett, 1987; Mackenzie, 1991; Mackenzie, unpublished). The presence of these nuclear genes in nature may have allowed the *pvs* mutation to be retained with a negligible impact on plant productivity. It has also been observed that some wild *Phaseolus* selections may, in nature, be cross-pollinating (D. Debouck, Cali, Colombia, personal communication). Consequently, it is possible that the presence of *pvs* in some *Phaseolus* lines facilitates this mode of reproduction. The extent of distribution of the *pvs* sequence in these wild *Phaseolus* lines is not yet known.

A particularly unusual feature of the cms system in common bean is the mechanism by which pollen fertility is restored. Several nuclear genes have been identified that effect fertility restoration to cms common bean; at least one of these genes, *Fr*, is distinct in its mode of action. The introduction of nuclear gene *Fr* to the cms line by crossing results in permanent restoration of full fertility (Mackenzie and Bassett, 1987). This restoration of wildtype phenotype is accompanied by the complete loss of *pvs* from the mitochondrial genome (Mackenzie and Chase, 1990). Loss of the mitochondrial sequence, effected in this case by restorer gene *Fr*, can also occur at very low frequency spontaneously. These spontaneous reversions to fertility appear to be identical in mitochondrial DNA rearrangement to the *Fr*-directed process. Comparative physical mapping of the mitochondrial genome of the cms line and a derived fertile revertant indicates that the mitochondrial genome of cms common bean probably exists as three inter-recombining autonomous mitochondrial molecules (Janska and Mackenzie, 1993). Only one of these autonomous molecules contains *pvs*. The introduction of *Fr* appears to direct elimination of this *pvs*-containing molecule, leaving the remaining genome intact (Janska and Mackenzie, 1993). A novel set of subsequent genetic experiments has allowed the development of a hypothesis for *Fr* action. According to this model, the mitochondrial population of the cms line exists as a mixture of pvs^+ and pvs^- forms. Consequently, the *Fr*-directed mitochondrial alteration occurs as the result of selection against a subpopulation of *pvs*-containing mitochondria during gamete formation (He *et al.*, 1994).

A second cms system has been observed in broad bean (*Vicia faba*). As in common bean, the cms broad bean line can be restored by a single dominant nuclear gene, and restoration of pollen fertility is stable and permanent. Also similar to common bean, the cms broad bean line undergoes spontaneous reversion to fertility. In broad bean, however, cms is associated with the presence of two large double-stranded RNA molecules within the cytoplasm (Grill and Garger, 1981; Scalla *et al.*, 1981). Virus-like RNA molecules are not associated with cms in common bean (Mackenzie *et al.*, 1988).

The '447' cytoplasm of *V. faba* with its unusual features of instability and permanent restoration, was first described by Bond (1966a,b). The virus-like particles observed in the cytoplasm of this cms line are visible as membranous structures and contain high molecular weight linear double-stranded RNAs of approximately 16.7 kb in size. These virus-like inclusions were associated with sterility by their disappearance from the cytoplasm of fertile revertants and their loss upon restoration of fertility by a nuclear fertility restorer gene. These large dsRNAs are not detectable in the mitochondria of the male sterile plants, nor do these RNAs demonstrate homology, by gel blot hybridization, to the mitochondrial or chloroplast genomes of the male sterile plant (Turpen *et al.*, 1988). Some hybridization is detected, however, when cDNA clones derived from the dsRNAs are used to probe blotted nuclear DNA from the male sterile and fertile lines (Turpen *et al.*, 1988). The membrane-bound RNAs are enclosed with an RNA-dependent RNA polymerase (Lefebvre *et al.*, 1990). However, their elimination upon introduction of a nuclear restorer gene suggests that their maintenance is not completely independent of nuclear control. This virus-like RNA is also apparently transcriptionally active, giving rise to at least one discreet transcript of 4.5 kb by an unusual process of strand-displacement (Pfieffer *et al.*, 1993). If these virus-like inclusions in *V. faba* cells are the result of virus infection, the particular class of plant virus has not yet been determined.

Cytoplasmic Mutants in Soybean

Although not yet well characterized, several additional mutants demonstrating maternal inheritance in legumes have been collected. The majority of these have been identified in soybean, owing largely to the concerted efforts of a small group of investigators. Most available mutants appear to influence chloroplast development or function, selected based on a yellow leaf phenotype (for example Palmer and Mascia, 1980; Shoemaker *et al.*, 1985; Cianzio and Palmer, 1992). The sites of mutation have not been determined in most of these. Some of the cytoplasmic mutants do, however, demonstrate genetic interaction with the nuclear genotype. In certain cases, this interaction is observed as lethality (Palmer and Cianzio, 1985; Palmer, 1992), suggesting that these cytoplasmically inherited mutations occur within genes important to nuclear-chloroplast coordination. This collection of genetic materials should be extremely useful for in-depth investigation of nuclear-chloroplast interactions.

Concluding Comments

Despite the enormous diversity within the legume family and the agricultural importance of many of its members, cytoplasmic genetic characterization of this family is sparse. This is surprising because soybean has long been a model system for photosynthesis and respiration studies, and legumes are most important for their ability to conduct nitrogen fixation. Nitrogen fixation requires an oxygen-free environment; consequently, mitochondrial gene regulation and function in nitrogen fixing tissues is of obvious intrigue. The genetic diversity available in this plant family has provided some of the more interesting evidence to date for the continuing evolutionary processes associated with endosymbiosis. It is clear that many members of the Fabaceae family will become increasingly important systems as some of these questions begin to be addressed.

References

Abad, A., Mehrtens, B. and Mackenzie, S. (1995) Specific expression in reproductive tissues and fate of a mitochondrial sterility-associated protein in cytoplasmic male-sterile bean. *Plant Cell* 7, 271–285.

Bailey-Serres, J., Leroy, P., Jones, S.S., Wahleithner, J.A. and Wolstenholme, D.R. (1987) Size distributions of circular molecules in plant mitochondrial DNAs. *Current Genetics* 12, 49–53.

Bond, D.A. Fyfe, J.L. and Toynbee-Clarke, G. (1966a) Male sterility in field beans (*Vicia faba* L.) III. Male sterility with a cytoplasmic type of inheritance. *Journal of Agricultural Science* 66, 359–367.

Bond, D.A., Fyfe, J. L. and Toynbee-Clarke, G. (1966b) Male sterility in field beans (*Vicia faba* L.) IV. Use of cytoplasmic male sterility in the production of F1 hybrids, and their performance in trials. *Journal of Agricultural Science* 66, 369–377.

Brown, G.G., Auchincloss, A.H., Covello, P.S., Gray, M.W., Menassa, R. and Singh, M. (1991) Characterization of transcription initiation sites on the soybean mitochondrial genome allows identification of a transcription-associated sequence motif. *Molecular and General Genetics* 228, 345–355.

Cannon, G.C. and Heinhorst, S. (1990) Partial purification and characterization of a DNA helicase from chloroplasts of *Glycine max*. *Plant Molecular Biology* 15, 457–464.

Chanut, F.A., Grabau, E.A. and Gesteland, R.F. (1993) Complex organization of the soybean mitochondrial genome: Recombination repeats and multiple transcripts at the atpA loci. *Current Genetics* 23, 234–247.

Chase, C.D. and Ortega, V.M. (1992) Organization of ATPA coding and 3' flanking sequences associated with cytoplasmic male sterility in *Phaseolus vulgaris* L. *Current Genetics* 22, 147–153.

Chiu, W.-L. and Sears, B.B. (1992) Electron microscopic localization of replication origins in *Oenothera* chloroplast DNA. *Molecular and General Genetics* 232, 33–39.

Chu, N.M., Oishi, K.K. and Tewari, K.K. (1981) Physical mapping of the pea chloroplast DNA and localization of the ribosomal RNA genes. *Plasmid* 6, 279–292.

Cianzio, S.R. and Palmer, R.G. (1992) Genetics of five cytoplasmically inherited yellow foliar mutants in soybean. *Journal of Heredity* 83, 70–73.

Close, P.S., Shoemaker, R.C. and Keim, P. (1989) Distribution of restriction site polymorphism within the chloroplast genome of the genus *Glycine*, subgenus *Soja*. *Theoretical and Applied Genetics* 77, 768–776.

Covello, P.S. and Gray, M.W. (1992) Silent mitochondrial and active nuclear genes for subunit 2 of cytochrome c oxidase (cox2) in soybean: evidence for RNA-mediated gene transfer. *EMBO Journal* 11, 3815–3820.

Doyle, J.J., Doyle, J.L. and Brown, A.H.D. (1990) A chloroplast-DNA phylogeny of the wild perennial relatives of soybean (*Glycine* subgenus *Glycine*): Congruence with morphological and crossing groups. *Evolution* 44, 371–389.

Fitter, J.T. and Rose, R.J. (1993) Investigation of chloroplast DNA heteroplasmy in *Medicago sativa* L. using cultured tissue. *Theoretical and Applied Genetics* 86, 65–70.

Flamand, M.C., Goblet, J.P., Duc, G., Briquet, M. and Boutry, M. (1992) Sequence and transcription analysis of mitochondrial plasmids isolated from cytoplasmic male sterile lines of *Vicia faba*. *Plant Molecular Biology* 19, 913–923.

Gepts, P. (1988) A middle American and an Andean common bean gene pool. In: Gepts, P. (ed.) *Genetic Resources of Phaseolus Beans*. Kluwer Academic Publ., Dordrecht, pp. 375–390.

Gold, B., Carrillo, N., Tewari, K. and Bogorad, L. (1987) Nucleotide sequence of a preferred maize chloroplast genome template for *in vitro* DNA synthesis. *Proceedings of the National Academy of Sciences USA* 84, 194–198.

Grabau, E.A. (1985) Nucleotide sequence of the soybean mitochondrial 18S rRNA gene: evidence for a slow rate of divergence in the plant mitochondrial genome. *Plant Molecular Biology* 5, 119–124.

Grabau, E.A. (1987) Cytochrome oxidase subunit II gene is adjacent to an initiator methionine tRNA gene in soybean mitochondrial DNA. *Current Genetics* 11, 287–293.

Grabau, E. and Gengenbach, B.G. (1989) Cytochrome oxidase subunit III gene from soybean mitochondria. *Plant Molecular Biology* 13, 595–597.

Grabau, E., Havlik, M. and Gesteland, R. (1988) Chimeric organization of two genes for the soybean mitochondrial ATPase subunit 6. *Current Genetics* 13, 83–89.

Grabau, E.A., Davis, W.H. and Gengenbach, B.G. (1989) Restriction fragment length polymorphism in a subclass of the 'Mandarin' soybean cytoplasm. *Crop Science* 29, 1554–1559.

Grabau, E.A., Davis, W.H., Phelps, N.D. and Gengenbach, B.G. (1992) Classification of soybean cultivars based on mitochondrial DNA restriction fragment length polymorphisms. *Crop Science* 32, 271–274.

Grill, L.K. and Garger, S.J. (1981) Identification and characterization of double-stranded RNA associated with cytoplasmic male sterility in *Vicia faba*. *Proceedings of the National Academy of Sciences USA* 78, 7043–7046.

He, S., Lyznik, A. and Mackenzie, S. (1994) Pollen fertility restoration by nuclear gene Fr in cms bean: Nuclear-directed alteration of a mitochondrial population. *Genetics* 139, 955–962.

Hedrick, L.A., Heinhorst, S., White, M.A. and Cannon, G.C. (1993) Analysis of soybean chloroplast DNA replication by two-dimensional gel electrophoresis. *Plant Molecular Biology* 23, 779–792.

Heinhorst, S., Cannon, G.C. and Wessbach, A. (1990) Chloroplast and mitochondrial DNA polymerases from cultured soybean cells. *Plant Physiology* 92, 939–945.

Herdenberger, F., Weil, J. H. and Steinmetz, A. (1988) Organization and nucleotide sequence of the broad bean chloroplast genes trnL-UAG, ndhF and two unidentified open reading frames. Current Genetics 14, 609–615.

Hervieu, F., Charbonnier, L., Bannerot, H. and Pelletier, G. (1993) The cytoplasmic male-sterility (cms) determinant of common bean is widespread in Phaseolus coccineus L. and Phaseolus vulgaris L. Current Genetics 24, 149–155.

Hervieu, F., Bannerot, H. and Pelletier, G. (1994) A unique cytoplasmic male sterility (CMS) determinant is present in three Phaseolus species characterized by different mitochondrial genomes. Theoretical and Applied Genetics 88, 314–320.

Horn, M.E., Sherrard, J.J. and Widholm, J.M. (1983) Photoautotrophic growth of soybean cells in suspension culture. Plant Physiology 72, 426–429.

Janska, H. and Mackenzie, S. (1993) Unusual mitochondrial genome organization in cytoplasmic male sterile common bean and the nature of cytoplasmic reversion to fertility. Genetics 135, 869–879.

Johns, C., Lu, M., Lyznik, A. and Mackenzie, S. (1992) A mitochondrial DNA sequence is associated with abnormal pollen development in cytoplasmic male sterile bean plants. Plant Cell 4, 435–449.

Johnson, L.B. and Palmer, J.D. (1989) Heteroplasmy of chloroplast DNA in Medicago. Plant Molecular Biology 12, 3–11.

Khairallah, M.M., Adams, M.W. and Sears, B.B. (1990) Mitochondrial DNA polymorphisms of Malawian bean lines: Further evidence for two major gene pools. Theoretical and Applied Genetics 80, 753–761.

Khairallah, M.M., Sears, B.B. and Adams, M.W. (1992) Mitochondrial restriction fragment length polymorphisms in wild Phaseolus vulgaris L.: insights on the domestication of the common bean. Theoretical and Applied Genetics 84, 915–922.

Kolodner, R. and Tewari, K. (1975) Presence of displacement loops in the covalently closed circular chloroplast deoxyribonucleic acid from higher plants. Journal of Biological Chemistry 250, 8840–8847.

Lee, D.J., Blake, T.K. and Smith, S.E. (1988) Biparental inheritance of chloroplast DNA and the existence of heteroplasmic cells in alfalfa. Theoretical and Applied Genetics 76, 545–549.

Lee, D.J., Caha, C.A., Specht, J.E. and Graef, G.L. (1993) Chloroplast DNA evidence for non-random selection of females in an outcrossed population of soybeans (Glycine max L.) Theoretical and Applied Genetics 85, 261–268.

Lee, D.J., Caha, C.A., Specht, J.E. and Graef, G.L. (1994) Analysis of cytoplasmic diversity in an outcrossing population of soybean (Glycine max, (L) Merr.). Crop Science 34, 46–50.

Lefebvre, A., Scalla, R. and Pfeiffer, P. (1990) The double-stranded RNA associated with the 447 cytoplasmic male sterility in Vicia faba is packaged together with its replicase in cytoplasmic membranous vesicles. Plant Molecular Biology 14, 477–490.

Levings, C.S., III, and Pring, D.R. (1979) Mitochondrial DNA of higher plants and genetic engineering. In: Setlow, J.K. and Hollaender, A. (eds) Genetic Engineering, Vol. 1. Plenum Press, New York, pp. 205–222.

Macfarlane, J.L., Wahleithner, J. A. and Wolstenholme, D. R. (1990) A gene for cytochrome c oxidase subunit III (coxIII) in broad bean mitochondrial DNA: structural features and sequence evolution. Current Genetics 17, 33–40.

Mackenzie, S. (1991) Identification of a sterility-inducing cytoplasm in a fertile accession line of Phaseolus vulgaris L. Genetics 127, 411–416.

Mackenzie, S. and Bassett, M. (1987) Genetics of fertility restoration in cytoplasmic male sterile *Phaseolus vulgaris* L. *Theoretical and Applied Genetics* 74, 642–645.

Mackenzie, S. and Chase, C.D. (1990) Fertility restoration is associated with loss of a portion of the mitochondrial genome in cytoplasmic male sterile common bean. *Plant Cell* 2, 905–912.

Mackenzie, S., Pring, D.R. and Bassett, M. (1988) Large double-stranded RNA molecules in *Phaseolus vulgaris* L. are not associated with cytoplasmic male sterility. *Theoretical and Applied Genetics* 76, 59–63.

Masoud, S.A., Johnson, L.B. and Sorensen, E.L. (1990) High transmission of paternal plastid DNA in alfalfa plants demonstrated by restriction fragment polymorphic analysis. *Theoretical and Applied Genetics* 79, 49–55.

Meeker, R., Nielsen, B. and Tewari, K. (1988) Localization of replication origins in pea chloroplast DNA. *Molecular and Cellular Biology* 8, 1216–1223.

Michalowski, C., Breunig, K.D. and Bohnert, H. J. (1987) Points of rearrangements between plastid chromosomes: location of protein coding regions on broad bean chloroplast DNA. *Current Genetics* 11, 265–274.

Morgens, P.H., Grabau, E.A. and Gesteland, R.F. (1984) A novel soybean mitochondrial transcript resulting from a DNA rearrangement involving the 5S rRNA gene. *Nucleic Acids Research* 12, 5665–5684.

Mubumbila, M., Gordon, K.H.J., Crouse, E.J., Burkard, G. and Weil, J.H. (1983) Construction of the physical map of the chloroplast DNA *Phaseolus vulgaris* and localization of ribosomal and transfer RNA genes. *Gene* 21, 257–266.

Nugent, J.M. and Palmer, J.D. (1991) RNA-mediated transfer of the gene coxII from the mitochondrion to the nucleus during flowering plant evolution. *Cell* 66, 473–481.

Palmer, J.D., Osorio, B., Aldrich, J. and Thompson, W. F. (1987) Chloroplast DNA evolution among legumes: loss of a large inverted repeat occurred prior to other sequence rearrangements. *Current Genetics* 11, 275–286.

Palmer, J.D., Singh, G.P. and Pillay, D.T.N. (1983) Structure and sequence evolution of three legume chloroplast DNAs. *Molecular and General Genetics* 190, 13–19.

Palmer, R.G. (1992) Conditional lethality involving a cytoplasmic mutant and chlorophyll-deficient malate dehydrogenase mutants in sobyean. *Theoretical and Applied Genetics* 85, 389–393.

Palmer, R.G. and Cianzio, S. (1985) Conditional lethality involving nuclear and cytoplasmic chlorophyll mutants in soybeans. *Theoretical and Applied Genetics* 70, 349–354.

Palmer, R.G. and Mascia, P.N. (1980) Genetics and ultrastructure of a cytoplasmically inherited yellow mutant in soybean. *Genetics* 95, 985–1000.

Pereira de Souza, A., Jubier, M.-F. and Lejeune, B. (1992) The higher plant nad5 mitochondrial gene: a conserved discontinuous transcription pattern. *Current Genetics* 22, 75–82.

Pfeiffer, P., Jung, J.L., Heitzler, J. and Keith, G. (1993) Unusual structure of the double-stranded RNA associated with the 447 cytoplasmic male sterility in *Vicia faba*. *Journal of General Virology* 74, 1167–1173.

Pring, D.R., Mullen, J.A. and Kempken, F. (1992) Conserved sequence blocks 5' to start codons of plant mitochondrial genes. *Plant Molecular Biology* 19, 313–317.

Scalla, R., Duc, G., Rigaud, J., Lefebvre, A. and Meignoz, R. (1981) RNA containing intracellular particles in cytoplasmic male sterile faba bean (*Vicia faba* L). *Plant Science Letters* 22, 269–277.

Schumann, C.M. and Hancock, J.F. (1989) Paternal inheritance of plastids in *Medicago sativa*. *Theoretical and Applied Genetics* 78, 863–866.

Shinozaki, K., Ohme, M.,Tanaka, M.,Wakasugi,T., Hayashida, N., Matsubayashi,T., Zaita, N., Chunwongse, J., Obokata, J.,Yamaguchi-Shinozaki, K., Ohto, C.,Torazawa, K., Meng, B.Y., Sugita, M., Deno, H., Kamogashira,T.,Yamada, K., Kusuda, J.,Takaiwa, F., Kato, A.,Tohdoh, N., Shimada, H. and Sugiura, M. (1986) The complete nucleotide sequence of the tobacco chloroplast genome: Its gene organization and expression. *EMBO Journal* 5, 2043–2049.

Shinozaki, L., Sun, C.R. and Sugiura, M. (1984) Gene organization of chloroplast DNA from the broad bean *Vicia faba*. *Molecular and General Genetics* 197, 363–367.

Shoemaker, R.C., Cody, A.M. and Palmer, R.G. (1985) Characterization of a cytoplasmically inherited yellow foliar mutant (cyt-Y3) in soybean. *Theoretical and Applied Genetics* 69, 279–283.

Shoemaker, R.C., Hatfield, P.M., Palmer, R.G. and Atherly, A.G. (1986) Chloroplast DNA variation in the genus *Glycine* subgenus *Soja*. *Journal of Heredity* 77, 26–30.

Singh, G.P.,Wallen, D.G. and Pillay, D.T.N. (1985) Positioning of protein-coding genes on the soybean chloroplast genome. *Plant Molecular Biology* 4, 87–93.

Smith, S.E., Bingham, E.T. and Fulton, R.W. (1986) Transmission of chlorophyll deficiencies in *Medicago sativa*. *Journal of Heredity* 77, 35–38.

Spielmann, A. and Stutz, E. (1983) Nucleotide sequence of soybean chloroplast DNA regions which contain the *psb* A and *trn* H genes and cover the ends of the large single copy region and one end of the inverted repeats. *Nucleic Acids Research* 11, 7157–7167.

Synenki, R.M., Levings, C.S., III and Shah, D. M. (1978) Physicochemical characterization of mitochondrial DNA from soybean. *Plant Physiology* 61, 460–464.

Turpen, T., Garger, S.J. and Grill, L.K. (1988) On the mechanism of cytoplasmic male sterility in the 447 line of *Vicia faba*. *Plant Molecular Biology* 10, 489–497.

Waddell, J.,Wang, X.-M. and Wu, M. (1984) Electron microscopic localization of the chloroplast DNA replicative origins in *Chlamydomonas reinhardtii*. *Nucleic Acids Research* 12, 3843–3856.

Wahleithner, J.A. and Wolstenholme, D.R. (1987) Mitochondrial plasmid DNAs of broad bean: nucleotide sequences, complex secondary structures and transcription. *Current Genetics* 12, 55–67.

Wahleithner, J.A. and Wolstenholme, D.R. (1988) Origin and direction of replication in mitochondrial plasmid DNAs of broad bean, *Vicia faba*. *Current Genetics* 14, 163–170.

Wintz, H., Chen, H.C. and Pillay, D.T.N. (1989) Partial characterization of a gene coding for subunit IV of soybean mitochondrial NADH dehydrogenase. *Current Genetics* 15, 155–160.

Wolfe, K.H. (1988) The site of deletion of the inverted repeat in pea chloroplast DNA contains duplicated gene fragments. *Current Genetics* 13, 97–99.

Wu, M., Lou, J., Chang, D., Chang, C. and Nie, Z. (1986) Structure and function of a chloroplast DNA replication origin of *Chlamydomonas reinhardtii*. *Proceedings of the National Academy of Sciences USA* 83, 6761–6765.

Plant Transposable Elements: Potential Applications for Gene Tagging in Soybean

L.O.Vodkin

Plant and Animal Biotechnology Laboratory, Department of Agronomy, University of Illinois, 1201 West Gregory, Urbana, Illinois 61801, USA.

Introduction

Transposable elements were first discovered in plant systems almost fifty years ago, and their physical isolation from plants was reported in the early 1980s. Understanding the effects of transposable elements on plant gene expression and variation has been an active area of research since that time (Peterson, 1987; Vodkin, 1989; Gierl and Saedler, 1992; Walbot, 1992). The isolation of transposable elements, elucidation of their effects on gene expression, and their use in gene tagging were first achieved by studying mutations of the anthocyanin pathway in several plant species.

Tagging genes with transposable elements is one way to isolate genes whose function is unknown. The finding that transposable elements can function in heterologous species also opens the process of gene tagging to species in which endogenous transposable elements have not yet been found. First, I will review the discovery of transposable elements and the major classes of elements in plant systems. Next, the recent advances in defining the transposition process and the function of elements in diverse species will be reviewed. Only a few systems are currently amenable for studying the function and use of transposable elements for gene tagging. Although soybean is a very important agronomic species, it is not generally an amenable genetic system for these purposes so advances in gene tagging in soybean will probably rely heavily on whether these approaches are found to be feasible in other systems. A discussion of soybean elements and some unusual variegation mutations in soybean and how they compare to other plant systems will be included.

Discovery of Genetic Units that Transpose

The first demonstration that genetic elements that could change their location in the genome was reported in the late 1940s by Barbara McClintock (reviewed in McClintock, 1984). She was mapping genes in maize by studying the effects of deletions caused by an inverted duplication on chromosome 9. The inverted duplication produces a chromosome break at a random position thus leading to fusion of the broken chromosomes after replication, an anaphase bridge, and subsequent breakage at a new position. As expected, McClintock found that the breakage-fusion-bridge (BFB) cycle led to the random loss of a number of kernel marker genes on chromosome 9 and she was able to map the relative positions of these genes. Unexpectedly, in one experiment she found that the site of chromosome breakage consistently mapped near the *waxy* locus toward the centromere leading to the loss of the *waxy* marker and all kernel markers distal to it.

McClintock coined the name Dissociation, or Ds, for the genetic factor that created non-random chromosome breakage near the *waxy* locus. Additionally, the genetic ratios of chromosome breakage demonstrated that a second genetic element was required for chromosome breakage by Ds. She named this genetic unit Ac (Activator) as it was necessary for activating the action of Ds. Transposition was first detected when McClintock found that the site of chromosome breakage caused by Ds had moved from proximal to *waxy* to a position near the *C* locus toward the end of chromosome 9.

The notion that genetic elements could move from one position in the chromosome to another was antithetical to the dogma that genes occupied fixed positions on the chromosomes. Thus, the studies of maize transposition represent a major discovery in genetics. The genetic discovery of transposable elements preceded their description at the molecular level by about 25 years. In the late 1960s, one class of polar mutations in bacteria was shown to be due to insertion sequences that are capable of their own transposition in the bacterial chromosome. The Ac and Ds elements described by McClintock were physically isolated in 1983 (Fedoroff *et al.*, 1983).

Although plant transposable elements were discovered by their ability to break chromosomes, most of the transposable elements in plants do not normally cause chromosome breakage. The presence of transposable elements is generally associated with genes that are mutated as a result of insertion of an element and display somatic variegation resulting in two phenotypes in a tissue. For example, purple and yellow sectors on a single kernel of corn or variable red and white sectors in flowers are often due to the action of a transposable element. The somatic mutability is caused by an element inserting into a gene and affecting the expression of the gene resulting in a null phenotype. The sectors of full colour arise from a clonal lineage of cells in which the element has exised from the gene and the normal function of the gene is restored. Excision of the element in premeiotic germ cells can lead to restoration of gene function and inheritance of the revertant phenotype in the next generation.

Major Classes and Characteristics of Plant Transposable Elements

The molecular analysis of plant transposable elements has been an active area of research since 1983 when the Ac and Ds elements were isolated from mutable alleles of the *waxy* locus (Fedoroff *et al.*, 1983; Shure *et al.*, 1983). The Ac and Ds elements represent the two-element controlling systems in which an autonomous element, like Ac, can mediate it own transposition and can also activate the transposition of nonautonomous elements like Ds. Ac is a 4.3 kilobase (kb) element that encodes a transposase, the enzyme that mediates excision and reintegration of the element from the DNA. The 5' and 3' termini of the Ac element are 11 base pair (bp) inverted repeat sequences and a duplication of 8 bp of the interrupted gene is found adjacent to the inverted repeats. The major feature that distinguishes Ds elements from Ac elements is that the former have internal deletions that remove or inactivate the transposase. The smallest Ds elements consist of little more than the 11 bp inverted repeats.

Although Ds elements cannot move on their own power, they can move if the transposase is supplied in *trans* from an Ac element in the genome. The transposase from an Ac element recognizes its own 11 bp inverted repeats (or those of the Ds elements) and creates a staggered nick at the ends of the 8 bp target duplication. The element is excised by the action of the transposase and can reinsert at another chromosomal site in which the transposase has generated an 8 bp staggered nick in the target DNA. There appears to be little specificity for the site of insertion in the target DNA. The original Ds elements discovered by McClintock, because of their ability to cause chromosome breakage, have a complicated structure. They are generally double Ds elements consisting of a Ds inserted into another Ds element in an inverted orientation (reviewed in Doring and Starlinger, 1984). The presence of four copies of the inverted repeat ends, some in direct orientation, is believed to lead to chromosome breakage because of aberrations in the transposition process created by the juxtaposition of direct rather than inverted repeat ends of the Ds elements. The transposase breaks sister chromatids in the process of cleaving as the direct repeats try to pair in inverted orientation. Direct molecular evidence that chromosome breakage is the result of aberrant transposition events was recently obtained using polymerase chain reaction to isolate intermediates in the process caused by chromosome-breaking Ds alleles of the *waxy* locus (Weil and Wessler, 1993). The chromosome-breaking properties of the Ds ends have also been re-created in tobacco (English *et al.*, 1993). Only a left and right end of a Ds element in direct orientation is sufficient for breakage. An inversion of one of the Ds ends occurs during the aberrant transposition and is detected as an intermediate by polymerase chain reaction in tobacco only when an active Ac element is present, indicating that formation of the intermediate is dependent on the transposase.

The En/Spm (Enhancer or Suppressor-mutator) element is another well-characterized, two-element family in maize. The autonomous element of this system is 8.3 kb long and contains 13 bp of inverted repeats followed by 3 bp of duplication of the target DNA (Pereira *et al.*, 1986). Nonautonomous, defective Spm (dSpm) elements are internal deletion derivatives. Thus, each transposable element family is defined by the unique sequence of its inverted repeats and the size of the target duplication. The transposase from Ac does not recognize the inverted repeats of the En/Spm system and vice versa. This explains the genetic observation that Ac cannot activate transposition of the defective Spm elements since the Ac transposase will not recognize the 13 bp inverted repeats. Likewise, the size of the the target duplication is characteristic for each element system. The Ac/Ds and the En/Spm families represent the best characterized transposons of the two element controlling systems. Nine different two-element systems have been described genetically in maize (Peterson, 1987).

The Mutator system in maize was identified genetically because it increases the frequency of mutation rates approximately 30-fold (Robertson, 1978). The first Mu1 sequence was isolated as an insertion in the *Adh1* gene (Bennetzen *et al.*, 1984). Nine different classes of mutator sequences have been identified that contain similar 220 bp terminal inverted repeats but have different internal sequences. None of these is responsible for the genetic activity of Mutator that causes somatic instability of Mu elements that have inserted into marker genes with easily observable phenotypes like the *bronze* locus. Mutator activity is normally inherited in 90% of the progeny of a testcross in which a Mutator line is crossed to a non-Mutator line (Lillis and Freeling, 1986). Thus, there appear to be multiple copies of the active element. Recently, unusual Mutator stocks have been found in which the Mutator activity segregated 1:1 in the progeny of a testcross. This indicated the presence of a single, active element which has been cloned from this line (Chomet *et al.*, 1991) and also from a *bronze* allele (Hershberger *et al.*, 1991). The MuR (or Mu9) element is 4.9 kb and cosegregates with the genetic Mutator activity.

The controlling element families like Ac and Spm replicate by a nonreplicative, loss/gain mechanism in which the element is excised from its location and moves to a new position. In contrast, the Mutator elements transpose by a replicative mechanism in which a copy of the element is made and moves to a new location.

Another major class of transposable elements, the retrotransposons, has characteristics similar to retroviral elements. Generally the retrotransposons contain a region of sequence similarity to reverse transcriptases and probably move via a RNA intermediate. They are divided into two subclasses, those that contain long terminal repeats (LTR) and those that do not. Both classes are flanked by small duplications of the target DNA. The non-LTR retrotransposons are often truncated at the 5' ends and contain a variable number of A residues at the 3' ends. Both of these features are consistent with movement via a RNA intermediate. Examples of retrotransposons are the Bs and Cin elements in maize (Johns *et al.*, 1985; Schwarz-Sommer *et al.*, 1987; Jin and Bennetzen, 1989).

Maize remains the premier genetic system for the study of plant transposable elements but transposable elements have been isolated from several other organisms. The Tam1 element in snapdragon appears to be an analogue of the maize En/Spm element with similar 13 bp inverted repeat termini (Nacken *et al.*, 1991) while Tam3 in snapdragon is an analogue of the Ac element (Hehl *et al.*, 1991). Several other small elements including Tpc1 of parsley (Herrmann *et al.*, 1988), Ips-r of pea (Bhattacharyya *et al.*, 1990), Tst1 of potato (Kikuchi *et al.*, 1991), dTph1 of petunia (Gerats *et al.*, 1990), and Tat1 of *Arabidopsis* (Peleman *et al.*, 1991) are defective elements flanked by 8 bp target duplications and have similarities in their terminal inverted repeats to the Ac/Ds family. The Tnt1 retrotransposon in tobacco was found during selection for nitrate reductase negative cell lines (Grandbastein *et al.*, 1989) and three new families of tobacco retrotransposons were found to be activated during tissue culture (Hirochika, 1993).

Transposable Element Effects on Gene Expression and Variation

Transposable elements can insert into the promoter area of the target gene, the coding region, or into introns. The site of insertion of the element has different effects on expression of the target gene. If the element resides in the promoter area, it may affect the quantitative expression of the gene or the tissue specificity by interfering with transcriptional factors. In the coding region, the transcripts may terminate prematurely. In some cases, the element termini mimic intron splicing signals leading to element excision from the gene by intron processing (Kim *et al.*, 1987; Wessler, 1988).

The excision of plant transposable elements is generally imprecise (Schwarz-Sommer *et al.*, 1985a). Often a portion of the target site duplication is left after the excision event. Thus, a 'footprint' remains in the gene and the function of the gene may be altered even after excision of the element. The footprints and other alterations like small deletions caused by excision are another source of genetic variation created by transposable element action. Thus, there is little doubt that the movement of transposable elements has provided raw material for evolution as have other mutagenic agents. An interesting example is the wrinkled trait character in *Pisum sativum*. Gregor Mendel examined this characteristic as one of the seven traits he used to determine the principles of genetic inheritance and segregation. The molecular basis for the wrinkled pea trait is the presence of a small insertion element with structural similarities to the Ac element in maize. The insertion interrupts the coding region of a starch branching enzyme expressed during seed development (Bhattacharyya *et al.*, 1990). The small insertion is stable leading to inheritance of the wrinkled phenotype as a simple recessive trait.

Autonomous transposable elements often self-destruct during their movement by the formation of internal deletions. Whereas deletions result in a permanent loss of autonomy, methylation of element sequences can result in a temporary silencing of active elements. There is a correlation of the methylation state of Mutator elements with the genetic activity of Mutator stocks (Chandler and Walbot, 1986). The changes in phase between active and inactive states of the Spm element have been correlated with changes in methylation of a GC-rich area in an upstream control region of the En/Spm element (Fedoroff, 1989). The activity of the Ac element is also correlated with methylation of element sequences (Chomet *et al.*, 1987). Likewise, cryptic elements which are heavily methylated can be activated if the methylation pattern is disturbed.

Transposable Element Action in Heterologous Species

Transposition in heterologous species was demonstrated first for the Ac element transformed into tobacco using *Agrobacterium tumefaciens* vectors (Baker *et al.*, 1986). The element also continues to excise in the first generation progeny (Taylor *et al.*, 1989). The primary Ac transcript appears to be processed to the same size message in tobacco as in maize (Finnegan *et al.*, 1988).

In order to select for excision events, the Ac element was placed between the promoter and the coding region of an antibiotic resistance gene such as neomycin phosphotransferase (NPT) or hygromycin phosphotranferase (HPT). The presence of the Ac element prevents expression of the antibiotic resistance gene. Upon excision of the element, expression of antibiotic resistance is restored (Baker *et al.*, 1987). Thus, antibiotic resistance reflects excision of the Ac element and its possible transposition to other locations in the genome. The frequency of excision in *Nicotiana plumbaginifolia* ranged from 34% to 100% in the progeny of first generation transformed plants and most plants exhibited from one to six Ac elements that had independently transformed to new locations (Marion-Poll *et al.*, 1993). Ac continued to transpose in the third and fourth generation progeny.

The Ac element has been shown to be active in rice calli derived from protoplasts transformed by electroporation with the Ac element (Jing-liu *et al.*, 1991) and in regenerated rice plants (Izawa *et al.*, 1991). Similarly, Ac functions in flax (Roberts *et al.*, 1990), carrot cell lines (Van Sluys and Tempe, 1989), and tomato plants (Yoder, 1990; Rommens *et al.*, 1991). In tomato, the Ac element amplified to more than 15 copies in only two generations (Yoder *et al.*, 1990).

It is important to study the mechanisms of transposition in order to determine the feasibility of developing efficient transposition systems for other plants. The frequency of Ac transposition was examined in tobacco using a streptomycin phosphotransferase gene (Jones *et al.*, 1989). Normal tobacco plants grown in the presence of streptomycin are white whereas transgenic plants carrying the streptomycin resistance gene are green. Variegated white and green sectors result from transposition of the Ac element out of the streptomycin phosphotransferase gene. This assay showed that increasing the dosage of Ac elements in the transgenic tobacco increased the somatic excision frequency. This finding contrasted with the natural situation in maize in which the somatic transposition frequency is decreased with increasing dosage of Ac elements. However, other studies show that very high levels of Ac transposase expression begin to inhibit Ds excision in tobacco cotyledons. These data indicate the mechanism may be similar to maize; however, higher threshold levels of Ac may be required in tobacco before inhibition of Ds transposition is observed (Scofield *et al.*, 1993).

In maize, germinal excision frequencies range from 1% to 10% and more than half of the germinal revertants contain a transposed Ac element, most of them to linked sites in the chromosomes. Studies of Ac action in tomato showed that germinal transposition frequencies were similar or slightly lower and that about 50% of the germinal revertants in tomato also carried a transposed Ac (Jones *et al.*, 1990). The transposed Ac elements also mapped to sites genetically linked to the streptomycin excision products. Some transposed Ac elements in tobacco also are tightly linked to the empty donor site while others are more dispersed (Dooner *et al.*, 1991). These results suggest that transposon tagging experiments are more likely to succeed if the Ac element resides near to the desired target gene.

The Ac element is capable of transactivating Ds elements that have been transformed into heterologous species. The complex double Ds elements that cause chromosome breakage in maize have also been introduced into tomato plants already transformed with the Ac element (Rommens *et al.*, 1991). As in maize, excision of the double Ds causes rearrangements and chromosome breakage in tomato and in tobacco rather than simple excision events (Rommens *et al.*, 1991; English *et al.*, 1993).

Other types of visible or selectable markers have been developed to assay for transposition. Petunia protoplasts transfected with a Ds element that interrupts the β-glucuronidase (GUS) gene will express GUS if the element excises in the presence of an Ac element (Houba-Herin *et al.*, 1990). Ac excision from the GUS gene was used to measure late excision events in tobacco leaves (Rommens *et al.*, 1991). The mouse dihydrofolate reductase gene (DHFR) that encodes methotrexate resistance was placed under control of the cauliflower mosaic virus (35S) promoter and located within a Ds element. The Ds-DHFR construct was transferred to tobacco calli that already contained an autonomous Ac element. The frequency of movement of the Ds-DHFR element was monitored by methotrexate resistance (Masterson *et al.*, 1989).

The En/Spm element also is active in heterologous species including tobacco plants (Masson and Fedoroff, 1989) and potato cultures (Frey *et al.*, 1989). By analogy to the Ac/Ds systems, the En/Spm element can transactivate defective Spm elements in other systems. Sequence analysis of the integration sites showed small target site duplications, and the excisions of the element were imprecise and characterized by footprints as in maize (Frey *et al.*, 1989).

The Tam3 element of snapdragon was proven to be an autonomous element when it was shown to transpose in tobacco (Martin *et al.*, 1989). Although there are no known defective Tam3 elements in snapdragon, an artificial, defective Tam3 that was created by internal deletion of part of the open reading frame was transactivated by the autonomous Tam3 element in tobacco (Haring *et al.*, 1991). On the other hand, the failure of Mu1 to transpose in *Arabidopsis* was an indication that it was not an active, autonomous element (Zhang and Somerville, 1987). To date, the transfer and function of retrotransposons in heterologous species have not been reported.

Although the flanking sequences are not hypermethylated, Tam3 rapidly becomes methylated at its ends in transgenic tobacco (Martin *et al.*, 1989). Transposition of Tam3 was not detected in the progeny of the transformed tobacco, possibly because of the methylation.

Agrobacterium tumefaciens is the most common means of introducing a transposable element into another species. However, *Agrobacterium* does not readily infect cereals. Electroporation has been used to introduce the Ac element into rice (Izawa *et al.*, 1991). The wheat dwarf virus has also been used to introduce Ac/Ds elements into cereal genomes (Laufs *et al.*, 1990).

The mechanisms of transposition by Ac and En/Spm are investigated by *in vitro* mutagenesis and reconstitution of functional transposition in tobacco or other heterologous systems. The 8.3 kb En/Spm element consists of four alternatively spliced transcripts designated tnpA, tnpB, tnpC, and tnpD. A tobacco transposition assay that detects the excision of a dSpm element from the GUS gene was used to test the function of each cDNA from the En/Spm element that was expressed under control of the CaMV 35S promoter. The tnpA and tnpD products were necessary for excision and transposition of the dSpm element and tnpB and tnpC are not required (Masson *et al.*, 1991). A low level of transposition is observed only with the tnpD protein so it is thought to be the transposase that binds to and cleaves at the ends of the 13 bp terminal inverted repeats. The tnpA product increases the frequency of transposition and may function to draw the element inverted repeat ends together in preparation for cleavage by tnpD or a combination of the tnpA and tnpD proteins. The tnpA protein has been expressed in *Escherichia coli* and binds a repetitive 12 bp motif present in the multiple copies within several hundred bases of the subterminal repetitive region. This sequence differs from the terminal 13 bp inverted repeats.

The binding of tnpA to the subterminal 12 bp repeats has also been shown to be responsible for the suppressor function of the Spm element.

McClintock had observed that some alleles containing dSpm elements are functional in the absence of an active Spm element elsewhere in the genome. She called these Spm-suppressible and postulated a second function for the Spm element in addition to transposition. Analysis of suppressible alleles has shown that they contain relatively small dSpm elements that are present within in the transcribed region of the gene. A molecular model for suppression proposes that the tnpA protein from an active Spm binds to the subterminal repeats and sterically blocks transcription through the gene (Schwarz-Sommer *et al.*, 1985b). In some cases, the dSpm element is capable of being removed by alternative splicing as its 13 bp terminal repeats resemble an intron splice site (Kim *et al.*, 1987). This explains the function of the gene in the absence of a transacting Spm element despite the presence of the dSpm element in the transcribed regions of the gene.

To recreate the suppressor effect, a transient GUS assay system in tobacco was developed. Plasmid constructs were developed with and without a dSpm sequence inserted into the untranslated sequences of the GUS gene driven by a plant promoter. These constructs were tested for their transient expression after transformation into control tobacco protoplasts or protoplasts isolated from tobacco plants that are transgenic for the tnpA protein (Grant *et al.*, 1990). Expression of tnpA resulted in suppression of GUS expression indicating that tnpA is responsible for the suppressor effect by interacting with the dSpm sequences in the GUS gene. Analysis of the DNA binding domain of the tnpA protein *in vitro* showed that dimers of the tnpA protein interacts tightly with a tail-to-tail dimer of the 12 bp motif in the subterminal region (Trentmann *et al.*, 1993). The binding domain of tnpA does not match any other known class of DNA binding proteins.

The Ac element contains only one transcript of 3.5 kb and an open reading frame (ORFa) that encodes a polypeptide of 807 amino acids (Fuswinkel *et al.*, 1991). The ORFa polypeptide has also been shown to bind to a repeating hexamer motif in the subterminal repeat region of the Ac element (Kunze and Starlinger, 1989). Deletion analysis of the Ac element revealed that high level of transposition in tobacco required about 200 bp at the end of each element indicating a functional requirement for the subterminal repeat region (Coupland *et al.*, 1989; Zhou *et al.*, 1991). Likewise, the binding domains of the ORFa protein from Ac are not similar to other recognized classes of DNA binding proteins (Feldmar and Kunze, 1991).

Gene Tagging

Transposable elements can be used to isolate genes of unknown function that are defined only by a mutant phenotype. The process of gene tagging requires an active transposable element that generates a transposon-induced mutation. The mutant gene containing the element can then be isolated using the cloned transposable element as a probe of a genomic library constructed from DNA of the mutant plant. The sequences flanking the element are likely to represent the mutant gene and can be used to probe DNA blots of wild type, mutant, and revertant plants in order to prove that the gene of interest has been isolated. The wild type gene can then be isolated from a genomic library.

Structural and regulatory genes of the anthocyanin pathway in maize and snapdragon were the first genes to be isolated by transposon tagging because of their easily visible phenotypes (reviewed in Vodkin, 1989; Gierl and Saedler, 1992; Walbot, 1992). Ac/Ds, En/Spm, and Mu elements have been used as tags. The *opaque-2* gene that regulates zein synthesis in maize was the first non-anthocyanin gene isolated by transposon tagging (Schmidt *et al.*, 1987; Motto *et al.*, 1988). A maize line containing Spm was used as the pollen donor to cross to a recessive *opaque-2* female. All of the progeny were normal except for several rare, *opaque-2* kernels in the F_1 population. A DNA fragment containing the Spm element and a small region of the *opaque-2* gene was identified using methylation sensitive enzymes to distinquish the active En/Spm element from the defective Spm elements elsewhere in the genome. Other genes that have been cloned include the *hcf-106* allele involved in chloroplast development (Martienssen *et al.*, 1989), the *knotted-1* gene that affects morphology (Hake *et al.*, 1989), and the *Y1* gene that is responsible for the production of β-carotene in endosperm in leaves of maize (Buchner *et al.*, 1990).

Since active transposable elements have been characterized from only a few plant species, gene tagging has been attempted in other species using the maize elements. The first such example was the tagging of a flower colour gene in petunia using the Ac element (Chuck *et al.*, 1993). An Ac element that interrupted the streptomycin phospotransferase gene was transformed into a purple-flowered variety of petunia using *Agrobacterium*. Transpositions of Ac are indicated by restored expression of the streptomycin resistance gene. The plants containing transposed Ac elements were self-pollinated and a new phenotype displaying variegated flower colour phenotype was found in the progeny. It segregated as a recessive gene and appeared to be allelic to the known mutation *Ph6*, one of several genes that modify pigmentation by affecting the pH of the corolla. A restriction fragment containing Ac sequences cosegregated with the mutant phenotype. Sequences flanking the Ac element were isolated and proven to represent the *Ph6* locus because germinal reversion to wild type was correlated with restoration of the wild type DNA restriction pattern using the flanking sequences as a probe.

Successful gene tagging requires that the element move frequently in the new host. As discussed in a previous section, the Ac element is active in many other species and is capable of activating Ds elements. Ac and Ds remain active and unmethylated in tobacco for several generations (Hehl and Baker, 1990). The use of a two-element system in heterologous species has several advantages. First, high level expression of the Ac transposase can be obtained by placing the Ac coding region under control of the CAMV 35S promoter. Increased frequency of Ds mobility is obtained in this manner. High rates of transposition of Ds from selectable marker genes leads to a 27% germinal transposition rate as measured by restoration of the resistance phenoptye specified by the marker gene (Grevelding *et al.*, 1992). The frequencies of transposition systems developed for both Ac and En/Spm in *Arabidopsis* are thought to be sufficient to allow transposon tagging for this species (Grevelding *et al.*, 1992; Fedoroff and Smith, 1993; Cardon *et al.*, 1993).

Another advantage of a two-element system for tagging in heterologous species is that the immobile Ac construct can be separated from the Ds-induced mutations by selfing or crossing (Fedoroff and Smith, 1993). As in maize, this allows the Ds-induced mutation to be stabilized. An advantage of cloning the Ds-induced mutations in heterologous species, as compared to maize, is that there are generally fewer of the Ds elements since they are new to the genome into which they are introduced. Thus, screening a library and identifying the Ds-tagged gene that corresponds to the phenotype is not as difficult as in maize. Polymerase chain reaction technology, specifically, inverse PCR techniques, is also aiding the isolation of genomic sequences flanking transposable elements (Earp *et al.*, 1990).

A gene tagging system was developed using a vector containing (i) a non-mobile En/Spm transposase gene under control of the CaMV promoter; (ii) a dSpm element; and (iii) a selectable hygromycin phophotransferase gene (Aarts *et al.*, 1993). After transformation with *Agrobacterium*, a male sterile plant was found in the third generation of self-fertilized progeny from a single plant that contained a transposing dSpm element. Using the dSpm probe, the gene designated *ms2* was isolated and represents the first nuclear male sterile gene isolated from plants. Another new mutation, *drl1*, generated by Ac tagging in *Arabidopsis* shows abnormal development of roots, shoots, and leaves (Bancroft *et al.*, 1993).

The Ac element has also been introduced into soybean tissue cultures (Zhou and Atherly, 1990). Soybean regeneration from culture is still not routine. However, recent progress in transformation of embryogenic cultures with the DNA gun has produced transgenic plants (McCabe *et al.*, 1988; Finer and McMullen, 1991). One of the difficulties in developing gene tagging systems in soybean will be to regenerate large numbers of transformed plants. The development of highly active elements that continue to be active in subsequent generations would seem to be the most feasible approach. This would reduce the number of primary transformants that need to be produced.

Soybean DNAs with Structural Features of Transposable Elements

The soybean *Le1* gene encodes a lectin protein that is specifically found in the cotyledons. Lectin accounts for about 2.5% of the total seed protein except in a few varieties that are homozygous for the *le1* allele and are null for lectin protein (Orf *et al.*, 1978). Isolation of the two alleles revealed that the lectin gene is present in *le1* genotypes but that its expression is prevented by a 3.5 kb insertion with structural properties of transposable elements (Goldberg *et al.*, 1983; Vodkin *et al.*, 1983). The element, designated Tgm1, has 13 bp terminal inverted repeats and is flanked by a duplication of 3 bp of the target lectin sequence. The Tgm1 element is part of the 'superfamily' of transposable elements that exist in diverse plant species and have terminal inverted repeats similar to the Spm element in maize (Vodkin, 1989). The maize En/Spm element, soybean Tgm1, and snapdragon Tam1 are often referred to as CACTA elements as they all begin with this five-base sequence. The soybean Tgm element also has a subterminal repetitive region with the motif ACATCGG (Rhodes and Vodkin, 1985). Interestingly, this sequence is different from the motif found in the En/Spm that is bound by the tnpA protein from maize En/Spm. In maize, the repetitive region is recognized by the tnpA product. Thus, it may be that a DNA binding protein of the En/Spm, Tgm, and Tam1 elements is species specific.

The Tgm1 element appears to be a deletion derivative of larger elements that exist in the soybean genome, some of which are over 12 kb in size and contain an ORF with homology to the tnpD region of En/Spm (Rhodes and Vodkin, 1988). A zinc finger domain has also been found in this ORF area that is conserved between soybean and maize (Vodkin and Vodkin, 1989). As discussed previously, the maize tnpD protein is necessary for transposition and is thought to be the transposase that binds to the terminal 13 bp inverted repeats. There are about 50 sequences in soybean that hybridize to internal regions of the Tgm1 element. Another repetitive sequence family with 500 to 800 members in soybean has weak structural homology to retrotransposons (Laten and Morris, 1993). Another soybean sequence, Tgmr, was found as a RAPD marker (random amplification of polymorphic DNA) and also has weak structural homology to the reverse transcriptase of the Tnt1 elements of tobacco (M. K. Bhattacharyya, personal communication).

Mutable Alleles and Anthocyanin Pathway Genes in Soybean

The soybean Tgm1 elements may represent a family of sequences that were once active. Whether any of the elements are currently active or capable of being activated is unknown. Chimeric phenotypes of several mutable alleles have been described, but none of these have been identified at the molecular level. One is the *w4*-mutable allele that results in a variegated purple and white flower (Groose *et al.*, 1988; 1990). This mutable allele has been shown to be an allele of the *W4* locus. Stable purple and stable white revertants have been derived from the *w4*-mutable stock and putatively represent element excisions. Some new mutants have been found in these populations but it is not known whether they represent transpositions from the *w4* locus (Palmer *et al.*, 1989).

In addition to the *W4* gene, there are four other genes known to affect flower colour. Soybean flowers are generally purple due to the action of the *W1* gene, which is thought to be a 5′, 3′ hydroxylase (Buzzell *et al.*, 1987). The recessive allele *w1* results in white flowers. Several unlinked genes appear to modify or overide the action of the dominant *W1* allele. These include the recessive *wm* allele that results in a magenta flower colour (Palmer and Kilen, 1987) and *wp* that results in pink flowers rather than purple (Stephens *et al.*, 1993). In the presence of the *W1* and recessive *w4* alleles, the *W3* allele results in a pattern phenotype in which the purple colour is restricted to a narrow band at the base of the flower on an otherwise white flower (Palmer and Kilen, 1987).

In parallel to the situation in maize, analysis of the anthocyanin pathway may be a way to identify additional transposable elements in soybean. To date, however, none of these flower colour genes has been isolated but several of the genes that encode enzymes in the pathway have been isolated. These include phenylalanine ammonia lyase (PAL), chalcone synthase (CHS), and dihydroflavonol reductase (DFR) (Frank and Vodkin, 1991; Akada *et al.*, 1991; Wang *et al.*, 1994). A restriction fragment length polymorphism of a dihydroflavonol reductase gene has been shown to be closely linked to the *W3* gene that specifies the purple throat pattern (Fasoula *et al.*, 1995). It is possible that the *W3* locus encodes the DFR enzyme. The *A1* gene in maize and the *pallida* locus in snapdragon encode DFR. Some of the *pallida* alleles show flower patterns due to the presence of a transposon footprints left by the action of the Tam3 element (Coen *et al.*, 1986).

Another mutable allele in soybean affects seed coat colour. The *r-m* allele of the *R* locus confers black stripes or spots on a brown background. Normally black (*R*) is dominant to brown (*r*). Black (*R**) and brown (*r**) seed can be derived at high frequency from the *r-m* stock. Some of these are somatic reversions and others are germinal reversions. However, the germinal revertants are unusual in that they are not completely stable. Instead they can switch back to the mutable phenotype at a high frequency (Chandlee and Vodkin, 1989). The *r-m* allele is also dominant to the revertant *R** allele (Wang *et al.*, in preparation).

Another major gene that affects anthocyanin pigmentation in the soybean seed coat is the *I* locus. Four alleles are known including the dominant *I* allele that inhibits pigmentation in the seed coat resulting in yellow seed. The i^i allele results in pigment only in the hilum area, the i^k alleles restricts pigment to a saddle-shaped region, and the *i* allele allows pigmentation over the entire seed coat. Spontaneous mutations from the dominant *I* to the recessive *i* allele have occurred independently in a number of varieties. Compared to the pigmented, homozygous recessive *i* genotypes, immature seed coats with the homozygous dominant *I* allele have increased levels of a mRNA and polypeptides for a proline rich cell wall protein, PRP1 (Lindstrom and Vodkin, 1991). Soybean seeds that also carry the recessive *t* alleles have defective seed coats. The *t* allele is thought to encode a 3′ flavonoid hydroxylase (Buzzell *et al.*, 1987). It is postulated that an epigenetic interaction between the flavonol pathway and the proline rich proteins affects the structural integrity of the seed coat (Nicholas *et al.*, 1993). An effect of anthocyanin pathway genes on cell wall proteins and tissue structure is not typical of mutants in better studied systems as maize and snapdragon. The PRP1 and PRP2 genes consist of tandem repeats of five amino acids, Pro-Pro-Val-Tyr-Lys. Genetic variation in the size of the PRP1 and PRP2 polypeptides has been shown to be due to precise deletions (or insertions) of even numbers of the protein repeat unit structure (Schmidt *et al.*, 1994).

Immature seed coats with the dominant *I* genotype (yellow) contain only 10% of the chalcone synthase mRNA found in seed coats with the homozygous recessive *i* genotype that results in anthocyanin pigmentation (Wang *et al.*, 1994). Similarly, the activity of CHS was only about 10% in the yellow seed coats. These data suggest that the biochemical basis of inhibition of anthocyanin in the seed coats is through reduced CHS mRNA levels leading to reduced CHS activity in seed coats with the dominant *I* allele. The dominant *I* allele also inhibits the production of primitive defence chemicals called proanthocyanidins that are present in pigmented, recessive *i* genotypes (Todd and Vodkin, 1993).

There are at least seven CHS genes in soybean but little is known about their tissue or developmental expression patterns. The nature of the dominant *I* allele is unknown but some changes in the restriction fragment pattern of CHS genomic sequences cosegregate with the *I* locus (Wang *et al.*, 1994; Todd and Vodkin, 1995). In addition, there appear to be large structural changes, perhaps including duplications and deletions of CHS genes, in the *I* allele and its spontaneous mutants (Todd and Vodkin, 1995). Either the *I* locus is a transacting regulatory gene or a genomic change in the CHS genes is responsible for the reduction of CHS mRNA(s) expressed in seed coats with the dominant *I* alleles. Suppression of endogenous CHS genes in a transacting manner by CHS transgenes has been described in several systems and has been termed cosuppression (Napoli *et al.*, 1990; van der Krol *et al.*, 1990). The molecular nature of the cosuppression phenomenon is not known but is postulated to involve a direct interaction between the transgene and the endogenous CHS loci.

Summary

Soybean has a number of unusual genetic resources that display variegated or pattern-specific mutations. Their investigation should aid in understanding the nature of gene expression and the regulation of the flavonoid and anthocyanin pathway in soybean. Whether any of these mutations will be due to transposable elements that can be used for gene tagging remains to be determined. Efforts to transform soybean with the well-characterized elements from maize is an alternative approach to develope a gene tagging system for soybean. Recent progress in soybean transformation and regeneration should make this goal feasible in the future. Both gene tagging with a heterologous element and gene identification by chromosome walking using restriction fragment length polymorphisms are theoretically feasible but are technically difficult at the present time. These goals may be routine in the future as technical improvements continue to develop.

References

Aarts, M.G.M, Dirkse, W.G., Stiekema, W.J. and Pereira, A. (1993) Transposon tagging of a male sterility gene in *Arabidopsis*. *Nature* 363, 715–717.

Akada, S., Kung, S.D. and Dube, S.K. (1991) The nucleotide sequence of gene 1 of the soybean chalcone synthase multigene family. *Plant Molecular Biology* 16, 751–752.

Baker, B., Schell, J., Lorz, H. and Fedoroff, N. (1986) Transposition of the maize controlling element 'Activator' in tobacco. *Proceedings of the National Academy of Sciences, USA* 83, 4844–4848.

Baker, B., Coupland, G., Fedoroff, N., Starlinger, P. and Schell, J. (1987) Phenotypic assay for excision of the maize controlling element Ac in tobacco. *EMBO Journal* 6, 1547–1554.

Bancroft, I., Jones, J.D.G. and Dean, C. (1993) Heterologous transposon tagging of the DRL1 locus in *Arabidopsis*. *Plant Cell* 5, 631–638.

Bennetzen, J.L., Swanson, J., Taylor, W.C. and Freeling, M. (1984) DNA insertion in the first intron of maize *Adh1* affects message levels: Cloning of progenitor and mutant *Adh1* alleles. *Proceedings of the National Academy of Sciences, USA* 81, 4125–4128.

Bhattacharyya, M.D., Smith, A.M., Ellis, T.H.N., Hedley, C. and Martin, C. (1990) The wrinkled-seed character of pea described by Mendel is caused by a transposon-like insert in a gene encoding starch-branching enzyme. *Cell* 60, 115–120.

Buchner, B., Kelson, T.L. and Robertson, D.S. (1990) Cloning of the y1 locus of maize, a gene involved in the biosynthesis of carotenoids. *Plant Cell* 2, 867–876.

Buzzell, R.I., Buttery, B.R. and MacTavish, D.C. (1987) Biochemical genetics of black pigmentation of soybean seed. *Journal of Heredity* 78, 53–54.

Cardon, G.H. Frey, M. Saedler, H. and Gierl, A. (1993) Mobility of the maize transposable element En/Spm in *Arabidopsis thaliana*. *Plant Journal* 3, 773–784.

Chandlee, J.M and Vodkin, L.O. (1989) Unstable expression of a soybean gene during seed coat development. *Theoretical and Applied Genetics* 77, 587–594.

Chandler, V.L. and Walbot, V. (1986) DNA modification of a maize transposable element correlates with loss of activity. *Proceedings of the National Academy of Sciences, USA* 83, 1767–1771.

Chomet, P.S., Wessler, S. and Dellaporta, S.L. (1987) Inactivation of the maize transposable element Activator (Ac) is associated with its DNA modification. *EMBO Journal* 6, 295–302.

Chomet, P. Lisch, D., Hardeman, K.J., Chandler, V.L. and Freeling, M. (1991) Identification of a regulatory transposon that controls the Mutator transposable element system in maize. *Genetics* 129, 261–270.

Chuck, G., Robbins, T., Nijjar, C., Ralston E., Courtney-Gutterson, N. and Dooner, H.K. (1993) Tagging and cloning of a petunia flower color gene with the maize transposable element activator. *Plant Cell* 5, 371–378.

Coen, E.S., Carpenter, R. and Martin, C. (1986) Transposable elements generate novel spatial patterns of gene expression in *Antirrhinum majus. Cell* 47, 285–296.

Coupland, G., Plum, C. Chatterjee, S., Post, A. and Starlinger, P. (1989) Sequences near the termini are required for transposition of the maize transposon Ac in transgenic tobacco plants. *Proceedings of the National Academy of Sciences, USA* 86, 9385–9388.

Dooner, H.K., Keller, J., Harper, E. and Ralston, E. (1991) Variable patterns of transposition of the maize element Activator in tobacco. *Plant Cell* 3, 473–482.

Doring, H.-P. and Starlinger, P. (1984) Barbara McClintock's controlling elements: now at the DNA level. *Cell* 39, 253–259.

Earp, D.J., Lowe, B. and Baker, B. (1990) Amplification of genomic sequences flanking transposable elements in host and heterologous plants: a tool for transposon tagging and genome characterization. *Nucleic Acids Research* 18, 3271–3279.

English, J., Harrison, K. and Jones, J.D.G. (1993) A genetic analysis of DNA sequence requirements for Dissociation State I activity in tobacco. *Plant Cell* 5, 501–514.

Fasoula, D.A., Stephens, P.A., Nickell, C.D. and Vodkin, L.O. (1955) Cosegregation of purple-throat flower color with a dihydroflavonol reductase (DFR) polymorphism in soybean. *Crop Science* 35, 1028–1031.

Fedoroff, N., Wessler, S. and Shure, M. (1983) Isolation of the transposable maize controlling elements Ac and Ds. *Cell* 35, 235–242.

Fedoroff, N.V. (1989) About maize transposable elements. *Cell* 56, 181–191.

Fedoroff, N.V. and Smith, D.L. (1993) A versatile system for detecting transposition in *Arabidopsis. Plant Journal* 3, 273–289.

Feldmar, S. and Kunze, R. (1991) The ORFa protein, the putative transposase of maize transposable element Ac, has a basic DNA binding domain. *EMBO Journal* 10, 4003–4010.

Finer, J.J. and McMullen, M.D. (1991) Transformation of soybean via particle bombardment of embryogenic suspension culture tissue. *In Vitro Cellular and Developmental Biology* 27, 175–182.

Finnegan, E. J., Taylor, B.H., Dennis, E.S. and Peacock, W.J. (1988) Transcription of the maize transposable element Ac in maize seedlings and in transgenic tobacco. *Molecular and General Genetics* 212, 505–509.

Frank R.L. and Vodkin L.O. (1991) Sequence and structure of a phenylalanine ammonia-lyase gene from *Glycine max. DNA Sequence* 1, 335–346.

Frey, M., Tavantzis, S.M. and Saedler, H. (1989) The maize En-1/Spm element transposes in potato. *Molecular and General Genetics* 217, 172–177.

Fuswinkel, H., Schein, S., Courage, U., Starlinger, R. and Kunze R. (1991) Detection and abundance of mRNA and protein encoded by transposable element Activator (Ac) in maize. *Molecular and General Genetics* 225, 186–192.

Gerats, A.G.M. Huits, H., Vrijlandt, E., Marana, C., Souer, E. and Beld, M. (1990) Molecular characterization of a nonautonomous transposable element (dTph1) of petunia. *Plant Cell* 2, 1121–1128.

Gierl, A. and Saelder, H. (1992) Plant-transposable elements and gene tagging. *Plant Molecular Biology* 19, 39–49.

Goldberg, R.B., Hoschek, G. and Vodkin, L.O. (1983) An insertion sequence blocks expression of a soybean lectin gene. *Cell* 33, 465–475.

Grant, S.R., Gierl, A. and Saedler, H. (1990) En/Spm encoded tnpA protein requires a specific target sequence for suppression. *EMBO Journal* 9, 2029–2035.

Grandbastien, M., Spielmann, A. and Caboche, M. (1989) Tnt1, a mobile retroviral like transposable element of tobacco isolated by plant cell genetics. *Nature* 337, 376–380.

Grevelding, C., Becker, D., Kunze, R., Von Menges, A., Fantes V., Schell, J. and Masterson, R. (1992) High rates of Ac/Ds germinal transposition in *Arabidopsis* suitable for gene isolation by insertional mutagenesis. *Proceedings of the National Academy of Sciences, USA* 89, 6085–6089.

Groose, R.W., Weigelt, H.D. and Palmer, R.G. (1988) Somatic analysis of an unstable mutation for anthocyanin pigmentation in soybean. *Journal of Heredity* 79, 263–267.

Groose, R.W., Schulte, S.M. and Palmer, R.G. (1990) Germinal reversion of an unstable mutation for anthocyanin pigmentation in soybean. *Theoretical and Applied Genetics* 79, 161–167.

Hake, S. Volbrecht, E.V., Freeling, M. (1989) Cloning of knotted, the dominant morphological mutant in maize using Ds2 as a transposon tag. *EMBO Journal* 8, 15–22.

Haring, M.A., Teeuwen-de Vroomen, M.J., Nijkamp, J.J. and Hille, J. (1991) Trans-activation of an artificial dTam3 transposable element in transgenic tobacco plants. *Plant Molecular Biology* 16, 39–47.

Hehl, R. and Baker B. (1990) Properties of the maize transposable element Activator in transgenic tobacco plants: a versatile inter-species genetic tool. *Plant Cell* 2, 709–721.

Hehl, R., Nacken, W.K.F., Krause, A., Saedler, H. and Sommer, H. (1991) Structural analysis of Tam3, a transposable element from *Antirrhinum majus*, reveals homologies to the Ac element from maize. *Plant Molecular Biology* 16, 369–371.

Herrmann, A., Schulz, W. and Hahlbrock, K. (1988) Two alleles of the single copy chancone synthase gene in parsley differ by a transposon-like element. *Molecular and General Genetics* 212, 93–98.

Hershberger, R.J., Warren, C.A. and Walbot, V. (1991) Mutator activity in maize correlates with the presence and expression of the Mu transposable element Mu9. *Proceedings of the National Academy of Sciences, USA* 88, 10198–10202.

Hirochika, H. (1993) Activation of tobacco retrotransposons during tissue culture. *EMBO Journal* 12, 2521–2528.

Houba-Herin, N., Becker, D., Post, A., Larondelle, Y. and Starlinger, P. (1990) Excision of a Ds-like maize transposable element (Ac >) in a transient assay in *Petunia* is enhanced by a trancated coding region of the transposable element Ac. *Molecular and General Genetics* 224, 17–23.

Izawa, T., Miyazaki, C., Yamamoto, M., Terada, R., Iida, S. and Shimamoto K. (1991) Introduction and transposition of the maize transposable element Ac in rice (Oryza sativa L.). Molecular and General Genetics 227, 391–396.

Jin, Y.-K. and Bennetzen, J.L. (1989) Structure and coding properties of Bs1, a maize retrovirus-like transposon. Proceedings of the National Academy of Sciences, USA 86, 6235–6239.

Jing-liu, Z., Xiao-ming, L., Rui-zhu, C. Rui-xin, H. and Meng-min, H. (1991) Transposition of maize transposable element Activator in rice. Plant Science 73, 191–198.

Johns, M.A., Mottinger, J. and Freeling, M. (1985) A low copy number, copia-like transposon in maize. EMBO Journal 4, 1093–1102.

Jones, J.D.G., Carland, F.C., Maliga, R. and Dooner, H.K. (1989) Visual detection of transposition of the maize element Activator (Ac) in tobacco seedlings. Science 244, 204–207.

Jones, J.D.G., Carland, F., Lim, E., Ralston, E. and Dooner, H.K. (1990) Preferential transposition of the maize element Activator to linked chromosomal locations in tobacco. Plant Cell 2, 701–707.

Kikuchi, S., Liu, X., Frommer, W.B., Koster-Topfer, M. and Willmitzer, L. (1991) Identification and structural characterization of further DNA elements in the potato and pepper genomes homologous to the transposable element-like insertion Tst1. Molecular and General Genetics 230, 494–498.

Kim, H.-Y., Schiefelbein, J.W., Raboy, V., Furtek, D.B. and Nelson, O.E. (1987) RNA splicing permits expression of a maize gene with a defective Suppressor-mutator transposable element insertion in an exon. Proceedings of the National Academy of Sciences, USA 84, 5863–5867.

Kunze, R. and Starlinger, P. (1989) The putative transposase of transposable element Ac from Zea mays L. interacts with subterminal sequences of Ac. EMBO Journal 3177–3185.

Laten, H.M. and Morris, R.O. (1993) Sire-1, a long interspersed repetitive DNA element from soybean with weak sequence similarity to retrotransposons: initial characterization and partial sequence. Gene 134, 153–159.

Laufs, J., Wirtz, U., Kammann, M., Matzeit, V., Schaefer, S., Schell, J., Dzernilofshy, A.R., Baker, B. and Gronenborn, B. (1990) Wheat dwarf virus Ac/Ds vectors: Expression and excision of transposable element introduced into various cereals by a viral replicon. Proceedings of the National Academy of Sciences, USA 87, 7752–7756.

Lillis, M. and Freeling, M. (1986) Mu transposons in maize. Trends in Genetics, July, 183–188.

Lindstrom, J.T. and Vodkin, L.O. (1991) A soybean cell wall protein is affected by seed color genotype. Plant Cell 3, 561–571.

Marion-Poll, A., Marin, E., Bonnefoy, N. and Pautot, V. (1993) Transposition of the maize autonomous element Activator in transgenic Nicotiana plumbaginifolia plants. Molecular and General Genetics 238, 209–217.

Martienssen, R.A., Barkan, A., Freeling, M. and Taylor, W.C. (1989) Molecular cloning of a maize gene involved in photosynthetic membrane organization that is regulated by Robertson's Mutator. EMBO Journal 8, 1633–1639.

Martin, C., Prescott, A., Lister, C. and MacKay, S. (1989) Activity of the transposon Tam3 in Antirrhinum and tobacco: possible role of DNA methylation. EMBO Journal 8, 997–1004.

Masson, P. and Fedoroff, N.V. (1989) Mobility of the maize Suppressor-mutator element in transgenic tobacco cells. *Proceedings of the National Academy of Sciences, USA* 86, 2219–2223.

Masson, P., Strem, M. and Fedoroff, N. (1991) The *tnpA* and *tnpD* gene products of the Spm element are required for transposition in tobacco. *Plant Cell* 3, 73–85.

Masterson, R.V., Furtek, D.B., Grevelding, C. and Schell, J. (1989) A maize Ds transposable element containing a dihydrofolate reductase gene transposes in *Nicotiana tabacum* and *Arabidopsis thaliana*. *Molecular and General Genetics* 219, 461–466.

McCabe, D.E. Swain, W.F., Martinell, B.J. and Christou, P. (1988) Stable transformation of soybean (*Glycine max*) by particle acceleration. *Bio/Technology* 6, 923–926.

McClintock, B. (1984) The significance of responses of the genome to challenge. *Science* 226, 792–801.

Motto, M., Maddaloni, M., Ponziani, G., Brembilla, M., Marotta, R., Di Fonzo, N., Soave, C., Thompson, R. and Salamini F. (1988) Molecular cloning of the 02-M5 allele of *Zea Mays* using transposon marking. *Molecular and General Genetics* 212, 488–494.

Nacken, W.K.F., Piotrowiak, R., Saedler, H. and Sommer, H. (1991) The transposable element Tam1 from *Antirrhinum majus* shows structural homology to the maize transposon En/Spm and has no sequence specificity of insertion. *Molecular and General Genetics* 228, 201–208.

Napoli, C., Lemieux, C. and Jorgensen, R. (1990) Introduction of a chimeric chalcone synthase gene into petunia results in reversible co-suppression of homologous genes *in trans. Plant Cell* 2, 279–289.

Nicholas, C.D., Lindstrom, J.T. and Vodkin, L.O. (1993) Variation of proline rich cell wall proteins in soybean lines with anthocyanin mutations. *Plant Molecular Biology* 21, 145–156.

Orf, J.H., Hymowitz, T., Pull, S.P. and Pueppke, S.G. (1978) Inheritance of soybean seed lectin. *Crop Science* 18, 899–900.

Palmer, R.G. and Kilen, T.C. (1987) In: Wilcox, J.R. (ed.) *Soybean: Improvement, Production, and Uses*, 2nd edn. American Society of Agronomy, Madison, Wisconsin, pp. 135–209. Palmer, R.G., Hedges, B.R., Benavente, R.S. and Groose, R.W. (1989) The w-4 mutable line in soybean. *Developmental Genetics* 10, 542–551.

Peleman, J., Cottyn, B., Van Camp, W., Van Montagu, M. and Inze, D. (1991) Transient occurrence of extrachromosomal DNA of an *Arabidopsis thaliana* transposon-like element, Tat1. *Proceedings of the National Academy of Sciences, USA* 88, 3618–3622.

Pereira, A., Cuypers, H., Gierl, A., Schwarz-Sommer, S. and Saedler, H. (1986) Molecular analysis of the En/Spm transposable element system of *Zea mays*. *EMBO Journal* 5, 835–841.

Peterson, P.A. (1987) Mobile elements in plants. In: *CRC Critical Reviews in Plant Sciences* 6, 105–208.

Rhodes, P.R. and Vodkin, L.O. (1985) Highly structured sequence homology between an insertion element and the gene in which it resides. *Proceedings of the National Academy of Sciences, USA* 82, 493–497.

Rhodes, P.R. and Vodkin, L.O. (1988) Organization of the Tgm family of transposable elements in soybean. *Genetics* 120, 597–604.

Roberts, M.R., Kumar, A., Scott, R. and Draper, J. (1990) Excision of the maize transposable element Ac in flax callus. *Plant Cell Reports* 9, 406–409.

Robertson, D.S. (1978) Characterization of a mutator system in maize. *Mutation Research* 58, 21–28.

Rommens, C.M.T., van der Biezen, E.A., Ouwerkerk, P.B.F., Nijkamp, J.J. and Hille, J. (1991) Ac-induced disruption of the double Ds structure in tomato. *Molecular and General Genetics* 228, 453–458.

Schmidt, R.J., Burr, F.A. and Burr, B. (1987) Transposon tagging and molecular analysis of the maize regulatory locus *opaque2*. *Science* 238, 960–963.

Schmidt, J.S., Lindstrom, J.T. and Vodkin, L.O. (1994) Genetic length polymorphisms create size varition in proline rich proteins of the cell wall. *Plant Journal* 6, 177–186.

Schwarz-Sommer, S., Gierl, A., Cuypers, H., Peterson, P.A. and Saedler, H. (1985a) Plant transposable elements generate the DNA sequence diversity needed in evolution. *EMBO Journal* 4, 591–597.

Schwarz-Sommer, S., Gierl, A., Berndtgen, R. and Saedler, H. (1985b) Sequence comparison of states of a1-m1 suggests a model of Spm/En action. *EMBO Journal* 4, 2439–2443.

Schwarz-Sommer, S., Leclercqu, L., Goebel, E. and Saedler, H. (1987) Cin4, an insert altering the structure of the *A1* gene in *Zea mays*, exhibits properties of nonviral retrotransposons. *EMBO Journal* 6, 3873–3880.

Scofield, S.R., English, J.J. and Jones, J.D.G. (1993) High level expression of the Activator transposase gene inhibits the excision of Dissociation in tobacco cotyledons. *Cell* 75, 507–517.

Shure, M., Wessler, S. and Fedoroff, N.V. (1983) Molecular identification and isolation of the *Waxy* locus in maize. *Cell* 35, 225–233.

Stephens, P.A., Nickell, C.D. and Vodkin, L.O. (1993) An association of increased protein and seed size with pink flower color in soybean. *Crop Science* 33, 1135–1137.

Taylor, B.H., Finnegan, E.J., Dennis, E.S. and Peacock, W.J. (1989) The maize transposable element Ac excises in progeny of transformed tobacco. *Plant Molecular Biology* 13, 109–118.

Todd, J.J. and Vodkin, L.O. (1993) Pigmented soybean (*Glycine max*) seed coats accumulate proanthocyanidins during development. *Plant Physiology* 102, 663–670.

Todd, J.J. and Vodkin, L.O. (1995) Molecular genetics of the soybean *I* locus which inhibits pigment synthesis in a spatial manner. *Journal of Cellular Biochemistry* 21A, 460.

Trentmann, S.M., Saedler, H. and Gierl, A. (1993) The transposable element En/Spm-encoded TNPA protein contains a DNA binding and a dimerization domain. *Molecular and General Genetics* 238, 201–208.

van der Krol, A.R., Mur. L.A., Beld, M. and Mol, J.N.M. (1990) Flavonoid genes in petunia: Addition of a limited number of gene copies may lead to a suppression of gene expression. *Plant Cell* 2, 291–299.

Van Sluys, M.A. and Tempe, J. (1989) Behavior of the maize transposable element Activator in *Daucus carota*. *Molecular and General Genetics* 219, 313–319.

Vodkin, L.O., Rhodes, P.R. and Goldberg, R.B. (1983) A lectin gene insertion has the structural features of a transposable element. *Cell* 34, 1023–1031.

Vodkin, L.O. (1989) Transposable element influence on plant gene expression and variation. In: Marcus, A. (ed.) *The Biochemistry of Plants: Plant Molecular Biology,* Vol. 15, Academic Press, New York, pp. 83–132.

Vodkin, M.H. and Vodkin, L.O. (1989) A conserved zinc finger domain in higher plants. *Plant Molecular Biology* 12, 593–594.

Walbot, V. (1992) Strategies for mutagenesis and gene cloning using transposon tagging and T-DNA insertional mutagenesis. *Annual Review of Plant Physiology: Plant Molecular Biology* 43, 49–82.

Wang, C.-S., Todd, J.J. and Vodkin, L.O. (1994) Chalcone synthase mRNA and activity are reduced in yellow seed coats with dominant *I* alleles, *Plant Physiology* 105, 739–748.

Weil, C.F. and Wessler, S.R. (1993) Molecular evidence that chromosome breakage by Ds elements is caused by aberrant transposition. *Plant Cell* 5, 515–522.

Wessler, S.R. (1988) Phenotypic diversity mediated by the maize transposable elements Ac and Spm. *Science* 242, 399–405.

Yoder, J.I. (1990) Rapid proliferation of the maize transposable element Activator in transgenic tomato. *Plant Cell* 2, 723–730.

Zhang, H. and Sommerville, C.R. (1987) Transfer of the maize transposable element Mu1 into *Arabidopsis thaliana. Plant Science* 48, 165–173.

Zhou, J.H. and Atherly, A.G. (1990) In situ detection of transposition of maize controlling element (Ac) in transgenic soybean tissues. *Plant Cell Reports* 8, 542–546.

Zhou, J.H., Myers, A. and Atherly, A.G. (1991) Functional analysis of the 3'-terminal sequence of the maize controlling element (Ac) by internal replacement and deletion mutagenesis. *Genetica* 84, 13–21.

Limitations and Potentials of Genetic Manipulation of Soybean*

J.E. Specht and G.L. Graef

Department of Agronomy, University of Nebraska, Lincoln, Nebraska 68583, USA

Introduction

USA soybean [*Glycine max* (L.) Merr.] yields have steadily increased by about 21 kg ha^{-1} per year. This is the regression coefficient obtained when national soybean yields, estimated by the USDA since 1924, are regressed on production year. The upward trend in on-farm soybean yield is essentially attributable to the continuous adoption by most USA soybean producers of yield-enhancing agricultural technologies. The research and development activities of public and proprietary organizations are the primary source of these technologies.

Technological enhancement of on-farm soybean yields is obtained in two basic ways: genetic improvement, which results when farmers adopt new cultivars that have a higher yield potential, and agronomic improvement, which results when farmers adopt cultural practices that ameliorate the yield-reducing factors in the production environment (Specht and Williams, 1984). Genetic technology is very rapidly adopted by the USA soybean producer, primarily because: (i) it is relentlessly recurrent, which makes the timely replacement of obsolete cultivars an economic necessity; (ii) it is easily introduced into a soybean production system, usually without any need to implement a coordinate change in cultural practices; and (iii) it is inexpensive, or offers a coincident reduction of some other production cost, e.g. a new pest-resistant cultivar obviates the need for the pesticide.

* Contribution from Hatch Research Projects 12-194 and 12-184 of the Nebraska Agricultural Research Division, Lincoln, Nebraska 68583, USA.

The Impact of Past and Current Soybean Breeding Efforts

In this chapter, we intend to focus on the contributions of genetic technology to soybean improvement, with particular emphasis on the limitations and potentials of biotechnology. Before doing so, however, we believe it is instructive to reflect upon the impact of the past and current soybean breeding efforts.

Genetic improvement in the soybean has been estimated to be about 10 to 18 kg ha^{-1} annually, based on the yield potential of several 'landmark' cultivars (Luedders, 1977; Boerma, 1979; Wilcox *et al.*, 1979). Specht and Williams (1984), using data from a 3-year yield trial of 240 northern USA cultivars, regressed cultivar mean yield (which ranged from 650 to 4400 kg ha^{-1}) on cultivar release year (which ranged form 1902 to 1977). They found that genetic improvement in yield had averaged 18.8 kg ha^{-1} per year, but also noted that this figure was biased upward by the nonrecurrent 25% advance in yield potential that occurred in the 1940s – the result of a change in breeding method. Prior to that time, breeders evaluated the existent and incoming plant introductions and then simply released the best ones to farmers as cultivars. Breeders now search for potential cultivar releases by evaluating the innumerable recombinant lines that originate from the selfed progeny of hybridized parents.

To obtain a more contemporary estimate of genetic improvement in yield, Specht and Williams (1984) limited their regression analysis to just the 137 cultivars of hybrid origin that were released between 1939 and 1977. They considered the resultant regression coefficient of 12.5 kg ha^{-1} per year to be a conservative estimate of the annual genetic gain that northern USA soybean breeders were actually achieving in soybean yield potential. Specht and Williams (1984) contended that genetic technology probably accounted for about one-half of the 23.7 kg ha^{-1} annual improvement in on-farm soybean yields in the 19 northern states of the USA, the area of adaptation of these 137 cultivars. They noted that the 23.7 kg ha^{-1} figure reflected the realized impact of all farmer-adopted genetic and agronomic technologies.

Criteria for Evaluating New Methods of Genetic Manipulation

There are three points from the preceding discussion of past and current genetic gains in yield that are relevant in the assessment of any new method for the genetic manipulation of the soybean. The first point is that, irrespective of your perspective on the potential of biotechnology in soybean breeding, the number that must be targeted for enhancement is the 12.5 kg ha^{-1} annual genetic gain in yield that current breeding procedures are apparently generating. Any proposed modification of an existing breeding procedure, or any proposed new breeding method, must be evaluated with that number in mind. In this regard, it should be noted that the annual genetic gain obtained with any recurrently applied breeding method is estimable (Allard, 1960; Baenziger and Peterson, 1992) using this equation:

$$G_y = (k)(\sigma_{g'}^2)/(\sigma_p)(y)$$

where G_y is the annual genetic gain obtainable with recurrent selection; k is the standardized selection differential, effectively the selection intensity (i.e. q/n, where q = lines selected and n = lines evaluated) expressed in phenotypic standard deviation units; $\sigma_{g'}^2$ is the 'fixable' portion of the total genetic variance among the evaluated lines, which in the soybean is mainly of the additive portion of the genetic variance and possibly some additive × additive epistasis; σ_p is the phenotypic standard deviation of the means of the n lines; and y is the number of years required to complete one breeding cycle.

The genetic gain equation intuitively reveals the four ways by which a recurrent breeding method can be modified to improve annual genetic gain: (i) increase k via techniques that reduce the resource and labour cost of phenotypic evaluation, thus allowing a cost-effective increase in q relative to n; (ii) enhance $\sigma_{g'}^2$, preferably by methods that avoid a concurrent reduction in the phenotypic mean of the population of n lines; (iii) decrease σ_p via improvements in the precision of trait measurement, thus reducing the nongenetic portion of the phenotypic variance and concurrently increasing heritability; and/or (iv) reduce y by shrinking the time required for the completion of one breeding cycle. As we discuss the various biotechnological breeding methods, we will often refer to these four components of annual genetic gain when evaluating the limitations and potentials of those methods.

The second point of relevance in assessing a new method for genetic improvement of soybean is the nonrecurrent burst in genetic yield improvement that occurred in the 1940s. Instead of selection among and within plant introductions, soybean breeders began using hybridization, which allowed for recombination of parental genes and exploitation of a previously untapped source of genetic variability. Does this historical precedent suggest that some new breeding method could bring about a similar-sized burst in yield improvement? It is difficult for us to envision this happening, but it is certainly not beyond the realm of possibility if some heretofore unknown source of genetic variability, suitable for yield improvement, could be identified and accessed.

The third point of relevance is that the genetic improvement of yield potential *per se* is not the only objective in a breeding programme. Genetic gains in yield potential can quickly evaporate if the existent germplasm happens to be susceptible to some newly emergent biotic or abiotic stress. To protect genetic gains in yield potential, a substantial amount of a breeder's time and effort is periodically expended on searching the germplasm pools for genes that will provide resistance to these stresses. In addition to yield enhancement and yield protection, most breeders have a third objective – genetic improvement in the quality of the seed or forage product. In all three instances, yield, protection, and quality, biotechnological methods of plant improvement deserve careful consideration, simply because the needed genetic variability, or the specific genes, may not always be accessible or creatable with traditional breeding procedures.

The Basis for the Biotechnological Approaches to Plant Breeding

Since the 1940s, the genetic improvement of the soybean has depended upon two primary instruments of genetic manipulation – sexual hybridization and selection. Sexual hybridization is used to create genetic variability, and to procure and deploy specific genes. Selection, based on a measurement of the phenotype, is used to identify the superior recombinant inbred individuals (the plants) or their progeny (the lines). The various techniques of biotechnology, many of which are discussed in the foregoing chapters of this book, now offer plant breeders the prospect of freedom from the limitations imposed by the above dependencies. This became possible because of three developments:

1. Plant regeneration – the ability to recover plants, not only from micropropagated meristematic tissue, but also from *in vitro* cultures of protoplasts, cells, undifferentiated plant tissue (callus), pollen, ovules, embryos, and other explant tissue (Moore and Collins, 1983; Evans, 1989; Lal and Lal, 1990). In the soybean, plant regeneration can be induced from zygotic embryo tissue and some other explant tissues (hypocotyls, leaf discs, etc.), but a reproducible method of plant regeneration from protoplasts has not yet been reported (see Chapters 7 and 9).

2. Genetic transformation – the ability to integrate engineered DNA sequences into the genomes of regenerable plant cells, such that the regenerated plants or their progeny transmit the introduced DNA, and the phenotype it encodes, in a Mendelian fashion (Wu, 1989; Draper and Scott, 1991). In the soybean, transgenic plants have been derived by using *Agrobacterium* to deliver and insert an engineered alien gene into the genome of regenerable cells (Hinchee *et al.*, 1990), and by bombardment of organogenic host tissue with particles coated with alien DNA. Both methods lead to integration of alien DNA segments into the genomic DNA of the host cells (Christou *et al.*, 1993; Chapter 7). Transformation of soybean protoplasts is also possible via DNA uptake mediated by electroporation or PEG treatment, but as noted above, plant regeneration from protoplast cultures has not been successful.

3. Gene mapping – the ability to track, within segregating populations and across generations, any DNA sequence that encodes, or is tightly linked to, a desirable phenotype. With molecular markers, selection can then be based on the genotype as well as the phenotype. Thus, both qualitative or quantitative genes can be tracked (Tanksley, 1983; Stuber, 1990, 1992). Much of the gene mapping effort in the soybean, including efforts directed at the detection of quantitative trait loci (QTL), was completed only recently (see Chapter 2). Therefore, the use of marker-facilitated selection and QTL tagging in the soybean is still in its infancy. However, the work of Concibido *et al.* (1994) is an excellent example of the kind of resolution that is possible when presumably polygenic soybean traits are subjected to a marker-facilitated genetic analysis.

The Application of Biotechnological Methods in Soybean Breeding

Regardless of how it is accomplished, the strategy of plant breeding remains: create genetic variability, evaluate variants and select superior phenotypes. A basic challenge that breeders face is to identify superior genotypes from evaluation of phenotypes. Furthermore, it is not the genotypes, but the individual genes that are passed on to the progeny.

Biotechnology can contribute to the basic strategy of plant breeding in two ways.

1. The creation of genetic variability is no longer limited by sexual compatibility considerations.
2. Using genetic maps and marker-facilitated selection, the soybean breeder can identify a recombinant individual on the basis of its genotype as well as its phenotype.

With these two points in mind, let us briefly discuss specific biotechnological applications that have been, or can be, used in the soybean.

Methods used to create genetic variability

The creation of genetic variability in a breeding programme is generally achieved by hybridization, occasionally by induced mutation. For soybean cultivar development, the majority of crosses are intraspecific, elite × elite crosses. A review of soybean registration articles in *Crop Science* during the past decade reveals that the majority of soybean cultivars result from single crosses between cultivars or elite breeding lines. Three-way crosses, backcrosses, and intermated populations account for most of the remainder of the registered cultivars. In fact, the same is true for all modern domestic soybean cultivars released from public institutions from 1947 to 1991 (Bernard *et al.*, 1988, 1991). A few cultivar releases have originated from the interspecific mating of *G. max* with its wild relative *G. soja* Sieb. and Zucc., but only after one or more backcrosses were made to the adapted parent (Fehr *et al.*, 1990a,b).

Mutation breeding is generally successful when there is a need to create variability of a range or type that is not available in the existing germplasm (Borojevic, 1990). For example, chemical mutagenesis of seeds resulted in several soybean germplasm lines with altered fatty acid profiles (Hammond and Fehr, 1983 a,b; Wilcox and Cavins, 1986, 1990). On the other hand, induced mutation may not always be necessary. The fatty acid profile of soybean seeds was also successfully altered by selection within intermated populations (Burton *et al.*, 1989, 1994).

In what ways can biotechnology contribute to the creation of genetic variability?

1. Asexual methods like somatic hybridization and genetic transformation can provide the soybean breeder with access to the genetic variability and genes of any organism, even newly engineered genes.
2. Wide hybridization and chromosomal engineering would enable the breeder to introduce variability to the cultivated soybean from its close relatives.
3. Molecular genetic maps and DNA fingerprints could facilitate a better pairing of parents for mating purposes, because parental selection would be based on both genotypic and phenotypic data.

These applications of biotechnology to the creation of genetic variability could presumably improve genetic gain by increasing the frequency of lines in the desired phenotypic and genotypic range, without increasing $\sigma_{g'}^2$ *per se.* Although y could potentially be decreased by applications of different biotechnologies, the time required for line development may actually be lengthened in many instances. Regulatory requirements, for example, may lengthen the testing and evaluation of genetically engineered products. However, the value of a specific transformed trait, along with patent or other legal protection, may make that time investment worthwhile.

Genetic transformation is a useful means of producing novel genetic variants, which are commonly known as transgenic plants (Lal and Lal, 1990; Draper and Scott, 1991). Two methods have been used successfully to create transgenic plants.

One method is *Agrobacterium*-based, wherein the bacterium mediates the delivery of alien DNA to regenerable cells of cultured host tissue, with the *Ti* plasmid mediating the subsequent integration of that DNA into the host cell genome. An excellent discussion of this method can be found in Draper and Scott (1991). Hinchee *et al.* (1990) used a cotyledon regeneration system and reported that, of the 128 soybean plantlets they regenerated from kanamycin-selected, *Agrobacterium*-inoculated cotyledons of the cultivar Peking, six were demonstrably transgenic. This procedure was used to successfully produce glyphosate-tolerant transgenic soybean plants. A major limitation to the method is that only a few, mainly obsolete, soybean cultivars (e.g. Peking) are susceptible to *Agrobacterium* infection. Thus, an alien gene introduced into Peking must be moved to other cultivars by backcrossing, which can cut into the short average marketable lifetime of most new cultivars.

The other method involves delivering the alien DNA into host cells via bombardment of regenerable plant tissue with small particles coated with the alien DNA (Christou *et al.*, 1993; Chapter 7). Because micropropagated soybean meristem tissue (such as embryonic axes dissected from mature seeds) can be used as the bombardment target, plant regeneration occurs via a *de novo* organogenic pathway, making it very frequent and less subject to somaclonal variation. The chimerism that occurs because transgenics regenerate from multicellular meristematic zones is a problem, but not a critical one. Proof that the DNA delivered by microprojectiles is inserted into host cell DNA is based on two lines of evidence: Mendelian segregation of the introduced foreign gene in the R_2 and later generations, and the presence of the foreign DNA in the putative transgenic soybean lines (Christou *et al.*, 1993). Little is known about the mechanism by which free DNA is recombinationally integrated into host DNA, but it seems to occur frequently enough to make particle bombardment a popular approach for gene transfer.

Three technical problems with gene transfer still need resolution or refinement:

1. The number of alien DNA copies in transgenic plants varies greatly, and the homologous recombination and unequal crossing over that might occur with multiple copies could lead to genetic instabilities (Draper and Scott, 1991).
2. Transgenic plants containing the same alien gene frequently exhibit widely variant levels in the expression of that gene. This phenomenon is often called a 'position effect' (i.e. the site of insertion usually differs in each transgenic plant), because the expression of the inserted gene seems to be greatly influenced by the 'context' of the surrounding host DNA. Thus, each transgenic plant must be assayed for expression.
3. It is possible for an alien gene to be inserted into an existing gene. Subtle mutations of this type might not be detectable in the laboratory, but could be costly if the mutation occurred, for example, in a pest resistance gene. Continued research on the mechanisms of alien DNA insertion into host DNA will hopefully result in a better understanding of this process and how the insertion position can be made more deterministic.

The most suitable genes for genetic transformation are, of course, those encoding traits with a well-characterized physiological, biochemical, molecular, and genetic basis (Snape *et al.*, 1990; Briggs, 1992). The first genes used in the genetic transformation of the soybean were plant and bacterial genes encoding resistance to broad-spectrum herbicides (Stalker, 1991). Transgenic soybean cultivars with glyphosate resistance are now in the field testing stage and could be made available to soybean producers within a year or two (Christou *et al.*, 1993). Other possible targets for gene transfer to the soybean include: (i) insect resistance, via the transfer of *Bacillus thuringiensis* genes that encode proteins toxic to insects or via the transfer of genes encoding protease inhibitors (Lal and Lal, 1993); (ii) viral resistance, via the transfer of genes engineered to produce viral coat proteins (Chapter 10; Sturtevant and Beachy, 1993); and (iii) altered seed storage proteins (Chapter 6; Krebbers *et al.*, 1993). Almost any cloned gene, irrespective of its source, can be inserted into the soybean, so it is quite possible that soybean cultivars can be specifically engineered to serve as 'biological factories' that can produce unique biological substances or useful industrial chemicals (Pen *et al.*, 1993; Hiatt and Mostov, 1993).

The primary limitation of gene transfer technology is, of course, that it only works with genes that can be cloned. Even then, the genes most likely to be transferred will be those whose phenotypic effect is large enough to warrant the effort. Therefore, the genes most likely to be utilized for the genetic transformation of the soybean will continue to be those that offer yield protection (herbicide tolerance, pest resistance), or modify product quality (alter the protein and oil constituents of the soybean seed).

Chromosome-mediated gene transfer can complement the transfer of specific genes. Wide hybridization is of interest to soybean breeders because of the genetic variation that may exist at the subgeneric level. The genus *Glycine* has two subgenera: *Soja*, which consists of *G. max* and its wild annual relative *G. soja*, and *Glycine*, which includes 12 wild perennial species. All behave as diploids (2n = 40), but other evidence (e.g., duplicate gene pairs) suggests that the 40-chromosome *Glycine* species are diploidized allotetraploids with a basic chromosome number of x = 10. Genomic relationships among various *Glycine* species have been postulated on the basis of the success of interspecific and intersubgeneric hybridizations, the degree of F_1 viability, and the meiotic chromosome associations that have been observed in F_1 plants (Hymowitz *et al.*, 1991).

Intersubgeneric hybridization between *G. max* and its wild perennial relatives has been attempted, and viable F_1 plants have been obtained, often via *in vitro* seed culture (Hymowitz *et al.*, 1991), but the hybrids are nearly always sterile. Singh *et al.* (1993) report on fertile backcross-derived plants from *G. max* × *G. tomentella* intersubgeneric hybrids. Perennial species have phenotypes of interest to breeders such as greater degrees of drought and salt tolerance, and apparent immunity to some common soybean diseases. An example of successful intergeneric gene transfer is the disease resistance found in some wheat cultivars. In this instance, rye chromosome segments were introduced into the wheat genome (translocations 1A/1R and 1B/1R) as a result of wheat × rye crosses (Baenziger and Peterson, 1992). A wide hybridization method, one that is capable of generating *G. max* lines with specific chromosomes or chromosomal segments of a perennial species, would be a genetic advance welcomed by many soybean breeders. Chromosomal engineering might also be possible if hybrid and cybrid plants created by protoplast fusion or other methods of somatic hybridization could be regenerated (Moore and Collins, 1983; Lal and Lal, 1990). Certainly, aneuploid stocks (e.g. a complete set of viable trisomics) would be extremely useful for association of genes and molecular marker linkage maps to specific soybean chromosomes.

The production of doubled haploids from F_1 pollen or anther culture is a technique that can be used to access more quickly the potential genetic variability available in a mating. The doubled haploid (DH) procedure provides a means for the rapid creation of homozygous recombinant individuals from F_1 plants (Baenziger and Peterson, 1992). In traditional breeding methods, F_2-derived lines initially have equal amounts of among-line and within-line variance. With each generation of selfing before line derivation, the within-line additive genetic variance decreases by one half, corresponding to the decrease in heterozygotes each generation. The among-line additive genetic variance increases with each generation of selfing because of the increase in homozygous genotypes. With the doubled haploid procedure, the maximum among-line additive genetic variance is available for selection upon production of the DH lines. Whether this is advantageous or not depends upon the speed with which pollen from F_1 plants can be converted into field-testable lines, since three or four generations of inbreeding per year can be easily obtained in soybean breeding programmes that use off-season nurseries. In addition, by selfing to homozygosity, recombination is doubled relative to DH lines derives from F_1 plants. In any event, it is not currently possible to test this method, since there have been no reliable reports of the successful production of doubled haploids in the soybean (Hymowitz *et al.*, 1991).

Cell and tissue culture itself can serve as a useful mutagenic treatment, generating 'somaclonal variation', a term describing regenerated plants with phenotypes that are subtly or radically different from that of the initially cultured plant genotype. Some somaclonal variants have proven to be quite useful in the ornamental plant trade and in horticultural crop species (Lal and Lal, 1990). However, except for specific herbicide resistances and biochemical mutants, the successes of somaclonal variation have been mostly of the serendipitous kind. Most soybean breeders seeking an alteration in a given phenotype would use the mutagenic effects of cell culture only as a last resort, simply because a low frequency of plant regeneration makes the cost of mutant discovery prohibitively expensive. Somaclonal variation is a hindrance in those instances where cell or tissue culture is necessary, but genetic transformation is the objective.

Methods used to evaluate and select variants

A major bottleneck in any breeding and selection programme is the evaluation of the variants that are generated. Whether a breeder generates five hundred populations from hand crossing, 40,000 F_1 seeds using male sterility, or 150,000 M_1 seeds from chemical mutagenesis, the decisions on how to allocate resources for inbreeding, line development, and evaluation remain.

Development and application of more efficient field designs and sophisticated statistical procedures may improve the breeder's ability to identify superior phenotypes, and hopefully genotypes. Further developments in statistics, experimental design, and information technology will primarily impact the breeder's ability to precisely evaluate and select among larger numbers of genotypes, thereby increasing k in the numerator and decreasing σ_p in the denominator of the genetic gain equation. Decisions on when and how to evaluate phenotypes will be guided by the specific objectives and traits involved.

How can biotechnology contribute to evaluation of genetic variability and selection of superior genotypes?

1. Technologies now allow a characterization of diversity at the DNA level, using markers like RFLPs (Shoemaker *et al.*, 1992), RAPDs (Williams *et al.*, 1990), SSRs (Cregan *et al.*, 1994), and DAF (Caetano-Anolles *et al.*, 1991). The improved knowledge of allelic diversity at the molecular level may improve understanding and evaluation of different mating and selection systems and their effect at the gene level.

2. The development of high-density genetic maps and the integration of maps based on conventional and various molecular markers (Muehlbauer *et al.*, 1988; Shoemaker *et al.*, 1992), greatly improves the potential for marker-assisted selection in cultivar improvement programmes.

3. *In vitro* selection offers opportunities for certain traits, and may expand to other traits as knowledge and technologies improve.

All three applications of biotechnology to evaluation of variants and selection of superior individuals have the potential to greatly increase selection intensity (k) and reduce the time involved in evaluation and cultivar development (y), particularly with the automation of many laboratory procedures.

Cell culture can be useful as an *in vitro* means for the selection of traits whose expression is more easily evaluated at the cellular or molecular level (Moore and Collins, 1983; Lal and Lal, 1990). Because a large number of cells can be evaluated with minimal effort and cost, the amount of genetic variability that is potentially assayable is greatly increased, and n can be made very large. However, three procedural criteria must be met before an *in vitro* technique becomes useful in this regard: (i) the cells possessing the desirable phenotypes must be regenerable into plants; (ii) the selected cell phenotype must be expressible at the whole plant level; and (iii) the phenotypic expression must be the result of a stable alteration of the nuclear (or cytoplasmic) genome, so that traditional breeding methods can be used in the subsequent genetic manipulation of that phenotype. Although there are many reports of the 'successful' use of an *in vitro* technique in the literature, evidence that the latter two criteria have been met is frequently inconclusive or else absent in those reports (Lal and Lal, 1990).

Allelic and genotypic diversity at the molecular level have been evaluated in soybean (Shoemaker *et al.*, 1986; Close *et al.*, 1989; Grabau *et al.*, 1989, 1992). Lee *et al.* (1992, 1994) evaluated genotypic diversity in an outcrossing population of soybeans to assess the randomness of mating in the population and characterize the maintenance of cytoplasmic diversity during successive cycles of outcrossing. Shoemaker *et al.* (1992) indicated that a detailed molecular genetic map could increase our understanding of genome structure and evolution and, with pedigree information, could be used to determine ancestral sources of markers in the genome of a particular cultivar. Application of nuclear and cytoplasmic DNA markers to genetic diversity and its fate in different breeding schemes could be informative, regarding the efficiency of the method and creation and maintenance of desirable gene combinations.

Finally, the use of DNA markers to facilitate selection of superior genotypes receives much attention, particularly with the potential to identify and manipulate specific chromosome segments important in the expression of quantitative traits like yield. The marker-facilitated approaches of plant breeding are based on a simple concept: genetic linkage between a polymorphic marker locus and any locus of functional interest to the breeder (Stuber, 1990, 1992; Helentjaris, 1992; Lande, 1992). In effect, the allelic status detected at the marker locus in a recombinant individual is an indicator of the allelic status at the functional locus, with the reliability of that indication being a function of the tightness of the genetic linkage between the two loci. Ideally, each marker locus should be highly polymorphic (i.e. has multiple alleles whose codominant phenotypes are easily discernible from one another), and exhibit no epistatic interactions with other marker loci. The only other requirement is the need to have large numbers of marker loci. The loci used as markers in the soybean are mostly of the DNA type, such as RFLP, RAPD, and more recently, SSR. Isozymes and some pigmentation and morphological markers can also serve as useful markers.

The selection of parents is an important consideration in any breeding programme. Molecular genetic maps and DNA fingerprints may be useful to evaluate the genotypes of potential parents and their combining ability. The evidence accumulated to date, however, suggests that selection of parents based on molecular marker genotypes is not particularly useful in the identification of superior parents or parental combinations (Dudley *et al.*, 1992; Dudley, 1993). This is probably due to our present level of understanding of the relationship between molecular genotype and desirable phenotypic traits. Marker-assisted selection has the potential to be a powerful aid to increase genetic gain.

Dudley (1993) outlines the steps in marker-assisted selection as: (i) identifying associations between marker alleles and QTL; and (ii) using these associations to develop improved lines or populations. While a high-density genetic map would increase the probability of identifying marker–QTL linkages, one marker is effective if it is the right one. Most of the reports in the literature to date have moved to the second step in the process without first developing a saturated genetic map, and without establishing strong relationships between marker alleles and QTL by adequately measuring the phenotype.

The identification of associations between marker alleles and QTL involves two steps: calibration and verification. Firstly, the power of marker-facilitated selection is very dependent upon an accurate assessment of the phenotypes in the population where the marker–QTL associations are initially identified. It is interesting to note that a new cultivar is not released without several hundred replications of data over four or five years and more than 80 environments, yet QTL studies typically use less than four or five environments with minimal replication to develop linkages between markers and QTL. Sufficient resources need to be devoted to the calibration phase. Secondly, it is necessary to verify the marker–QTL associations identified in the calibration. Marker-facilitated selection can reduce the requirements for large-scale field evaluation and phenotypic measurement, but not during the calibration phase of its implementation.

After the marker–QTL associations are identified and confirmed, application to general breeding populations can be made. As mentioned in most studies and reviews, however, the repeatability between populations will depend on the segregation of the marker and QTL in the breeding population, whether the marker–QTL linkage is in coupling or repulsion and is the same in both populations, and the effect of epistasis (cf. Dudley, 1993). Application of marker-assisted selection to breeding populations could effectively increase heritabilities for traits like yield, and produce better responses, or similar responses with less effort, than traditional methods (Lande, 1992). Furthermore, a selection index that includes both phenotypic measurements and a molecular marker score can increase the selection response relative to phenotypic selection alone, particularly if much of the additive genetic variance in a character can be explained using the molecular markers (Lande, 1992).

For qualitatively inherited traits, a linked reporter locus is useful in many respects. For example, it can be used to sort the dominant phenotypes of the functional locus into homozygous and heterozygous classes. A marker-facilitated method for detecting specific genotypes in segregating populations is very useful during backcrossing, not only to expedite the introgression of a donor parent gene into the recurrent parent, but especially to accelerate the recovery of the recurrent parent genome. It is particularly effective in those cases where the measurement of the functional phenotype is either too expensive or too difficult, or where several genes with the same phenotype (e.g. unlinked duplicate genes for disease resistance) need to be incorporated into a single line.

Summary

Judging from the literature and paper sessions at scientific meetings, we are in the logarithmic phase of information generation relative to biotechnology applications to plant breeding. The new technologies that emerged over the past ten years are now being used to obtain correlative data for important agronomic traits, and to develop genetic maps. Map development should be a priority because the saturated genetic map is necessary, not only for effective marker-assisted selection, but also for map-based cloning.

Innovative application of the technologies and development of new approaches will improve our understanding of genome organization and gene regulation, the role of regulatory sequences in gene expression *in vivo* and their contributions to continuous variation, and genotype-environment interactions. After obtaining and assimilating the information from our current calibration phase, breeders will be in a position to apply transformation and marker-assisted selection routinely and with confidence in cultivar development programmes.

References

Allard, R.W. (1960) *Principles of Plant Breeding*. John Wiley & Sons, New York.

Baenziger, P.S. and Peterson, C.J. (1992) Genetic variation: Its origin and use for breeding self-pollinated species. In: Stalker, H.T. and Murphy, J.P. (eds) *Plant Breeding in the 1990s*. CAB International, Wallingford, UK, pp. 69–92.

Bernard, R.L., Juvik, G.A., Hartwig, E.E. and Edwards, Jr., C.J. (1988) Origins and pedigrees of soybean varieties in the United States and Canada. *US Department of Agriculture, Technical Bulletin* No. 1746. (Correction insert added in 1991).

Borojevic, S. (1990) *Principles and Methods of Plant Breeding*. Elsevier Science Publishers, B.V., Amsterdam.

Boerma, H.R. (1979) Comparison of past and recently developed soybean cultivars in maturity groups VI, VII, and VIII. *Crop Science* 19, 611–613.

Briggs, S.P. (1992) Identification and isolation of agronomically important genes from plants. In: Stalker, H.T. and Murphy, J.P. (eds) *Plant Breeding in the 1990s*. CAB International, Wallingford, UK, pp. 373–387.

Burton, J.W., Wilson, R.F., Brim, C.A. and Rinne, R.W. (1989) Registration of soybean germplasm lines with modified fatty acid composition of seed oil. *Crop Science* 29, 1583.

Burton, J.W., Wilson, R.F. and Brim, C.A. (1994) Registration of N79–2077–12 and N87–2122–4, two soybean germplasm lines with reduced palmitic acid in seed oil. *Crop Science* 34, 313–314.

Caetano-Anolles, G., Bassam, B.J. and Gresshoff, P.M. (1991) DNA amplification fingerprinting using very short arbitrary oligonucleotide primers. *Bio/Technology* 9, 553–557.

Christou, P., McCabe, D.E., Swain, W.F. and Russell, D.R. (1993) Legume transformation.

In: Verma, D.P.S. (ed.) *Control of Plant Gene Expression*. CRC Press, Boca Raton, Florida, pp. 547–564.

Close, P.S., Shoemaker, R.C. and Keim, P. (1989) Distribution of restriction site polymorphism within the chloroplast genome of the genus *Glycine*, subgenus *Soja*. *Theoretical and Applied Genetics* 77, 768–776.

Concibido, V.C., Denny, R.L., Boutin, S.R., Hautea, R., Orf, J.H. and Young, N.D. (1994) DNA marker analysis of loci underlying resistance to soybean cyst nematode (*Heterodera glycines* Ichinohe). *Crop Science* 34, 240–246.

Cregan, P.B., Akkaya, M.S., Bhagwat, A.A., Lavi, U. and Rongwen, J. (1994) Length polymorphisms of simple sequence repeat (SSR) DNA as molecular markers in plants. In: Gresshoff, P.M. (ed.) *Plant Genome Analysis*. CRC Press, Boca Raton, Florida, pp. 43–49.

Draper, J. and Scott, R. (1991) Gene transfer to plants. In: Grierson, D. (ed.) *Plant Genetic Engineering*. Blackie & Son, Glasgow and London, pp. 38–81.

Dudley, J.W., Saghai Maroof, M.A. and Rufener, G.K. (1992) Molecular marker information and selection of parents in corn breeding programs. *Crop Science* 32, 301–304.

Dudley, J.W. (1993) Molecular markers in plant improvement: Manipulation of genes affecting quantitative traits. *Crop Science* 33, 660–668.

Evans, D.A. (1989) Techniques in plant cell and tissue culture. In: Kung, S. and Arntzen, C.J. (eds) *Plant Biotechnology*. Butterworth Publishers, Boston, pp. 53–76.

Fehr, W.R., Cianzio, S.R. and Welke, G.A. (1990a) Registration of 'SS202' soybean. *Crop Science* 30, 1361.

Fehr, W.R., Cianzio, S.R., Welke, G.A. and LeRoy, A.R. III (1990b) Registration of 'SS201' soybean. *Crop Science* 30, 1361.

Grabau, E.A., Davis, W.H. and Gengenbach, B.G. (1989) Restriction fragment length polymorphism in a subclass of the 'Mandarin' soybean cytoplasm. *Crop Science* 29, 1554–1559.

Grabau, E.A., Davis, W.H., Phelps, N.D. and Gengenbach, B.G. (1992) Classification of soybean cultivars based on mitochondrial DNA restriction fragment length polymorphisms. *Crop Science* 32, 271–274.

Hammond, E.G. and Fehr, W.R. (1983a) Registration of A5 germplasm line of soybean. *Crop Science* 23, 192.

Hammond, E.G. and Fehr, W.R. (1983b) Registration of A6 germplasm line of soybean. *Crop Science* 23, 192–193.

Helentjaris, T.G. (1992) RFLP analyses for manipulating agronomic traits in plants. In: Stalker, H.T. and Murphy, J.P. (eds) *Plant Breeding in the 1990s*. CAB International, Wallingford, UK, pp. 357–372.

Hiatt, A. and Mostov, K. (1993) Assembly of multimeric proteins in plant cells: Characteristics and uses of plant-derived antibodies. In: Hiatt, A. (ed.) *Transgenic Plants: Fundamentals and Applications*. Marcel Decker, Inc. New York, pp. 221–237.

Hinchee, M.A.W., Newell, C.A., Connor-Ward, D.V., Armstrong, T.A., Deaton, W.R., Sato, S.S. and Rozman, R.J. (1990) Transformation and regeneration of non-solanaceous crop plants. In: Gustafson, J.P. (ed.) *Gene Manipulation in Plant Improvement II*. Plenum Press, New York, pp. 203–212.

Hymowitz, T., Palmer, R.G. and Singh, R.J. (1991) Cytogenetics of the genus *Glycine*. In: Tsuchiya, T. and Gupta, P.K. (eds) *Chromosome Engineering in Plants: Genetics, Breeding, Evolution. Part B*. Elsevier, New York, pp. 53–63.

Krebbers, E.,Van Rompaey, J. and Vandekerckhove, J. (1993) Expression of modified seed storage proteins in transgenic plants. In: Hiatt, A. (ed.) *Transgenic Plants: Fundamentals and Applications.* Marcel Decker, Inc. New York, pp. 37–60.

Lal, R. and Lal, S. (1990) *Crop Improvement Utilizing Biotechnology.* CRC Press, Boca Raton, Florida.

Lal, R. and Lal, S. (1993) *Genetic Engineering of Plants for Crop Improvement.* CRC Press, Boca Raton, Florida.

Lande, R. (1992) Marker-assisted selection in relation to traditional methods of plant breeding. In: Stalker, H.T. and Murphy, J.P. (eds) *Plant Breeding in the 1990s.* CAB International,Wallingford, UK, pp. 437–451.

Lee, D.J., Caha, C.A., Specht, J.E. and Graef, G.L. (1992) Chloroplast DNA evidence for non-random selection of females in an outcrossed population of soybeans (*Glycine max* (L.) Merr.). *Theoretical and Applied Genetics* 85, 261–268.

Lee, D.J., Caha, C.A., Specht, J.E. and Graef, G.L. (1994) Analysis of cytoplasmic diversity in an outcrossing population of soybeans [*Glycine max* (L.) Merr.]. *Crop Science* 34, 46–50.

Luedders, V.D. (1977) Genetic improvement in yield of soybeans. *Crop Science* 17, 971–972.

Moore, G.A. and Collins, G.B. (1983) New challenges confronting plant breeders. In: Tanksley, S.D. and Orton,T.J. (eds) *Isozymes in Plant Genetics and Breeding. Part A.* Elsevier Science Publishers B.V., Amsterdam, pp. 25–58.

Muehlbauer, G.J., Specht, J.E.,Thomas-Compton, M.A., Staswick, P.E. and Bernard, R.L. (1988) Near-isogenic lines – a potential resource in the integration of conventional and molecular marker linkage maps. *Crop Science* 28, 729–735.

Pen, J., Sijmons, P.C., van Ooijen, A.J.J. and Hoekema, A. (1993) Protein production in transgenic crops: Analysis of plant molecular farming. In: Hiatt, A. (ed.) *Transgenic Plants: Fundamentals and Applications.* Marcel Decker, Inc. New York, pp. 239–251.

Shoemaker, R.C., Hatfield, P.M., Palmer, R.G. and Atherly, A.B. (1986) Chloroplast DNA variation in the genus *Glycine* subgenus *Soja. Journal of Heredity* 77, 26–30.

Shoemaker, R.C., Guffy, R.D., Lorenzen, L.L. and Specht, J.E. (1992) Molecular genetic mapping of soybean: map utilization. *Crop Science* 32, 1091–1098.

Singh, R.J., Kollipara, K.P. and Hymowitz,T. (1993) Backcross (BC$_2$-BC$_4$)-derived fertile plants from *Glycine max* and *G. tomentella* intersubgeneric hybrids. *Crop Science* 33, 1002–1007.

Snape, J.W., Law, C.N.,Worland, A.J. and Parker, B.B. (1990) Targeting genes for genetic manipulation in crop species. In: Gustafson, J.P. (ed.) *Gene Manipulation in Plant Improvement II.* Plenum Press, New York, pp. 55–76.

Specht, J.E. and Williams, J.H. (1984) Contribution of genetic technology to soybean productivity: retrospect and prospect. In: Fehr, W.R. (ed.) *Genetic Contributions to Yield Gains of Five Major Crop Plants.* CSSA Special Publication No. 7, Crop Science Society of America, Madison,Wisconsin, USA, pp. 49–74.

Stalker, D.M. (1991) Developing herbicide resistance in crops by gene transfer technology. In: Grierson, D. (ed.) *Plant Genetic Engineering.* Blackie & Son, Glasgow and London, pp. 38–81.

Stuber, C.W. (1990) Molecular markers in the manipulation of quantitative characters. In: Brown, A., Clegg, M., Kahler, A. and Weir, B. (eds) *Plant Population Genetics,*

Breeding, and Genetic Resources. Sinauer Associates, Sunderland, Massachusetts, USA, pp. 334–350.

Stuber, C.W. (1992) Biochemical and molecular markers in plant breeding. *Plant Breeding Reviews* 9, 37–61.

Sturtevant, A.P. and Beachy, R.N. (1993) Virus resistance in transgenic plants: Coat protein-mediated resistance. In: Hiatt, A. (ed.) *Transgenic Plants: Fundamentals and Applications.* Marcel Decker, Inc. New York, pp. 93–112.

Tanksley, S.D. (1983) Gene mapping. In: Tanksley, S.D. and Orton, T.J. (eds) *Isozymes in Plant Genetics and Breeding. Part A.* Elsevier Science Publishers B.V., Amsterdam, pp. 25–58.

Wilcox, J.R. and Cavins, J.F. (1986) Registration of C1640 soybean germplasm. *Crop Science* 26, 209–210.

Wilcox, J.R. and Cavins, J.F. (1990) Registration of C1726 and C1727 soybean germplasm with altered levels of palmitic acid. *Crop Science* 30, 240.

Wilcox, J.R., Schapaugh, Jr., W.T., Bernard, R.L., Cooper, R.L., Fehr, W.R. and Niehaus, M.H. (1979) Genetic improvement of soybeans in the Midwest. *Crop Science* 19, 803–805.

Williams, J.G.K., Kubelik, A.R., Livak, K.J., Rifalski, J.A. and Tingey, S.V. (1990) DNA polymorphisms amplified by arbitrary primers are useful as genetic markers. *Nucleic Acids Research* 18, 6531–6535.

Wu, R. (1989) Methods for transforming plant cells. In: Kung, S. and Arntzen, C.J. (eds) *Plant Biotechnology.* Butterworth Publishers, Boston, pp. 35–51.

In vitro Selection and Culture-induced Variation in Soybean

6

J.M. Widholm

Plant and Animal Biotechnology Laboratory, Department of Agronomy, University of Illinois, 1201 West Gregory, Urbana, Illinois 61801, USA.

Introduction

Soybean cells grow readily when placed under culture conditions and have been studied as undifferentiated friable callus or suspension cultures for many years. These studies include those of Miller (1960) where a sensitive cytokinin assay was developed with callus of the cv. Acme. These friable cultures will not undergo differentiation to regenerate plants, however, so new approaches were developed within the last ten years to produce cultures capable of regeneration into fertile plants via either organogenesis or embryogenesis. These cultures usually consist of relatively large tissue masses which are not as ideal as single or small clumps of cells for *in vitro* selection. Thus only a few selection studies have been carried out with regenerable soybean cultures. However, since plants can now be routinely regenerated there have been descriptions of somaclonal variation in progeny of regenerated soybean plants from several laboratories as summarized below.

In this review I have kept the 'In vitro Selection' and 'Tissue Culture Induced Variation or Somaclonal Variation' sections separate since the former is a directed process and the latter is observed only in progeny of regenerated plants. Some feel that *in vitro* selection finds mutants which result from somaclonal variation but the *in vitro* selection frequency is usually much lower than that of somaclonal variation so this conclusion can be questioned. This will be discussed in more detail later in this review.

In the studies described here the regenerated plants will be denoted R_0, their selfed progeny R_1, and so on.

In vitro Selection

Biochemical traits

One of the first attempts to carry out selection with soybean tissue cultures was that of Limberg *et al.* (1979) where a suspension culture denoted SB-1 (cv. Mandarin) was grown in a medium with maltose rather than sucrose as the sole carbon source. Initially the cells grew very slowly with 60–70% viability, but after 20 cell doublings the culture had decreased its doubling time from 200 to 100 h and after 310 cell doublings the doubling time reached 24 h which is the same as the wildtype on the sucrose medium. This ability to grow with maltose as the sole carbon source was retained even after 100 cell doublings with sucrose as the carbon source thus demonstrating the stability of the trait. The variant cells were able to take up ^{14}C-maltose much more rapidly than the wildtype which might be the mechanism causing the increased growth rate.

Selection was also carried out with a feeder cell system devised by Weber and Lark (1979) where cells are plated on top of a filter placed on a sponge suspended over suspension cultured feeder cells in a Petri dish. The system can support the efficient growth of a low number of cells or protoplasts or a few survivors in a large number of dying cells during selection. An SB-1 line resistant to the purine analogue, 6-thioguanine, was selected by transfer to gradually increasing concentrations. Another line resistant to another purine analogue, 8-azaguanine, was also selected using, in this case, one-step selection with a high analogue concentration. To avoid the death of the few surviving cells caused by toxic substances released by the dying cells, the filters with the cells were transferred to fresh feeder plates every two days. The mechanism of resistance was not determined for either cell line.

The mutagenesis efficiency of several mutagens was tested by selecting for the ability of soybean suspension cultured cells to grow when plated on the feeder plate system (Weber and Lark, 1979) with maltose as the carbon source (Weber and Lark, 1980). Usually about 1 to 3.5×10^{-7} of the unmutagenized cells formed colonies within six to seven days on the maltose medium. Treatment with the mutagens decreased the overall colony forming ability but also increased the frequency of maltose utilizing colonies by about 10- to 10,000-fold. The frame shift mutagen hycanthone was most effective, while the alkylating agents, ethyl methanesulfonate, methyl methanesulfonate, N-methyl-N'-nitro-N-nitrosoguanidine and ultraviolet light radiation were less effective. The soybean cells did not show photoreactivation following ultraviolet light treatment upon a 24 h exposure to visible light, in contrast to carrot (Howland, 1975) and tobacco (Trosko and Mansour, 1968) cells, where photoreactivation occurred. Trosko and Mansour (1968) could not demonstrate photoreactivation with *Happlopappus* cultures, however.

The maltose-utilization trait selected for by Weber and Lark (1980) was stable and the frequency of maltose-utilizing colonies was increased by muta-gens, which indicates that the maltose utilization ability was caused by a mu-tation. However, since the exact mechanism was not elucidated and plants could not be regenerated to test if the trait was heritable in progeny, it is im-possible to conclude unequivocally that the selected lines are mutants. The feeder plate selection method has the advantage of allowing growth of a few cells even though most of the cells could be killed by the mutagenic or selec-tion treatments. Under normal plating or suspension culture conditions, cell growth might not be observed if the majority of the cells were killed by some toxic treatment. This phenomenon, which is probably due to the release of toxic compounds from dying cells, has been called cooperative death by We-ber and Lark (1979). The plating of the cells on a filter also allows one to move this filter and the cells to fresh or altered media easily.

Polacco (1979) devised a negative selection system using the phosphate analogue arsenate to kill growing but not nongrowing soybean cells (cv. Kanrich). The selective killing was demonstrated by showing that 1 mM sodium arsenate killed suspension cultured cells growing in the normal culture medium or in medium with urea as the sole nitrogen source. However, if citrate was added to chelate the Ni^{2+} in the medium, which is an essential cofactor for the urease activity needed to utilize urea, the cells stopped grow-ing and then, if treated with arsenate, less than 50% of the cells were killed. This negative selection system was applied to cells which were first mutagen-ized with ethyl methanesulfonate and then grown on a medium with urea as the sole nitrogen source, where the arsenate was applied to kill all cells which could use the urea for growth. Twelve colonies formed when 2.2×10^7 of the treated cells were plated on the normal nitrate- and ammonium-containing medium. Of these, only one did not grow rapidly on either the normal or urea media. The growth of this line was stimulated by the addition of extracellular proteins, bovine serum albumin or amino acids, but the specific lesion was not described and it did not appear to be urease, which was the intended tar-get.

Studies of urease-null mutants have led to the finding that soybean cul-tures can be contaminated by a slow-growing pink bacterium, *Methylobacter-ium mesophilicum*, which can grow with methanol as a carbon source (Holland and Polacco, 1992). These bacteria, which are often called pink-pig-mented facultative methylotrophic (PPFM) bacteria, are ubiquitous on plants and can be detected on most leaves, including soybean, by pressing the leaves onto a solid medium containing methanol as the sole carbon source (Corpe, 1985). Holland and Polacco (1992) also found these PPFM bacteria in their soybean callus cultures when callus homogenates were plated on the metha-nol-containing medium. These bacteria also contribute the urease enzyme activity found in tissue cultures of certain urease-null soybean mutants. Therefore biochemical studies with such contaminated cultures could give spurious results.

We have carried out a series of studies of ureide metabolism with a soybean suspension culture where contamination with PPFMs would adversely affect the conclusions (Stahlhut and Widholm, 1989 a,b,c). The cells contained the ureides allantoin and allantoic acid when grown with NH_4^+ and NO_3^- as nitrogen sources and could be adapted to grow on allantoate, allantoin, inosine monophosphate, or uric acid as sole nitrogen sources by repeated subculture. Growth was inhibited when the purine degradation inhibitors, allopurinol and allantoxanamide, were added to cultures grown with intermediates before the point of inhibition (xanthine dehydrogenase and uricase, respectively) and not with those after the point of inhibition. These inhibitors also prevented the incorporation of label from ^{14}C-inosine monophosphate into allantoin and allantoate. The soybean suspension cultures could grow with urea as the sole nitrogen source and this required the presence of Ni^{2+}, a urease cofactor, but were inhibited by the urease inhibitor phenylphosphoradiamidate (PPD) (Stahlhut and Widholm, 1989b). Both PPD or another urease inhibitor, acetohydroxamic acid (AHA), inhibited ^{14}C-urea utilization and $^{14}CO_2$ evolution. However, the growth of cells on allantoin as the sole nitrogen source did not require Ni^{2+} and was not inhibited by PPD and the urease inhibitors PPD or AHA, or the cofactor Ni^{2+} had little effect on $^{14}CO_2$ release from ^{14}C-allantoin by the cells. These results indicate that urea is not an intermediate in the metabolism of ureides by soybean cell cultures, as was also shown by the last study in this series (Stahlhut and Widholm, 1989c), where allantoin degradation occurs via the enzyme allantoinase to produce allantoic acid which is probably degraded by enzymes to glycolate, CO_2 and NH_3 without urea as an intermediate as found in soybean plants (Winkler *et al.*, 1988). These studies show that purine and ureide metabolism in the cultured soybean cells is similar to that of whole plants.

We feel that the soybean suspension cultures used in these selection and purine and ureide metabolism studies were not contaminated with PPFMs, since allantoin degradation did not utilize urease (Stahlhut and Widholm, 1989b) and urea is not an intermediate (Stahlhut and Widholm, 1989c), while it is known that the PPFM bacteria utilize a pathway with urea as an intermediate (Polacco and Holland, 1993). Thus the PPFMs could not be carrying out the ureide metabolism we were measuring. We also could not find any bacterial colonies after grinding the cells and plating on the methanol medium (Corpe, 1985).

We have also initiated two soybean suspension cultures (Corsoy and PI 437833) which can grow photoautotrophically (CO_2 as the sole carbon source and light as the sole energy source) (Horn *et al.*, 1983, Rogers and Widholm, 1988). Like most photoautotrophic plant cell cultures, these required a long period of adaptation from green heterotrophic callus to mixotrophic suspension cultures and finally to photoautotrophic suspensions. These cultures have gradually improved their photosynthetic characteristics (CO_2-fixation enzyme and chlorophyll levels) over a period of many years (reviewed by Widholm, 1992). The Corsoy line can now be reversibly bleached and greened by changing the light conditions and the sugar content of the medium which is quite in contrast to the original long term period needed for initiation of the culture. This new ability, along with the gradual improvement in photosynthetic characteristics, indicates that selection for cells with these altered characteristics has occurred during the evolution of these photoautotrophic soybean suspension cultures. That these cultures are indeed mutants would seem to be difficult to prove, however.

Iron deficiency chlorosis, which can be observed with certain soybean cultivars on high pH calcareous soils, is an agronomic problem which can cause serious yield losses. Sain and Johnson (1984) found that suspension cultures of a known iron-efficient cv. Hawkeye grew better on low iron concentrations than did one initiated from the iron-inefficient line PI 54619-5-1. Hawkeye, but not the PI cells, could continue to grow well for many transfers on 27 μM Fe. The short-term uptake rate of Fe from the medium was similar for both cell lines. When these cell lines were incubated in growth-limiting Fe concentrations, both lines adapted and could grow at the same rate as the control kept on the normal Fe level (108 μM). Thus this study showed that *in vitro* growth with low Fe in the medium could differentiate between efficient and inefficient genotypes and selection could produce cell lines from both genotypes which were more Fe-efficient than before selection.

In other studies of Fe nutrition, 19 soybean genotypes with varying levels of Fe-efficiency were evaluated for Fe chlorosis at five field locations and callus initiated from each was also grown on a medium with low Fe (10 μM instead of 100 μM) or low auxin (0.02 μM naphthaleneacetic acid [NAA] instead of 0.2 μM) to measure growth rates (Stephens *et al.*, 1990). The field evaluation showed that the genotypes used ranged from very Fe-efficient to very Fe-inefficient, and similar variation was seen when the callus growth was measured on the two test media in comparison with the control grown on the normal high-Fe and high-NAA medium. The correlation between field scores and callus growth response was $r^2 = 0.78$ for the low-auxin medium and $r^2 = 0.72$ for the low-Fe medium.

The callus growth Fe-efficiency test medium was improved by adding 10 mM NaHCO$_3$ by Graham *et al.* (1992). The presence of HCO$_3$ $^-$ in the nutrient solution had been shown previously by Byron and Lambert (1983), among others, to enhance the ability of the system to differentiate between Fe-efficient and Fe-inefficient soybean genotypes. When the improved medium was used to measure the callus growth of ten soybean genotypes which were also rated for Fe-efficiency at five field locations, a very high correlation ($r^2 = 0.92$) was obtained between callus growth reduction and field chlorosis ratings. In this case the callus was initiated from epicotyls of axenic soybean seedlings so the remainder of the plant survived and the axillary buds formed shoots to reconstitute a whole plant which will produce seed. This general method is being used in a commerical breeding programme by placing cotyledon pieces on the test medium where efficient genotypes form callus and inefficient ones turn brown and form no callus. This method is also nondestructive since only a cotyledon is removed from the plant.

Selection for resistance to the dihydrofolate reductase (DHFR) inhibitor, methotrexate (MTX, a folic acid analogue), was carried out by Weber *et al.* (1985) using a haploid soybean suspension culture. The cells were plated on feeder plates and irradiated with ultraviolet light to kill 90% of the cells. After 14 days (to allow recovery and mutation fixation) the filter was moved to fresh medium with 22 μM MTX to select for resistance in one step or to 0.22 μM MTX followed by transfer of the growing colonies to gradually increasing concentrations for gradual selection. Suspension cultures were initiated from the resistant colonies and ten of these were cloned through protoplasts for further study. The resistance was stable in the absence of MTX. Two of the ten lines studied showed very low rates of MTX uptake which would explain their resistance. The DHFR enzyme activity was not altered in any of the ten lines as far as MTX inhibition kinetics or activity levels. Thus the resistance mechanism for most of the resistant lines is unknown.

Disease resistance

Culture filtrates of the fungus *Phialophora gregata*, the causal organism of the important soybean disease brown stem rot, were incorporated into the soybean callus growth medium at different dilutions (Gray *et al.*, 1986). Callus of the susceptible cultivars was affected more than that of the resistant genotypes PI 437833 or PI 84946-2 as shown by growth inhibition, browning and viability as measured by triphenyltetrazolium chloride reduction. Culture filtrates from nonpathogenic *P. gregata* isolates had no effect on callus from susceptible or resistant genotypes. Tobacco callus was not affected by culture filtrates from the soybean pathogenic isolates.

Selection using growth inhibitory levels of the *P. gregata* culture filtrate with organogenic soybean cultures initiated and maintained by the methods of Barwale *et al.* (1986) produced several plants after several recurrent selection cycles (Y.Q. Guan, L.E. Gray and J.M. Widholm, unpublished). However, the progeny grown from seed produced by these plants were not resistant when challenged with the fungus in the greenhouse. Thus the regenerated plants were escapes or the resistance trait was epigenetic and was thus not passed to the progeny.

Addition of culture filtrates or purified cell walls of the soybean fungal pathogen *Phytophthora megasperma* to soybean suspension cultures caused the accumulation of the phytoalexin, glyceollin, and the loss of cell viability (Bhandal *et al.*, 1987). The additon of glyceollin to the medium also caused cell death. Some cells would remain viable if the culture filtrate was rinsed out within 24 h after treatment. In the case of *P. megasperma*, a toxin is probably not involved in symptom development, but elicitors consisting of cell wall degradation products (glucans) are (Ebel *et al.*, 1976). The elicitors induce phytoalexin (glyceollin) synthesis which can kill the cells as in a hypersensitive response. Thus, since the plants which are resistant are those that react more rapidly to elicitation, the corresponding cells would probably be the ones which die first and would be eliminated rather than survive the selection. Therefore an elicitor-based selection scheme would appear to not be useful for selecting resistant cells unless the most rapidly responding cells could somehow be identified and be recovered. The soybean cells were able to metabolize the added glyceollin in this study (Bhandal *et al.*, 1987).

Toxic culture filtrates from the fungus, *Septoria glycines*, which causes the soybean leaf disease, brown spot, were used to select for resistance using organogenic callus cultures (Song *et al.*, 1994). The culture filtrate has been shown to contain a high molecular weight polysaccharide with a high uronic acid content which is the toxic substance (Song *et al.*, 1993). The culture filtrate inhibited soybean callus growth but did not affect the growth of calli of the nonhost plants carrot, corn, cotton, *Datura innoxia* and tobacco. The organogenic callus cultures were selected by transfer many times on medium containing growth inhibitory levels of the *S. glycines* culture filtrates and a few plants were regenerated from the surviving cultures. The progeny of the selected regenerated plants segregated for resistance to the disease when inoculated in the field. The resistance was characterized by a much slower development of the disease symptoms, leaf browning, chlorosis and abscission. The resistance was initially found only in shorter, weaker plants. Unfortunately, the resistance continued to segregate after several self-fertilizations of resistant plants and is now observed rarely, indicating some sort of genetic instability. Resistance to this disease would be very valuable since screening of the germplasm collection has not identified any natural resistance.

Recently we have found that the culture filtrate of the fungal pathogen *Fusarium solani*, the causal organism of the sudden death syndrome disease in soybean, inhibits soybean callus growth (Jin *et al.*, 1993). The growth inhibition is more severe with genotypes known to be most susceptible to the disease in greenhouse tests where plants are inoculated with the live organism. Callus browning was also noted in the susceptible genotypes and this could be quantitated by the absorbency at 330 nm of acetone extracts. There appears to be two toxins which are proteins as determined by susceptibility to proteinase K. The culture filtrates also inhibit the growth of embryogenic suspension cultures.

Herbicide resistance

Selection for tolerance to the herbicide atrazine was carried out by culturing cotyledonary node plus epicotyl explants from soybean seedlings of the PI 438489B on a medium containing 48 mg l^{-1} (0.22 mM) atrazine (Wrather and Freytag, 1991). Sixty-six per cent of the explants died but the survivors regenerated a total of 150 plants via organogenesis. When a normally lethal level of atrazine was applied to the soil surrounding the 150 regenerated plants, ten plants survived. When the 100 plants grown from seeds produced by these ten tolerant plants were treated with atrazine only two plants survived and again only two of 78 seedlings grown from the seed produced by these two tolerant R_1 plants survived. This tolerance appears to not be maternally inherited in contrast with the usual case for most of the previously described selected triazine resistances. The low proportion of tolerance in the progeny would indicate that the resistance is controlled in a complex manner by more than one gene or that it is unstable.

Antibiotic resistance

Selection for kanamycin resistance with organogenic soybean cultures, inoculated with *Agrobacterium tumefaciens* harbouring an *nptII* gene in the T region of a Ti plasmid, did successfully produce transformed fertile plants (Hinchee *et al.* 1988). While only 6% of the regenerated shoots were transformed when the kanamycin selection was applied, none were obtained without selection. This low frequency of successful selection, due to a high frequency of escapes, illustrates one of the problems encountered in working with organogenic soybean cultures which are multicellular and are grown on solid medium. The inhibitor must be transported from the bottom through the tissue mass to affect all parts so the portion in contact with the medium would have higher concentrations than that at the top. Thus, the dose used is a compromise so that the top is inhibited without killing the rest of the callus mass completely. Escapes are likely to be found after selection for this reason.

The development of the continuously proliferating embryogenic soybean suspension cultures described by Finer and Nagasawa (1988) should provide a good system for *in vitro* selection since the cultures grow as masses of continuously proliferating, globular stage embryos. The embryos can be induced to regenerate plants when desired. The only successful selection that has been reported with this system has been in transformation studies like those of Finer and McMullen (1991) where the selectable marker gene *AphIV*, which codes for the hygromycin detoxification enzyme, hygromycin phosphotransferase, was used to transform embryogenic suspension cultures via particle bombardment. Transformed cells expressing hygromycin resistance were selected with normally lethal levels of the antibiotic. The resistant cells formed yellow-green outgrowths on the white-brown tissues killed by hygromycin within four to six weeks. These growing clumps could be manually removed and grown further. Transformed plants were regenerated from these selected cultures and some transformed progeny could also be produced upon self-pollination. However, the regenerated plants (R_0 generation) often showed abnormalities as described below.

Tissue Culture-induced Variation or Somaclonal Variation

The term somaclonal variation, which was coined by Larkin and Scowcroft (1981), can be defined as tissue culture-induced genetic variation (mutations). With this definition it is of importance to remember that mutations are heritable through meiosis unlike epigenetic variation which might be observed in cultures or in regenerated plants but will not be passed to progeny (Meins, 1983). The plants of abnormal appearance often regenerated from soybean cultures are not usually a result of somaclonal variation since the next generation plants are usually normal in size and morphology. Thus only the progeny of the R_0 plants should be inspected for somaclonal variation.

Once plant regeneration from soybean tissue cultures became possible, a number of reports describing somaclonal variation began appearing. We were able to identify low levels of variation in progeny of 263 plants regenerated by both organogenesis and embryogenesis from six different genotypes (Barwale and Widholm, 1987, 1990). The mutants recovered included wrinkled leaf, multiple shoots, chlorophyll-deficient, dwarf and partial and complete sterility. Some phenotypes such as twin seeds, abnormal leaf morphology, abnormal leaflet number and some chlorophyll deficiency were not always stably inherited. Significant changes in seed oil and protein levels were found when about 700 R_3 families were analysed, but these changes were not found in the next generation.

One of the two wrinkled leaf mutants identified was crossed reciprocally with normal plants and the trait was found to be inherited only from the maternal parent (Stephens *et al.*, 1991a). This result indicates that the trait is due to a mutation carried by either the mitochondrial or plastid DNA. We have recently identified another wrinkled leaf somaclonal variant in progeny produced by a plant regenerated from Clark 63 protoplasts. Thus, three independent wrinkled leaf mutants have been found in our work.

Field evaluation for two years at two locations of 86 lines produced by repeated self-pollination of progeny of normal appearance from nine plants regenerated via organogenesis (Barwale *et al.*, 1986) from the commercially important cultivar A 3127 did not find detrimental effects on agronomic traits in comparison with the control (Stephens *et al.*, 1991b). No statistically significant variation ($P < 0.05$) was found in seed quality, seed weight or seed yield but there was variation in maturity, lodging, height, seed protein and oil. All of the variation noted was beneficial but was small so that real improvements by the use of this variation by breeders appears to be unlikely. However, the fact that detrimental effects were not observed indicates that selection and transformation using tissue culture will not necessarily lead to impaired performance.

A total of about 1300 lines produced by self-pollinating individual R_0 plants and their progeny were also evaluated for resistance to infection by the *Septoria* brown spot causal fungus, *S. glycines*, in the field (S.M. Lim and

P.A. Stephens, unpublished). The R_0 plants were regenerated by both organogenesis and embryogenesis from several cultivars in the studies described by
Barwale *et al.* (1986). None of the lines showed any resistance however.

Graybosch *et al.* (1987) also carried out field evaluation of progeny of
plants regenerated from soybean tissue cultures. In this case 89 plants of the
cultivars, Calland, Funman and Wayne were regenerated by organogenesis by
the cotyledonary node regeneration system of Wright *et al.* (1986). Male sterility was found in progeny from two of the 89 lines observed in the R_1 generation. No changes in six marker genes involved in flower, pubescence, hilum or
pod colour were observed in about 7000 R_1 plants. There were two lines with
decreased yield and two with altered plant height but no variability in lodging
or maturity was noted. Here again most lines showed no detrimental effects
from the tissue culture procedure.

When a total of 44 families derived from plants of four different genotypes
regenerated by direct embryogenesis from immature cotyledons by the methods of Ranch *et al.* (1985) were evaluated in the R_1 and R_2 generations, five of
them showed variation in fertility, maturity, height or variegation in both
generations (Ranch and Palmer, 1987). Another variant, which had large
thick leaves, low and variable seed set and green leaves until harvest, was
found to be a tetraploid (80 chromosomes).

Ten plants were also regenerated from embryogenic cultures of the cultivar McCall using the methods of Lazzeri *et al.* (1985) with 15 mg l^{-1} (81 μM)
NAA or with 10 mg l^{-1} (45 μM) 2,4-dichlorophenoxyacetic acid (2,4-D) (Hildebrand *et al.*, 1989). There was some morphological and seed fatty acid composition variation in the R_1 generation but this was not observed in the next
generation. Male sterility was found in one of the lines in the R_2 and R_3 generations.

Plants were regenerated from cotyledonary node plus epicotyl explants
from germinating seeds of the genotypes Bedford, PI 88788 and PI 438489B
(Freytag *et al.*, 1989). Progeny segregated for albinism, lanceolate leaf shape,
leaf variegation, pod variegation, large thick leaves and indeterminate versus
determinate growth habit. Apparently a total of 12 heritable phenotypes were
identified in the progeny from the 131 regenerated plants which would give a
somaclonal variation frequency of about 9%.

In studies where different 2,4-D concentrations were used to induce embryogenesis with immature zygotic soybean cotyledons, using the methods of
Ranch *et al.* (1985), the frequency of somaclonal variation decreased with increasing 2,4-D concentrations (Shoemaker *et al.*, 1991). This increase in genetic fidelity paralleled the normality of the adventitious embryos formed and
their ability to form complete plants. At the lowest 2,4-D concentration used,
22.5 μM (5.0 mg l^{-1}), abnormal cotyledonary stage embryos were formed
and only 10% or less developed into plants. With the highest concentration
used, 200 μM (44 mg l^{-1}), from 80–100% of the embryos formed could convert to form plants. About 40% of the plants regenerated from embryos induced with the low 2,4-D level produced progeny with heritable variation

while only 3% did in the case of the high 2,4-D level. The variant phenotypes observed included partial and complete sterility, curled and wrinkled leaves, dwarfism, chlorophyll deficiency and chimerism, change from indeterminate to determinate growth, lack of unifoliates, yellow-edged cotyledons and variation in the isozymes, aconitase, diaphorase and malate dehydrogenase.

Further study of the somaclonal variants produced by Shoemaker *et al.* (1991) followed the variants produced by the 475 R_0 plants obtained from nine different genotypes (Amberger *et al.*, 1992a). One of two wrinkled leaf variants showed segregation in the R_2 generation indicating that it was controlled by a nuclear recessive allele. The other wrinkled leaf variant showed segregation which did not fit any predictable genetic ratio so that maternal inheritance is a possibility as was noted with another wrinkled leaf trait by Stephens *et al.* (1991a). A chlorophyll deficiency trait also segregated in a 1 : 3 ratio indicating that it likewise is controlled by a recessive, nuclear allele. A curled leaf trait segregated in a ratio of one (dwarf with curled leaves) to two (intermediate height with curled leaves) to one (normal wild type) indicating that the curled leaf trait is dominant. Two of the chlorophyll deficient variants produced chimeric plants along with normal green, viable chlorophyll deficients and/or lethal chlorophyll deficients in segregating populations. This genetic instability is being studied further to attempt to find the cause which could be active transposable elements.

The overall frequency of somaclonal variation in this study (Amberger *et al.*, 1992a) was about 8%. The new variants which had not been identified before as somaclonal variants include malate dehydrogenase and aconitase nulls, curled leaves, lethal chlorophyll deficiencies, no unifoliates and yellow-edged cotyledons. The isozyme variants were found when the patterns of a total of nine isozymes were examined in three R_2 seed from each of 500 individual R_1 plants produced by 185 of the R_0 plants.

The aconitase and malate dehydrogenase-null somaclonal variants were studied further by Amberger *et al.* (1992b) and both were inherited as single, recessive, nuclear alleles. The aconitase mutant which lacked the *Aco2*-b isozyme had not been identifed before. The malate dehydrogenase-null was originally identified as a non-lethal chlorophyll deficient variant which was then found to lack two malate dehydrogenase isoforms. The malate dehydrogenase-null trait and the chlorophyll deficiency traits always segregate together. The new malate dehydrogenase-null was found to be allelic to four known malate dehydrogenase-nulls in the *Mdh*1 gene, three of which have w4-m as a parent which is homozygous for mutable alleles at the w4 locus (Groose *et al.*, 1990).

Field evaluation of a total of 113 lines derived from different regenerated plants of three cultivars was carried out at two locations for two years (Hawbaker *et al.*, 1993). The R_0 plants were regenerated by direct embryogenesis using 2,4-D in the studies described by Amberger *et al.* (1992a). The plants showed no obvious phenotypic changes in comparison with the uncultured controls. However, significant variation for all ten quantitative traits examined was found. The traits evaluated included days to beginning bloom, beginning seed, beginning maturity, full maturity, height, lodging, seed yield, seed weight, protein content and oil content. Both favourable and detrimental changes were found with most of the traits. The changes found were considered to be relatively small and to not be very useful for improving soybean if used in a breeding programme.

Mechanisms of somaclonal variation

Somaclonal variation is caused by the occurrence of some sort of genetic instability which apparently occurs during the tissue culture process. It appears as if several different mechanisms can be involved including transposable element activation, chromosome breakage and rearrangement, DNA amplification and single base changes (reviewed by Lee and Phillips, 1988). These occurrences have all been observed in some plant species.

The herbicide 2,4-D has often been implicated as being a mutagen since high concentrations cause chromosomal aberrations (Sikka and Sharma, 1976) and mutations (Kumari and Vaidyanath, 1989) but the study of Shoemaker et al. (1991) shows that very high 2,4-D levels are associated with less somaclonal variation than are lower levels. There was also an association in these studies of low somaclonal variation with the normality of the embryos formed and their ability to convert into plants. Studies of how normal development and the lack of induced mutations are connected could lead to a better understanding of the cause of somaclonal variation, i.e. genome instability.

Studies with Minsoy and Noir 1 soybean suspension cultures grown for over 1000 generations have shown that there can be changes in the RFLP alleles in DNA of cultures initiated from roots but not from those initiated from cotyledons or stems (Roth et al., 1989). The changes noted are almost always made to match those found previously in other soybean genotypes. The authors suggest that the production of particular alleles could result from specific recombinational events which can occur frequently. No matter what the mechanism, the changes in RFLP alleles clearly indicate that DNA base sequences are changing in tissue culture. Why this only happens in cultures initiated from roots is unclear. Most soybean cultures which can regenerate plants are initiated from immature cotyledons or cotyledonary node tissues, so apparently would not be as unstable.

The DNA methylation levels of the 5S RNA genes of suspension cultures initiated from roots of Noir 1, Minsoy and an F1 hybrid between the two were measured after various times in culture using the restriction enzymes HpaII and MspI (Quemada et al., 1987). The methylation levels decreased from almost 100% in the intact plant to about zero in cultures grown for eight months, and then increased again in the same cultures grown for four years. These studies show clearly that DNA methylation patterns can be altered in tissue cultures. It is known that DNA methylation is important in the regulation of gene expression and also in the activation of transposable elements. The decreased methylation found in the soybean cultures, if widespread, could possibly lead to the activation of transposable elements which can cause mutations.

One of the isozyme nulls studied by Amberger *et al.* (1992b), which lacks two nuclear encoded malate dehydrogenase bands, is allelic to four previously identified mutants. Three of these mutants were found in progeny of a line *w4-m* which has been shown to be homozygous for mutable alleles (Groose *et al.*, 1990). The high rate of change of anthocyanin pigmentation in this line controlled by the *w4* locus and the high rate of mutations at other loci suggest that transposable elements are active and may have caused the three malate dehydrogenase nulls. This could indicate that this chromosomal region is a hot-spot for transposable element movement which might be triggered by tissue culture as has been observed previously in maize and alfalfa in particular (reviewed by Lee and Phillips, 1988).

We (Barwale and Widholm, 1990) regenerated plants from the lectin-null mutants, Sooty and Columbia, to determine if this process might activate the excision of the transposable element residing in the lectin gene known to cause the null phenotype (Goldberg *et al.*, 1983). Correct excision of transposable elements can allow the resumption of normal gene expression. However, all 168 R_1 seeds examined still lacked lectin as determined immunologically while the control normal lines were all lectin-positive. More lines need to be screened, however, to adequately test this hypothesis with this system.

Conclusions

The studies discussed here have demonstrated that friable soybean cultures can be used in simple and complicated selection experiments, but since plants cannot be regenerated, the results are only of basic and not applied research value. However, studies of fungal resistance using fungal culture filtrates and Fe-chlorosis evaluations *in vitro* can be used as practical germplasm screening tools.

The soybean cultures do seem to carry out purine and ureide metabolism and react to Fe stress and the fungal culture filtrates in a fashion similar to whole plants so should be ideal systems for basic studies. It should also be possible to select for Fe efficiency and resistance to the culture filtrates using regenerable cultures so that plant improvement should be possible. The development of more friable regenerable cultures or a reproducible protoplast system from which plants can be regenerated readily would expedite the *in vitro* selection work.

In the cases where *in vitro* selection was carried out with regenerable soybean cultures and plants were then regenerated, the selected trait was either not expressed, as in the case of selection with *P. gregata* culture filtrates (Y.Q. Guan, L.E. Gray and J.M. Widholm, unpublished), or was unstable in progeny, as with selection with *S. glycines* culture filtrates (Song *et al.*, 1994) or atrazine (Wrather and Freytag, 1991). However, selection for resistance to the antibiotics, kanamycin (Hinchee *et al.*, 1988) or hygromycin (Finer and McMullen, 1991) did apparently produce stably resistant plants and progeny. Why there are problems in selecting for useful traits is not clear. There have been examples with other species where selection for desirable traits such as herbicide resistance in corn (Newhouse *et al.*, 1991) and amino acid analogue resistance in *Datura innoxia* (Ranch *et al.*, 1983) has produced stable traits which are inherited by progeny. In fact, the corn herbicide resistance has now been backcrossed into commercial inbreds and hybrids are being sold to farmers for commercial production.

Even though genetic transformation has become the most popular tool to attempt to improve plants, in many cases the genes needed to do this have not been isolated so *in vitro* selection should still have a place. Good examples of the possibilities are the disease resistant selection systems described here, if stable resistance can be obtained.

The observations of Holland and Polacco (1992) that PPFMs contaminate their soybean tissue cultures raises a serious question about the use of soybean cultures in many experiments, especially those involving biochemical studies. We have not found these bacteria in any of our cultures when we homogenized them and plated the homogenate on the methanol-containing medium (Corpe, 1985). The reason we do not see PPFM contamination could be due to our use of chlorine gas for sterilizing seed by the method of Gamborg *et al.* (1983) in one case or to the use of different tissues in other cases. We did detect PPFMs on the surfaces of leaves of 40 different greenhouse grown plant species including soybean using the leaf press method of Corpe (1985) so indeed the bacteria are present in the environment.

Somaclonal variation is clearly evident in soybean plants regenerated via both embryogenesis and organogenesis and affects both nuclear and cytoplasmic genes. The ploidy level can also be increased (Ranch and Palmer, 1987). In most of the studies summarized above, the authors state that the variant phenotypes are never found in the original seed maintained line so the tissue culture experience must have induced the changes. The overall frequency of the somaclonal variation is relatively low in soybean where, in all studies combined, a total of about 6.5% (Table 6.1) of the R_0 plants produced variant progeny, except in the case described by Shoemaker *et al.* (1991) where frequencies of 40% were found when low 2,4-D was used to induce embryogenesis. The somaclonal variation frequencies (Table 6.1) do not seem to be markedly affected by the genotype or regeneration method, except as mentioned, the 2,4-D level can be important. The soybean culture-induced mutation rate is low especially in comparison with *Zea mays* (maize) where as many as 1.2 (Edallo *et al.*, 1981) and 0.71 (Zehr *et al.*, 1987) simply inherited mutations per regenerated plant can be observed. Not all species show such high frequencies, however, so soybeans are not necessarily unusual.

Table 6.1. Frequency of somaclonal variation found in progeny of regenerated soybean plants. Certain numbers are estimates due to the data presentation methods of the original publications.

References and regeneration method	No. of R_0 plants	No. of independent mutants	(%)	No. of chlorophyll mutants
Barwale and Widholm (1987, 1990) Organogenesis and embryogenesis	263	8	3	1
Graybosch *et al.* (1987) Organogenesis	89	2	2	0
Ranch and Palmer (1987) Embryogenesis	44	5	11	1
Freytag *et al.* (1989) Organogenesis	131	12	9	3
Hildebrand *et al.* (1989) Embryogenesis	10	1	10	0
Amberger *et al.* (1992a) Embryogenesis	475	38	8	17
Totals	1012	66	6.5	22

Most of the mutations appear to be under nuclear control except for the wrinkled leaf trait where at least one of the several which has been identified is maternally inherited (Stephens *et al.*, 1991a). While we have not determined if the maternally inherited wrinkled leaf trait is caused by a mutation in the plastid or mitochondrial genomes, the mitochondrial DNA has been found to be especially unstable in some tissue cultures such as *Brassica campestris* where large inversions and a duplication can occur (Shirzadegan *et al.*, 1991).

Most of the soybean somaclonal variants produced so far will not be useful for soybean improvement through plant breeding. This has also been observed with somaclonal variation observed with other crops as well, although some disease resistances and improved horticultural traits might be useful enough to be used (Evans, 1989). One somaclonal variation derived commercial soybean variety will be marketed soon which matures earlier than the cultivar from which it was regenerated. If this earlier variety has the same or better characteristics and yield as the parent it would be considered to be improved so could be commercially viable.

It is important to note, as shown by Stephens *et al.* (1991b), Graybosch *et al.* (1987) and Hawbaker *et al.* (1993), that progeny of plants regenerated from soybean tissue cultures do not necessarily have detrimental characters and can perform under field conditions as well as control plants. This of course is important where genetic engineering or *in vitro* selection is applied and plants are regenerated for commercial use. Soybeans appear to be different from another legume, alfalfa, where Bingham and McCoy (1986) indicate that backcrossing may always be necessary to eliminate the undesired somaclonal variation found in the regenerated plants.

Sometimes *in vitro* selected mutants have been called somaclonal variants, but the selected mutants are usually recovered at a very low frequency near that found for spontaneous mutations (about 10^{-6}) or lower so that one could argue that the selection agent is simply picking out the normally occurring spontaneous mutants. In contrast, somaclonal variation is found at a higher frequency than normal background mutations, indicating a different mechanism. That somaclonal variation occurs at a higher frequency cannot be determined directly from the total frequency noted, (mean of 6.5%) with soybean (Table 6.1), since this frequency is for all heritable phenotypic variation which can be caused by mutations in many different genes. It has been estimated that chlorophyll levels are controlled by a total of about 300 different genes in barley (von Wettstein *et al.*, 1971). In the case of soybean somaclonal variation, there seems to be a total of about 22 mutations affecting chlorophyll levels in progeny of about 1012 total plants (Table 6.1). This would give a frequency of 2.2×10^{-2} per plant. If there are 300 different genes which can be mutated to cause these phenotypes then the frequency is 7.2×10^{-5} per gene. This is still almost 100 times higher than the usual spontaneous mutation frequency of about 10^{-6}. The *in vitro* selection frequencies found in the studies described in this review were generally near 10^{-7}.

The embryogenic suspension culture system seems to be very useful for transformation and for *in vitro* selection. However, the initiation requires skill and several months of time and the plants regenerated from the culture after it has grown for a year or two can all be sterile. This sterility problem means that new cultures need to be initiated regularly. One solution to this problem would be to freeze the young culture and thaw some at regular intervals to always maintain lines which will regenerate fertile plants. Soybean suspension cultured cells (Weber *et al.*, 1983) and callus (Engelmann, 1992) have been successfully cryopreserved in liquid nitrogen but no reports of success with the embryogenic suspension or other regenerable soybean cultures have been published.

In general, the *in vitro* selection and somaclonal variation results obtained thus far with soybean have not been very important for improving the plant, although certain screening methods can be applied to breeding programmes. Likewise, the somaclonal variants do not include many useful traits which is not unexpected. Now that we have good regeneration systems with such an important crop plant we can expect to see more soybean improvement result from the use of tissue cultures.

Acknowledgements

The unpublished results presented here were obtained using funds from the Illinois Agricultural Experiment Station, the American Soybean Association, the Illinois Soybean Program Operating Board, the US Department of Agriculture and Agrigenetics Research Associates.

References

Amberger, L.A., Palmer, R.G. and Shoemaker, R.C. (1992a) Analysis of culture-induced variation in soybean. *Crop Science* 32, 1103–1108.

Amberger, L.A., Shoemaker, R.C. and Palmer, R.G. (1992b) Inheritance of two independent isozyme variants in soybean plants derived from tissue culture. *Theoretical and Applied Genetics* 84, 600–607.

Barwale, U.B., Kerns, H.R. and Widholm, J.M. (1986) Plant regeneration from callus cultures of several soybean genotypes via embryogenesis and organogenesis. *Planta* 167, 473–481.

Barwale, U.B. and Widholm, J.M. (1987) Somaclonal variation in plants regenerated from cultures of soybean. *Plant Cell Reports* 6, 365–368.

Barwale, U.B. and Widholm, J.M. (1990) Soybean: plant regeneration and somaclonal variation . In: Bajaj, Y.P.S. (ed.) *Biotechnology in Agriculture and Forestry, Vol. 10, Legumes and Oilseed Crops I.* Springer-Verlag, Berlin, pp. 114–133.

Bhandal, I.S., Paxton, J.D. and Widholm, J.M. (1987) *Phytophthora megasperma* culture filtrate and cell wall preparation stimulate glyceollin production and reduce cell viability in suspension cultures of soybean. *Phytochemistry* 26, 2691–2694.

Bingham, E.T. and McCoy, T.J. (1986) Somaclonal variation in alfalfa. *Plant Breeding Reviews* 4, 123–152.

Byron, D.F. and Lambert, J.W. (1983) Screening soybeans for iron efficiency in the growth chamber. *Crop Science* 23, 885–886.

Corpe, W.A. (1985) A method for detecting methylotrophic bacteria on solid surfaces. *Journal of Microbiological Methods* 3, 215–221.

Ebel, J., Auers, A.R. and Albersheim, P. (1976) Host-pathogen interactions XII. Response of suspension-cultured soybean cells to the elicitor isolated from *Phytophthora megasperma* var. *sojae*, a fungal pathogen of soybeans. *Plant Physiology* 57, 775–779.

Edallo, S., Zucchinali, C., Perenzin, M. and Salamini, F. (1981) Chromosomal variation and frequency of spontaneous mutation associated with *in vitro* culture and plant regeneration in maize. *Maydica* 26, 39–56.

Engelmann, F. (1992) Effects of freezing in liquid nitrogen on the properties of a soybean (*Glycine max* L. var. *acme*) callus strain used as a bioassay for cytokinin activity. *Cryo-Letters* 13, 331–336.

Evans, D.A. (1989) Somaclonal variation – genetic basis and breeding applications. *Trends in Genetics* 5, 46–50.

Finer, J.J. and McMullen, M.D. (1991) Transformation of soybean via particle bombardment of embryogenic suspension culture tissue. *In Vitro Cell and Developmental Biology* 27P, 175–182.

Finer, J.J. and Nagasawa, A. (1988) Development of an embryogenic suspension culture of soybean [*Glycine max* (L.) Merrill]. *Plant Cell Tissue Organ Culture* 15, 125–136.

Freytag, A.H., Rao-Arelli, A.P., Anand, S.C., Wrather, J.A. and Owens, L.D. (1989) Somaclonal variation in soybean plants regenerated from tissue culture. *Plant Cell Reports* 8, 199–202.

Gamborg, O.L., Davis, B.P. and Stahlhut, R.W. (1983) Somatic embryogenesis in cell cultures of *Glycine* species. *Plant Cell Reports* 2, 209–212.

Goldberg, R.B., Horschek, G. and Vodkin, L.O. (1983) An insertion sequence blocks the expression of a soybean lectin gene. *Cell* 33, 465–475.

Graham, M.J., Stephens, P.A., Widholm, J.M. and Nickell, C.D. (1992) Soybean genotype evaluation for iron deficiency chlorosis using sodium bicarbonate and tissue culture. *Journal of Plant Nutrition* 15, 1215–1225.

Gray, L.E., Guan, Y.Q. and Widholm, J.M. (1986) Reaction of soybean callus to culture filtrates of *Phialophora gregata*. *Plant Science* 47, 45–55.

Graybosch, R.A., Edge, M.E. and Delannay, X. (1987) Somaclonal variation in soybean plants regenerated from the cotyledonary node tissue culture system. *Crop Science.* 27, 803–806.

Groose, R.W., Schulte, S.M. and Palmer, R.G. (1990) Germinal reversion of an unstable mutation for anthocyanin pigmentation in soybean. *Theoretical and Applied Genetics* 79, 161–167.

Hawbaker, M.S., Fehr, W.R., Mansur, L.M., Shoemaker, R.C. and Palmer, R.G. (1993) Genetic variation for quantitative traits in soybean lines derived from tissue culture. *Theoretical and Applied Genetics* 87, 49–53.

Hildebrand, D.F., Adams, T.R., Dahmer, M.L., Williams, E.G. and Collins, G.B. (1989) Analysis of lipid composition and morphological characteristics in soybean regenerants. *Plant Cell Reports* 7, 701–703.

Hinchee, M.A.W., Conner-Ward, D.V., Newell, C.A., McDonnell, R.E., Sato, S.J., Gasser, C.S., Fischhoff, D.A., Re, D.B., Fraley, R.T. and Horsch, R.B. (1988) Production of transgenic soybean plants using *Agrobacterium*-mediated DNA transfer. *Bio-Technology* 6, 915–922.

Holland, M.A. and Polacco, J.C. (1992) Urease-null and hydrogenase-null phenotypes of a phylloplane bacterium reveal altered nickel metabolism in two soybean mutants. *Plant Physiology* 98, 942–948.

Horn, M.E., Sherrard, J.H. and Widholm, J.M. (1983) Photoautotrophic growth of soybean cells in suspension culture. I. Establishment of photoautotrophic cultures. *Plant Physiology* 72, 426–429.

Howland, G.P. (1975) Dark-repair of ultraviolet-induced pyrimidine dimers in the DNA of wild carrot protoplasts. *Nature* 254, 160–161.

Jin, H., Widholm, J.M. and Hartman, G.L. (1993) Characterization of phytotoxin(s) involved in sudden death syndrome of soybean. *Phytopathology* 83, 1343.

Kumari, T. and Vaidyanath, K. (1989) Testing of genotoxic effects of 2,4-D using multiple genetic assay systems of plants. *Mutation Research* 226, 235–238.

Larkin, P.J. and Scowcroft, W.R. (1981) Somaclonal variation – a novel source of variability from cell cultures. *Theoretical and Applied Genetics* 60, 197–214.

Lazzeri, P.A., Hildebrand, D.F. and Collins, G.B. (1985) A procedure for plant regeneration from immature cotyledon tissue of soybean. *Plant Molecular Biology Reporter* 3, 160–167.

Lee, M. and Phillips, R.L. (1988) The chromosomal basis of somaclonal variation. *Annual Review of Plant Physiology and Plant Moleculcar Biology* 39, 413–437.

Limberg, M., Cress, D. and Lark, K.G. (1979) Variants of soybean cells which can grow in suspension with maltose as a carbon-energy source. *Plant Physiology* 63, 718–721.

Meins, F. (1983) Heritable variation in plant cell culture. *Annual Review of Plant Physiology* 34, 327–346.

Miller, C.O. (1960) An assay for kinetin-like materials. *Plant Physiology* Suppl. 35, XXVI.

Newhouse, K., Singh, B., Shaner, D. and Stidham, M. (1991) Mutations in corn (*Zea mays* L.) conferring resistance to imidazolinone herbicides. *Theoretical and Applied Genetics* 83, 65–70.

Polacco, J.C. (1979) Arsenate as a potential negative selection agent for deficiency variants in cultured plant cells. *Planta* 146, 155–160.

Polacco, J.C. and Holland, M.A. (1993) Roles of urease in plant cells. *International Review of Cytology* 145, 65–103.

Quemada, H., Roth, E.J. and Lark, K.G. (1987) Changes in methylation of tissue cultured soybean cells detected by digestion with the restriction enzymes *Hpa*II and *Msp*I. *Plant Cell Reports* 6, 63–66.

Ranch, J.P. and Palmer, R.G. (1987) A ploidy variant regenerated from embryogenic tissue cultures of soybean. *Soybean Genetics Newsletter* 14, 161–163.

Ranch, J.P., Rick, S., Brotherton, J.E. and Widholm, J.M. (1983) The expression of 5-methyltryptophan-resistance in plants regenerated from resistant cell lines of *Datura innoxia*. *Plant Physiology* 71, 136–140.

Ranch, J.P., Oglesby, L. and Zielinski, A.C. (1985) Plant regeneration from embryo-derived tissue cultures of soybean. *In Vitro Cell and Developmental Biology* 21, 653–658.

Rogers, S.M.D. and Widholm, J.M. (1988) Comparison of photosynthetic characteristics of two photoautotrophic cell suspension cultures of soybean. *Plant Science* 56, 69–74.

Roth, E.J., Frazier, B.L., Apuya, N.R. and Lark, K.G. (1989) Genetic variation in an inbred plant: variation in tissue cultures of soybean [*Glycine max* (L.) Merrill]. *Genetics* 121, 359–368.

Sain, S.L. and Johnson, G.V. (1984) Selection of iron-efficient soybean cell lines using plant cell suspension cultures. *Journal of Plant Nutrition* 7, 389–398.

Shirzadegan, M., Palmer, J.D., Christey, M. and Earle, E.D. (1991) Patterns of mitochondrial DNA instability in *Brassica campestris* cultured cells. *Plant Molecular Biology* 16, 21–37.

Shoemaker, R.C., Amberger, L.A., Palmer, R.G., Oglesby, L. and Ranch, J.P. (1991) Effect of 2,4-dichlorophenoxyacetic acid concentration on somatic embryogenesis and heritable variation in soybean [*Glycine max* (L.) Merr.]. *In Vitro Cell and Developmental Biology* 27P, 84–88.

Sikka, K. and Sharma, A.K. (1976) The effects of some herbicides on plant chromosomes. *Proceedings of the Indian Academy of Science* 42, 299–307.

Song, H.S., Lim, S.M. and Clark, J.M. (1993) Purification and partial characterization of a host-specific pathotoxin from culture filtrates of *Septoria glycines*. *Phytopathology* 83, 659–661.

Song, H.S., Lim, S.M. and Widholm, J.M. (1994) Selection and regeneration of soybeans resistant to the pathotoxic culture filtrates of *Septoria glycines*. *Phytopathology* 84, 948–951.

Stahlhut, R.W. and Widholm, J.M. (1989a) Growth on and catabolism of purine degradation intermediates by soybean [*Glycine max* (L.) Merr.] cell suspension cultures. *Journal of Plant Physiology* 133, 649–653.

Stahlhut, R.W. and Widholm, J.M. (1989b) Ureide catabolism by soybean [*Glycine max* (L.) Merrill] cell suspension cultures I. Urea is not an intermediate in allantoin degradation. *Journal of Plant Physiology* 134, 85–89.

Stahlhut, R.W. and Widholm, J.M. (1989c) Ureide catabolism by soybean [*Glycine max* (L.) Merrill] cell suspension cultures II. Assimilation of allantoin. *Journal of Plant Physiology* 134, 90–97.

Stephens, P.A., Barwale-Zehr, U.B., Nickell, C.D. and Widholm, J.M. (1991a) A cytoplasmically inherited, wrinkled-leaf mutant in soybean. *Journal of Heredity* 82, 71–73.

Stephens, P.A., Nickell, C.D. and Widholm, J.M. (1991b) Agronomic evaluation of tissue-culture-derived soybean plants. *Theoretical and Applied Genetics* 82, 633–635.

Stephens, P.A., Widholm, J.M. and Nickell, C.D. (1990) Iron-deficiency chlorosis evaluation of soybean with tissue culture. *Theoretical and Applied Genetics* 80, 417–420.

Trosko, J.E. and Mansour, V.H. (1968) Response of tobacco and *Happlopappus* cells to ultraviolet irradiation after posttreatment with photoreactivating light. *Radiation Research* 36, 333–343.

von Wettstein, D., Henningsen, K.W., Boynton, J.E., Kannangara, G.C. and Nielsen, O.F. (1971) The genic control of chloroplast development in barley. In: Boardman, N.K., Linnane, A.W. and Smillie, R.M. (eds) *Autonomy and Biogenesis of Mitochondria and Chloroplasts*. North-Holland, Amsterdam, pp. 205–223.

Weber, G. and Lark, K.G. (1979) An efficient plating system for rapid isolation of mutants from plant cell suspension. *Theoretical and Applied Genetics* 55, 81–86.

Weber, G. and Lark, K.G. (1980) Quantitative measurement of the ability of different mutagens to induce an inherited change in phenotype to allow maltose utilization in suspension cultures of the soybean, *Glycine max* (L.) Merr. *Genetics* 96, 213–222.

Weber, G., Roth, E.J. and Schweiger, H.-G. (1983) Storage of cell suspensions and protoplasts of *Glycine max* (L.) Merr., *Brassica napus* (L.), *Datura innoxia* (Mill.), and *Daucus carota* (L.) by freezing. *Zeitschrift für Pflanzenphysiologie* 109, 29–39.

Weber, G., deGroot, E. and Schweiger, H.-G. (1985) Methotrexate-resistant somatic cells of *Glycine max* (L.) Merr.: selection and characterization. *Journal of Plant Physiology* 117, 339–353.

Widholm, J.M. (1992) Properties and uses of photoautotrophic plant cell cultures. *International Review of Cytology* 132, 109–175.

Winkler, R.G., Blevins, D.G. and Randall, D.D. (1988) Ureide catabolism in soybeans III. Ureidoglycolate amidohydrolase and allantoate amidohydrolase are activities of an allantoate degrading enzyme complex. *Plant Physiology* 86, 1084–1088.

Wrather, J.A. and Freytag, A.H. (1991) Selection of atrazine tolerant soybean calli and expression of that tolerance in regenerated plants. *Plant Cell Reports* 10, 44–47.

Wright, M.S., Koehler, S.M., Hinchee, M.A. and Carnes, M.G. (1986) Plant regeneration by organogenesis in *Glycine max*. *Plant Cell Reports* 5, 150–154.

Zehr, B.E., Williams, M.E., Duncan, D.R. and Widholm, J.M. (1987) Somaclonal variation in the progeny of plants regenerated from callus cultures of seven inbred lines of maize. *Canadian Journal of Botany* 65, 491–499.

Soybean Seed Composition $\boxed{7}$

N.C. Nielsen

*USDA-ARS, Plant Production and Pathology Research Unit,
Department of Agronomy, Purdue University, West Lafayette, Indiana
47907, USA.*

Introduction

Efforts to improve the quantity and quality of food seed grains has long been an objective of mankind. Early efforts were undoubtedly confined to gathering and storing desirable landraces to provide a reliable food source. Later, as sophistication increased, genetic methodology evolved that permitted generation and selection of beneficial gene combinations. These genetic techniques, although supremely successful, relied on a random association of desirable gene combinations. Although breeding techniques increased the probability of finding and selecting useful gene combinations, they were dependent on maintenance of large populations of plants. With the advent of molecular biology, a means has arisen to intervene directly in the combination of specific desirable genes from diverse sources, to eliminate undesirable genes, and to carry out gene engineering to improve the quantity or quality of the primary constituents of seeds. In the limit, application of the directed approaches of biotechnology have the potential to reduce both the time and population sizes required for the recovery of elite germplasm. Because the successful application of molecular biology to crop improvement is quite dependent upon knowledge of the basic processes being manipulated, and information about these processes is invariably limited, it is clear that simultaneous application of both classical and molecular methodologies to seed crop improvement will be the rule for the foreseeable future. It also seems evident that the most immediate impact of molecular methodologies will be in the development of germplasm destined for niche markets.

The purpose of this chapter will be to examine current efforts to improve the quantity and quality of soybean seed constituents. Because of the apparent interdependence of classical breeding and molecular biological methodologies on seed improvement in the near term, the statistical interrelationships between inheritance of seed protein, oil and non-structural carbohydrate will be reviewed briefly. Reference will also be made to the rapidly growing body of knowledge about genes whose products comprise the prevalent proteins found in seeds at physiological maturity. Rapid advances are being made to engineer new oil compositions of soybeans, and because these are summarized elsewhere in this volume they will not be repeated here. However, a few successful and unsuccessful attempts to alter oil and protein composition have now been reported in the literature, and will be summarized in this chapter. Finally, an attempt will be made to assess the potential for successfully modifying seed composition. The hope is that this information will serve as a platform from which the interested reader can assess future limitations and potentials to modify seed composition.

General Features of Seed Development

Depending on the maturity group, genotype and growth conditions, soybean plants require 108 to 144 days from seed germination to the recovery of mature seeds (Gay *et al.*, 1980). Seed fill, from anthesis to physiological maturity, can take place within this period in a time ranging from 18 days at one extreme to about 70 days at the other (Reicosky *et al.*, 1982). During seed fill, embryo development progresses through a number of characteristic morphologically discernible stages. For a detailed description of this process, the interested reader should consult the carefully prepared article by Carlson and Lersten (1987). Briefly, however, the early stages in development are characterized by a period of rapid cell division. When seed fill occurs over a period of 60 days, rapid cell division will be completed by about 15 days after fertilization (DAF). Examination of the cell for organelles at this point in development will reveal a nucleus, mitochondria, plastids that contain starch grains, Golgi, smooth and rough endoplasmic reticulum, and a large central vacuole (Adams *et al.*, 1983). Rapid cell elongation in cotyledons, elevated metabolic activity and a swift increase in seed dry weight are characteristic of the next 35- to 40-day period in seed development. This stage in development is characterized by the rapid proliferation of oil and protein bodies until these vacuoles hinder microscopic visualization of other subcellular organelles. Generally the onset of oil accumulation can be demonstrated to slightly precede that of protein, and then oil accumulation plateaus before protein deposition ceases. The final week of development is characterized by rapid seed dehydration, and the rate of protein accumulation decreases rapidly. At physiological maturity, a typical soybean will contain about 20% oil, 40% protein and 12% non-structural carbohydrate on a dry weight basis, although cultivars with less than 18% oil or more than 50% protein can be found.

Genetic factors strongly affect seed composition. In general, the seed fill period for a particular soybean cultivar will remain constant from one generation to another if environmental growth conditions are reproducible. As even a casual examination will reveal, however, there is substantial genetically controlled variation in seed size. Such differences are not associated with only the duration of seed fill because cultivars of the same maturity vary considerably in the size of seeds they produce. Nonetheless, it has been established that the rate of seed growth is correlated statistically with seed size (Egli *et al.*, 1978; Zeiher *et al.*, 1982; Gent, 1983). The increased seed size can be attributed to greater rates of dry matter accumulation when large seeded varieties are compared to small seeded cultivars of similar maturities (Egli and Leggett, 1973; Beaver and Cooper, 1982). The genotypic differences in seed growth rate are not the consequence of short-term photosynthate production (Egli and Leggett, 1976). Instead, Egli *et al.* (1981) associated the high seed growth rates to greater numbers of cells per seed. For example, they showed the variety Essex had mature seeds averaging 113 mg per seed that grew at an average rate of 4.3 mg per seed per day and contained 6.6×10^6 cells per seed. Similar data for a second variety, Emerald, were 262 mg, 9.6 mg and 10.2×10^6 cells, respectively. Interestingly, the mechanism responsible for this increase in seed cell content and the genetics underlying this important phenomena remain to be explored.

As one might anticipate, changes in environmental growth conditions can also elicit changes in seed yield and composition. For example, it is possible to create a 30–40% increase in the number of pods per plant by increasing light intensity 60% late in flowering and early in pod development (Schou *et al.*, 1978). Although total seed weight per plant is increased by such a treatment, the increase in pods per node results in a decline in seed size and a lower seed growth rate (Kollman *et al.*, 1974; Openshaw *et al.*, 1979). The increase in total seed weight does not come directly from a light induced increase in the production of photosynthate in leaves, because not all of this carbohydrate is mobilized for the increased synthesis of protein, oil and carbohydrate (Streeter and Jeffers, 1979). Rather, the substrates for synthesis of seed storage components are withdrawn from the nearest source – pods, stems and petioles – rather than as newly synthesized substrate from leaves. Compositional changes also occur when environmental conditions result in either an extension or premature termination of seed fill compared with the interval that is normal for a given soybean genotype. The result of extended seed development is a larger seed with increased protein and decreased oil and carbohydrate. The opposite tends to occur when environment conditions, such as drought, hasten seed maturation (Egli *et al.*, 1978; Sato and Ikeda, 1979).

Because of its economic importance, substantial breeding efforts have been expended to alter the fatty acid composition of soybean oil. Early studies revealed that oil content and fatty acid composition are not simply inherited traits (Brim *et al.*, 1968; Howell *et al.*, 1972). Heritability estimates of per cent oil (Burton and Brim, 1981) and linoleate are low (Burton *et al.*, 1983), principally because of environmental interactions. Temperature effects in particular are recognized to profoundly affect oil composition (Collins and Howell, 1957; Gupta and Dhindsa, 1982). In general, low temperatures during seed development result in a greater accumulation of unsaturated fatty acids and starch, lower total oil content and a decreased accumulation of 18:1. The fatty acid composition of cellular membranes undergoes parallel changes. It has been suggested that increased polyunsaturation of the membrane triacylglycerols increases membrane fluidity, and thereby influences cell function at low temperature (Graham and Patterson, 1982). Indeed, Miquel *et al.* (1993) have provided convincing evidence that polyunsaturated fatty acids are required for the growth of higher plants at low temperatures. The latter report also questions the commonly held notion that low temperature damage is the result of disruption of membrane integrity, and points to the need for more definitive information about this important point.

Breeders have also expended considerable effort trying to increase soybean seed protein content. Several important generalizations need to be made concerning this work. The first is that increased protein content in seeds is inversely correlated with yield (Brim and Burton, 1979). The basis for this yield depression remains to be resolved. Second, an inverse correlation exists between oil and protein, and between carbohydrate and protein accumulation. Thus, carbohydrate and oil appear to be positively related to one another genetically, but both are negatively related to protein. Although there have been reports of yield and protein accumulation increasing together, these reports are rare (Brim and Burton, 1979; Burton and Brim, 1983). They have been used by some investigators to argue that oil and protein can be increased simultaneously with appropriate genetic recombination and selection (Wilson, 1987), although the obvious difficulty of this approach suggests it is unlikely to bear useful genotypes in the near term. Rather, an understanding of the basis for the linkage of oil and protein synthesis as it related to the synthesis and accumulation of these compounds in seeds, and the direct biotechnological manipulation of this process, may provide a more efficient approach.

It seems likely that the flux of carbon into carbohydrate, oil and protein is tightly regulated in the developing seed, and that at least part of the control at these steps is based on feedback regulatory mechanisms. Knowledge of the rate-limiting steps concerning the movement of carbon into protein and carbohydrate is speculative. However, results reported by Post-Beittenmiller *et al.* (1991) imply that acetyl CoA carboxylase carries out the rate-limiting step in fatty acid biosynthesis. At least two distinct forms of this enzyme are recognized to exist in many plants. The cytoplasmic form of the enzyme resembles animal acetyl CoA carboxylase in that it is a high molecular weight (220–270 kDa) multifunctional enzyme (Egin-Buhler and Ebel, 1983). Because most fatty acid biosynthesis occurs in plastids, the contributions of this enzyme to the synthesis of storage triglycerides is unclear. A second acetyl CoA carboxylase is located in chloroplasts and, in at least some plants, it has physical characteristics that resemble prokaryotic forms of the enzyme. The three functional components of these chloroplast enzymes, biotin carboxylase, biotin carboxyl carrier protein and carboxyl transferase, are on separate proteins (Kannangara and Stumpf, 1972; Sasaki *et al.*, 1993). In the case of animals, their acetyl CoA carboxylase is subject to regulation by citrate, phosphorylation and feed-back inhibition by long chain acyl CoAs. The limited information about regulatory aspects of the plant enzymes suggests that they may not be similar to those used in the animal system. Because the plant enzyme is known to be the target of aryloxyphenoxypropionate and cyclohexanedione herbicides (Parker *et al.*, 1990a,b), and has been cloned (Roesler *et al.*, 1994), new tools are available with which to analyse the enzyme. Manipulation of the regulatory functions of acetyl CoA carboxylase in seeds may permit the tight inverse linkage between oil and protein to be manipulated. Similar considerations might also be applied to regulated steps in the carbohydrate and protein biosynthetic pathways.

Seed Carbohydrate

Approximately 12% of the seed dry weight is nonstructural carbohydrate at physiological maturity. At early stages in the seed developmental process, when much of the synthetic activity is associated with pod development, the sugar content reflects the high fructose and glucose contents found in leaves. Later, during the active stage of seed development, when protein and oil are being deposited in enormous amounts, the sucrose and starch contents also increase rapidly. Starch content peaks at 11–12% of the seed dry weight toward the end of this period, and begins to decrease with a concomitant increase in soluble sugars as seed maturity approaches. ADPG-starch synthetase, which is located in chloroplasts, is apparently the predominant pathway of starch biosynthesis (Pavlinova and Turkina, 1978). At physiological maturity, starch typically accounts for 1–3% of seed dry weight (Yazdi-Samadi *et al.*, 1977; Wilson, 1987). The majority of the carbohydrate at seed maturity is either sucrose (41–68%), stachyose (12–35%) or raffinose (5–16%). The accumulation of stachyose and raffinose typically occurs late in seed development, immediately prior to and during seed dehydration (Konno, 1979). Even though raffinose and stachyose are translocatable sugars, evidence indicates that the majority of these compounds are synthesized in seeds *de novo* (Kandler and Hopf, 1980). Although sucrose, stachyose and raffinose account for the majority of non-structural carbohydrate in seeds at maturity, fructose, glucose, galactose, mannotriose, pinitol, galactopinitol and myo-inositol are present in lesser amounts (Schweizer *et al.*, 1978).

The relatively large amounts of raffinose and stachyose in soybeans at physiological maturity are problematic. Because the human digestive tract lacks α-galactosidase, these sugars are not metabolized effectively. Instead, they are fermented by bacterial flora in the gut with a resulting release of carbon dioxide as flatulence. In addition to causing an odour problem during animal production, the presence of these compounds in poultry feed have a negative effect on growth and development (Leske *et al.*, 1993). Variability exists among soybean cultivars for the content of oligosaccharide (Hymowitz *et al.*, 1972), although reports about the development of soybean varieties that lack stachyose and raffinose are apparently absent from the literature.

Raffinose, and the group of oligosaccharides related to it, are formed by the successive addition of galactose to a sucrose 'primer' at C-6 of glucose (Axelrod, 1965). Raffinose contains one such molecule and stachyose two. An enzyme preparation from soybeans capable of forming raffinose from sucrose and UDP-galactose has been described (Gomyo and Nakamura, 1966), which might lead one to expect that these oligosaccharides are produced by the successive addition of the galactose from the UDP-sugar. However, Schweizer *et al.* (1978) have provided evidence that raffinose synthesis requires sucrose and galactinol, and Handley *et al.* (1983) demonstrated the presence of galactinol synthetase activity in developing soybeans. Additional research directed toward isolation of these enzymes and evaluation of their importance to stachyose synthesis is important.

Genetic and biochemical studies concerning the inheritance of carbohydrates have been limited to the accumulation of the major components, and have typically not delved into the regulatory mechanisms of the genes which encode enzymes that carry out these processes. Openshaw and Hadley (1981) have shown that there is a high degree of heritability for total sugar content. This trait is apparently maternally inherited and governed by additive gene action (Openshaw and Hadley, 1978). These same authors reported that changes in total sugar content occur independently of stachyose and raffinose content.

The principle enzymes involved in starch metabolism are α- and β-amylase, and starch phosphorylase. Adams *et al.* (1981a,b) demonstrated the presence of both α- and β-amylase in developing, mature and germinating seeds, but starch phosphorylase was not found. Chloroplasts, which are the principle site for starch biosynthesis, contain the α-amylase, but the β-amylase is located elsewhere in cells of developing seeds (Preiss and Levi, 1980). The α-amylase activity peaks during mid-seed fill and then decreases in parallel with starch synthesis and degradation during seed development (Adams *et al.*, 1981b). The same authors reported that β-amylase activity increases during the late stages of seed fill. Of these three enzymes, only the *Sp* genetic system that encodes β-amylases is well characterized. Four alleles of this gene were described by Hildebrand and Hymowitz (1980a,b); Sp_1^{an} (Chestnut variety), Sp_1^a (Amsoy), Sp_1^b and sp_1 (Altona contains a mixture of the last two alleles).

Protein Composition

Although soybeans are the richest source of protein among crops that are commonly grown commercially, only a small proportion of the total protein produced each year finds its way into the food industry in European and North American markets. In contrast, soybeans are a primary source of protein in Oriental societies, where they are used in a number of traditional foods that are not well accepted in Europe, Canada and the United States. In Western societies, soy protein is usually used in combination with other foods, often as a protein extender. A primary impediment to its acceptance is the strong beany flavour associated with products derived from soy protein. It does, however, play an important role in the livestock industry in animal feeds. For this reason, soy protein is considered a co-product of soybean, and increasing attention is being paid in the industry to the seed protein content.

The average cultivated soybean contains about 40% protein, but there are accessions in the world soybean collection that have about 50% protein. The latter generally have low yields, and consequently are not used in the United States for commercial agriculture. Attempts have been made to integrate the high protein trait into agronomically acceptable cultivars. Such varieties typically have protein contents of 40–45%. Provar (Probst *et al.*, 1971) and Protana (Weber and Fehr, 1970) are early examples of such varieties, but they were not successful because of the yield depression typically associated with increased protein content. Backcrossing, a breeding method used to introgress simply inherited traits into a selected parent, has been used with limited success to improve seed protein concentration (Hartwig and Hinson, 1972; Wehrman *et al.*, 1987). A recent example of this approach is reported by Wilcox and Cavins (1995). The high protein from the donor parent Pando (49.8% protein) was backcrossed into Cutler 71 (40.8% protein). Lines were identified from successive backcross populations that had protein contents in excess of 47%, and which approached Cutler 71 in yield. A line from the backcross-3 population was identified that had a protein content of 47.2%, and a higher yield than Cutler 71. Nonetheless, the typical strong inverse statistical relationship between protein content and yield remained in the population from which this line was selected, and the oil content of the selected line was 17% as compared to about 20% in the case of Cutler 71. These results show that, although difficult, it is possible to transfer the high protein trait to agronomically acceptable soybean lines even in the face of the well recognized, strong negative statistical relationship between seed yield and protein content. A better understanding of the genetic and physiological mechanisms operative in controlling the relative amounts of oil and protein may facilitate the development of superior high protein varieties for the soybean industry.

Although many different enzymes are present in soybean seeds during development and at maturity, only a relatively small number of them accumulate in amounts that exceed 1% of the total seed protein. Of these, 70–80% are the 11S and 7S storage proteins known as glycinin and β-conglycinin, respectively. These are typical seed storage proteins in that they have no catalytic function, are rich in nitrogen by virtue of their high content of amide amino acids, and are accumulated within membrane-bounded vacuoles called protein bodies. Because of their prevalence in the seed, glycinin and β-conglycinin play a primary role in determining the functional and nutritional properties of foods made from soybeans. As with homologous storage proteins found in other legumes, they are nutritionally limiting in sulphur amino acids. Although methionine is considered an essential amino acid, when cysteine is limiting it contributes to the deficiency because it is synthesized from methionine.

In addition to these two storage proteins, the seeds contain 1–8% of several other proteins that do have catalytic activity and amino acid compositions that are more balanced from a nutritional point of view. In soybeans, these include lipoxygenases, the Kunitz trypsin inhibitor, Bowman-Birk and related protease inhibitors, lectin and urease. Although these proteins may contribute a selective advantage such as by making the seed less attractive to insect and animal predators, they have evolved to play a storage role by virtue of the proportion of the total seed protein they contribute. Many of them also have antinutritional properties when the seeds are used as a food.

Research concerning soybean seed proteins has a long history that stretches back to at least the turn of the century (Osborne and Campbell, 1898). Early work concerned their extraction, characterization in terms of their gross chemical features, and an assessment of solubility characteristics. Danielsson (1949) contributed important data about the sedimentation characteristics of seed proteins following invention of the ultracentrifuge. Catsimpoolas, Wolf, Shibasaki and many others subsequently provided information about purification and properties of individual protein components from the soybean seeds [see Circle and Smith (1972) for reviews of this early work]. Work in the past decade by plant breeders, biochemists and molecular biologists have focused on finding ways to correct nutritional inadequacies, eliminate antinutritional factors, deal with flavour limitations, as well as identify and exclude epitopes that elicit allergic responses when the soybean seed proteins are consumed. To discuss issues related to the limitations and potential for modifying these proteins, each of the groups of proteins identified earlier will be discussed individually.

Storage proteins

Structural features

The 11S proteins from dicot species are all quite similar (Plietz and Damaschum, 1986). They are isolated from seeds as hexamers with molecular masses that are reported to be between $320-375 \times 10^3$ (Wolf and Briggs,

1958; Badley *et al.*, 1975; Utsumi *et al.*, 1981). Each hexamer is composed of two trimers that are rotated 60 degrees with respect to one another to form a trigonal antiprism (Reichelt *et al.*, 1980; Plietz and Damaschun, 1986; Plietz *et al.*, 1988). The 11S hexamers can be made to undergo reversible dissociation into 3S monomers and 8S trimers (Wolf and Briggs, 1958). Under appropriate conditions, 15S aggregates of hexamers can also be produced. These structural changes probably represent part of the assembly mechanism that occurs *in vivo*, and which will be described more completely shortly.

Products from five genes are known to contribute glycinin subunits, although data from genomic southern hybridization experiments suggest other subunit genes could be present (Nielsen *et al.*, 1989). The nomenclature we have developed to identify these genes are summarized in Table 7.1. Each gene contains three introns, and although differences in intron length occur among the genes, they occur at similar relative positions (Nielsen *et al.*, 1989). If additional genes do exist, amino acid sequence studies carried out with purified subunits suggest that they probably do not make a major contribution to the total seed protein content, at least in those cultivars studied thus far. It should be noted that amino acid sequence data did reveal heterogeneity at selected residues in subunit A_2B_{1a} from CX635-1-1-1, the subunit that has

Table 7.1. Nomenclature identifying primary glycinin and β-conglycinin subunits of soybean.

Gene	Subunit	CX635-1-1-1 peptide designation[1]	Clone described
Gy1	G1	A1aB1b	Nielsen, 1989
Gy2	G2	A2B1a	Marco *et al.*, 1984
			Kim and Choi, 1989
			Kitamura *et al.*, 1990
			Momma *et al.*, 1985a
Gy3	G3	A1bB1a	Cho *et al.*, 1989
Gy4	G4	A5A4B3	Scallon *et al.*, 1985
			Momma *et al.*, 1985b
Gy5	G5	A3B4	Fukazawa *et al.*, 1985
β-Conglycinin			
Cgy1	α'	α'	Schuler *et al.*, 1982
			Lelievre *et al.*, 1992b
Cgy2	α	α	Sebastiani *et al.*, 1990
			Lelievre *et al.*, 1992b
Cgy3	β	β	Harada *et al.*, 1989
			Lelievre *et al.*, 1992b

[1] Nomenclature refers only to subunits characterized in CX635-1-1-1; because subunits from other soybean cultivars may have different primary sequences, it may lead to confusion if this nomenclature is used for cultivars other than CX635-1-1-1. See Moreira *et al.* (1979) Staswick *et al.* (1981) and Staswick *et al.* (1984a,b) for details about CX635-1-1-1.

been characterized most completely by protein sequence analysis (Staswick *et al.*, 1984a). Because all seeds used for these experiments did not originate from the same plant in the cultivar population, and only the five major genes were identified in this cultivar by DNA hybridization, the sequence heterogeneities in A_2B_{1a} probably reflect products of multiple alleles of the same gene which were present in the population of seeds used for the studies. These results are important because they imply that studies directed toward the identification of new glycinin genes must take into account whether any differences observed are derived from alleles of one of the five known genes, or are from a previously unidentified genetic locus. It is therefore necessary for any variant gene DNA sequences discovered in the future to be distinguished from the previously identified five major genes by a genetic segregation analysis.

The initial product from each glycinin gene is a nonglycosylated, pre-proglobulin that undergoes cleavage at a conserved Asn-Gly peptide bond. After cleavage, each subunit consists of two polypeptide chains connected by a single intra-chain disulphide bond (Staswick *et al.*, 1984b). One chain ($M_r \sim 40$ kDa) has an acidic isoelectric point and the other ($M_r \sim 20$ kDa) a basic isoelectric point. The acidic chain also contains a second conserved inter-chain disulphide bond. Unlike the other four glycinin subunits, G4 is unique in that it has a second Asn-Gly bond near the N-terminal end of the acidic chain that is cleaved during maturation. Thus cleavage of G4 by an asparaginyl endopeptidase results in an acidic chain of about 10 kDa, an acidic chain of about 40 kDa and a basic chain of about 20 kDa.

The five glycinin subunits can be divided into two groups based on amino acid sequence homology (Nielsen, 1984). G1, G2 and G3 comprise the Group-1 subunits, and G4 and G5, Group-2. Sequence homologies among subunits in the same group are around 90%, but are only about 50% when members of different groups are compared. Two groups of 11S subunits can be identified in other legume species (Casey *et al.*, 1986). Because the two groups seem to be a conserved feature among those legume and non-legume species whose 11S subunits have been characterized extensively, it is possible that some type of a functional difference exists between them. However, no evidence to justify this supposition has been reported.

The most notable structural difference between the two groups of proteins occurs in a part of the subunits immediately preceding the conserved Asn-Gly cleavage site. This area was denoted the hypervariable region (HVR) by Argos *et al.* (1985). The Group-2 glycinin subunits contain glutamate/aspartate-rich insertions in the HVR of about 100 amino acids when compared with those that belong to Group-1. Computer assisted predictions indicate that the HVR has extensive α-helical conformation, and is exposed to the surface of the subunit. Because of the extreme natural variability that occurs in this region, the HVR has been suggested as a location for modification to increase sulphur amino acid content in the glycinin subunits (Nielsen *et al.*, 1990).

The β-conglycinin complexes are isolated from seeds as trimers whose molecular masses are $150-175 \times 10^3$ (Thanh and Shibasaki, 1977, 1978a). Like the 11S proteins, manipulation of ionic strength causes the trimers to undergo reversible dissociation and reassociation (Thanh and Shibasaki, 1979). As isolated from seeds, the 7S trimers are made up of at least three major subunits, α', α and β (Thanh and Shibasaki, 1978b) with M_r 76,000, 72,000 and 53,000 (Sebastiani *et al.*, 1990), respectively. Both genomic and cDNA clones encoding each of the major subunit types have been reported (Coates *et al.*, 1985; Harada *et al.*, 1989; Sebastiani *et al.*, 1990). The coding regions in each of these genes contain five short introns (Harada *et al.*, 1989). Because 7S complexes with differing subunit compositions can be isolated (Thanh and Shibasaki 1976, 1978c), it is likely that the trimers are composed of a randomly assembled mixture of subunits. Unlike the glycinin subunits, each β-conglycinin subunit contains about 5% carbohydrate by weight. The glycosyl groups are of the high-mannose type that is typical of many plant glycoproteins, and are covalently linked to one or more asparagine residues (Yamauchi and Yamagishi, 1979).

In addition to the well known sulphur amino acid deficiency, soybeans have proved to be allergenic to some individuals. Shibasaki *et al.* (1980) associated allergenic activity with both the 7S and 11S proteins. Burks *et al.* (1988) used the sera of infants with soy-induced atopic dermatitis and identified the 7S fraction as the most allergenic. Pedersen and Djurtoft (1989) have focused on the glycinin fraction and have identified potential allergenic epitopes. Although the allergenic properties of soy-based foods has been recognized for some time, little work has been done with this potentially important aspect of soybean products. If a means can be identified to improve the acceptability of soybean products in food, additional attention must be devoted to the allergenic properties of these proteins. It is possible that the offending epitopes can be removed by gene engineering technology.

Utsumi (1992) has presented arguments that knowledge of storage protein three dimensional structure is a fundamental prerequisite for the design of structurally functional mutants with improved nutritional quality. Similar arguments can be advanced to eliminate allergenic epitopes. Although such information is presently not available for any of the soybean proteins, it has been reported for the homologous 7S proteins from both *Phaseolus vulgaris* (Lawrence *et al.*, 1990, 1994) and *Canavalia ensiformis* (Ng *et al.*, 1992; Ko *et al.*, 1992). Briefly, each 7S subunit features an internal repeat, each of which contains a β-barrel in a 'jelly-roll' type of folding motif. The two β-barrels interact to form the central core of the subunit, and are flanked on either side by α-helical domains in helix-turn-helix motifs. The N-terminal helix-turn-helix motif extends into another α-helical region that is followed by putatively extended portion of the backbone that connects to the C-terminal β-barrel. The putative extended region is invisible in the x-ray structure, probably because the amino acids in this region are not constrained in their movement. The N-terminal chain of the molecule lies across one face of the subunit, such that it interacts with the N-terminal β-barrel before connecting to the C-terminal β-barrel structure.

A number of workers have reported that there is structural similarity between the basic chain of the 11S subunits and the C-terminal portion of the 7S subunits (Argos *et al.*, 1985; Wright, 1988; Gibbs *et al.*, 1989), although relationships between the N-terminal portions of the two molecules have until now been obscure. With the availability of the refined 7S structure from *P. vulgaris*, Lawrence *et al.* (1994) aligned both the N-terminal and C-terminal portions of the molecules, and demonstrate that they bear a much closer relationship to one another than previously reported. The HVR of the 11S subunits described by Argos *et al.* (1985) correspond with the portion of the 7S structure discussed earlier that is in an extended motif and connects the N-terminal helix-turn-helix motif to the C-terminal β-barrel. Although this model will be of use in making decisions about sites in the molecules that should be modified to increase sulphur amino acid content, it is important that the three dimensional structure of 11S proteins also be obtained. In this regard, Utsumi *et al.* (1993) have recently described crystals of G3 glycinin trimers prepared in *E. coli*, and although unpublished, three different crystal forms have been generated from G4 hexamers produced in our laboratory (Beaman, unpublished results).

A third class of globulin can be isolated from soybeans that sediments at about 8S (Hu and Esen, 1982; Yamauchi *et al.*, 1984; Lilley and Nielsen, unpublished results). In most cultivars, this globulin accounts for 5–10% of the total seed protein. It has a molecular mass of about 158×10^3, and is a tetramer with subunits of about 42,000. Like the 11S subunits, each subunit contains two polypeptide chains; a large basic one of about 30,000 Da that is disulphide-linked to a smaller acidic one of about 16,000 Da (Lilley and Nielsen, unpublished results). This 8S globulin has a higher sulphur content than either the 7S or 11S globulins, but fewer acidic amino acids. It is this latter property that imparts a high isoelectric point (pH 9.5) on the protein (Lilley and Nielsen, unpublished results). Little work has been done on these proteins since their initial identification.

Synthesis and deposition of storage globulins

A model has emerged that accounts for the movement of storage protein precursors from their site of synthesis on rough endoplasmic reticulum to their site of deposition in protein storage vacuoles. Because much of the evidence supporting this proposal is summarized in detail elsewhere (Chrispeels *et al.*, 1982a,b; Shotwell and Larkins, 1989), only the essential features of this pathway will be mentioned here. As the storage proteins are synthesized into the lumen of the rough endoplasmic reticulum, short signal sequences are removed cotranslationally to form proglobulins (Sengupta *et al.*, 1981). Core glycosyl groups are added to most of the 7S precursors at this time, and then the conserved intra- and interchain disulphide chains that are typical of the mature 11S subunits are probably formed. Shortly after their arrival in the ER, trimers consisting of 7S and 11S proglobulin subunits are formed, and this occurs in an ATP dependent manner that probably requires the participation of molecular chaperones (Nielsen *et al.*, 1995). According to the present model, both 7S and 11S trimers transit through Golgi en route to the protein storage vacuoles, although other interpretations have been proposed (Robinson *et al.*, 1994).While in Golgi, the core glycosyl moieties on the 7S proglobulins undergo modification (Herman *et al.*, 1986; Sturm *et al.*, 1987). Once the trimers arrive in the protein bodies, they are apparently subjected to digestion by a number of proteases that also reside in these organelles. An asparaginyl endopeptidase cleaves the 11S proglycinin subunits in trimers at the conserved Asn-Gly peptide bond (Scott *et al.*, 1992), and this cleavage has been demonstrated to be a prerequisite for assembly of 8S trimers into 11S hexamers both *in vitro* and *in vivo* (Dickinson *et al.*, 1989; Nielsen *et al.*, 1995). The α' and α subunits of β-conglycinin also undergo cleavage about 40 amino acids from their N-terminals (Coates *et al.*, 1985), but it appears doubtful that this cleavage is of physiological significance. In this regard, 7S precursors from a wide variety of species undergo proteolytic cleavage post-translationally. Unlike the 11S subunit cleavage site which is highly conserved, the positions of those sites cleaved in the diverse 7S precursors are not.

Developing soybean seeds contain a small family of asparaginyl endopeptidases, members of which are potentially capable of cleaving 11S proglobulins post-translationally (Scott *et al.*, 1992; Muramatsu and Fukazawa, 1993; Hara-Nishimura *et al.*, 1995). A decision concerning which enzyme carries out cleavage of the conserved Asn-Gly bond that leads to hexamer formation in soybeans has not been resolved. One group of candidates are high molecular weight glycoproteins first identified by Scott *et al.* (1992). These enzymes apparently contain a sulphydryl group important for activity, and while slightly inhibited by E-64, are relatively insensitive to this specific sulphydryl protease inhibitor. The second candidates are sulphydryl proteases that have apparent molecular weights of 35–38 kDa (Hara-Nishimura *et al.*, 1991; Abe *et al.*, 1993; Muramatsu and Fukazawa, 1993) and do not bind to concanavalin-A affinity columns (Nielsen *et al.* 1995). Immunological cytological studies led to the conclusion that the low molecular weight nonglycosylated protease are in protein storage vacuoles (Hara-Nishimura *et al.*, 1995), although similar studies remain to be carried out with the high molecular weight glycosylated enzyme. SDS/gelatin activity gels revealed that activity due to the glycosylated high molecular weight protease was many fold greater than that of the nonglycosylated enzyme at the midmaturation stage of seed development. Upon seed germination, however, activity due to the low molecular weight nonglycosylated enzyme predominates (Nielsen *et al.*, 1995). Consistent with these results, the low molecular weight protease is encoded by a rare-class cDNA in a library prepared from mid-maturation stage soybeans.

As both enzymes have been proposed as active in assembly of proglycinin trimers into hexamers, additional data will be necessary before this point can be
settled.

Data concerning the organization of storage proteins in the protein storage vacuoles is sparse. Craig and colleagues have studied the deposition process of pea globulins by studying immunogold labelling of thin sections by
electron microscopy (Craig et al., 1980a,b), and it is reasonable to expect the
case in soybean to be identical. They showed that the storage proteins first
appeared as small clumped deposits at the periphery of the large central
vacuole in storage parenchyma cells. Interpretation of serial sections suggested
that the numerous protrusions of the central vacuole became more complex
as seed development proceeded, and that these protuberances eventually
fragmented into small, discrete, spherical vacuoles. Double immunogold labelling using gold particles of different sizes to mark 7S and 11S globulins revealed that both types of storage protein are localized in the same protein
bodies and appear to be distributed uniformly throughout the vacuole. Because the 11S and 7S subunits apparently share many common structural features (Lawrence et al., 1994), and are co-localized within the PSV, one might
suspect that the two kinds of subunits could co-assemble. However, Oliveira
(1994) has demonstrated that neither monomers and trimers of G4 glycinin,
nor monomers and trimers of the α subunit of β-conglycinin are capable of co-
assembly in vitro. If efforts to engineer soybean storage proteins are to be effective, additional information about how the storage globulins are packaged
with one another in the storage vacuoles would seem quite important.

Genomic organization of seed protein genes

Harada et al. (1989) have evaluated the genomic interrelationships of β-conglycinin genes in the Dare soybean variety. They concluded that the β-conglycinin gene family contains at least 15 members that are distributed among six
distinct regions of the genome. Together these regions encompass about 175
kilobase pairs of DNA. Three of these regions contain multiple genes (Region
A, five genes; Region B, two; Region C, three; Regions D, E, and F, one each).
Although the genes in Regions B and C all had the same transcriptional orientation, genes with the opposite transcriptional orientation were observed
in Region A. DNA that hybridizes with β-subunit gene probes are located in
each region, while those encoding α/α′-subunits are interspersed with the β-
subunit genes in only two of the three regions that contain clustered genes
[Region A, 2 α/α′ genes; Region C, α′ gene (Doyle et al., 1986)]. These results
are consistent with genetic linkage studies reported by Davies et al. (1985).
The linkage studies with subunit mobility variants indicated α and α′ are encoded by genes that assort independently, while the gene encoding the α subunit is linked to a gene encoding a β-subunit mobility variant. Moreover, data
obtained using the α′ null allele in Keburi indicate that it is encoded by a sin-

gle gene (Davies *et al.*, 1985). Together these data strongly suggest that the clusters of β-conglycinin genes are located on at least two chromosomes, but their precise location on one of the soybean genomic linkage maps is not yet available. Finally, the data of Harada *et al.* (1989) reveal that all of the β-conglycinin genes become transcriptionally activated at the same time, but that the α/α′ and β mRNAs accumulate and decay at different times during embryogenesis. Thus, like the glycinin genes (Nielsen *et al.*, 1989), both transcriptional and post-transcriptional mechanisms are involved in control of β-conglycinin gene expression. Unfortunately, the molecular basis that underlies neither the transcriptional nor post-transcriptional mechanisms operative during expression of these genes is well understood.

Similar chromosome walking experiments have been performed to understand the genomic organization of the glycinin genes in the variety Dare (Nielsen *et al.*, 1989). The three Group-1 genes are organized into two chromosomal domains, each about 45 kilobase pairs in length. The two domains have a high degree of homology, and each contain at least five other similarly placed genes that are expressed either in embryos or in mature plant leaves. *Gy1* and *Gy2* are found in a direct repeat separated by about five kilobase pairs of DNA in one domain, while *Gy3* occupies an equivalent position in the second domain. All three of these Group-1 glycinin genes are transcribed in the same direction relative to the homologous leaf and embryo coding regions around them. It is likely that the extra gene in the first domain arose by a gene duplication event. It is not known if the two Group-2 genes, *Gy3* and *Gy4*, occupy similarly organized domains in the genome. However, genetic linkage studies carried out with RFLPs for each gene showed that the *Gy1/Gy2*, *Gy3*, *Gy4* and *Gy5* loci all assorted independently of one another (Cho *et al.*, 1989). Mapping studies have located *Gy4* and *Gy5* on the public linkage map being developed by Shoemaker and colleagues (Diers *et al.*, 1994), and efforts are underway to identify where the Group-1 glycinin genes reside.

Because seed storage protein genes are abundant and tissue specific, they were among the first of the plant genes to be isolated. A number of the promoter regions in these genes have been studied and shown to contain the typical canonical eukaryotic promoter elements such as CAAT and TATA elements. Multiple polyadenylation sites are present in the 3′-noncoding regions. A 28 bp conserved sequence is located in the 5′-flanking region of all 11S legume genes, and denoted the 'legumin box' (5′-TCCATAGCCATGCATGCT-CAAGAATGTC-3′) (Bäumlein *et al.*, 1986). The 7S genes have an equivalent region called the 'vicilin-box' (5′-GCCACCTCAATTT-3′) (Gatehouse *et al.*, 1986). Also found in the promoter of legume genes are RY repeat elements (5′-CATGCAT-3′) and the CACA sequences (Goldberg, 1986). One of the RY elements generally located in the centre of the legumin box, and its removal, leads to a 30–50 fold decrease in gene activity (Lelievre *et al.*, 1992a; Bäumlein *et al.*, 1992). Chen *et al.* (1988) have characterized a 170 bp upstream element in the α′ subunit that is apparently involved in enhancing tissue specific and developmental expression when tested in heterologous plants (Allen *et al.*, 1989). In the promoter of the soybean lectin gene (Jofuku *et al.*, 1987) a 60 kDa nuclear protein was found to bind to two regions I: 12 bp, − 184 to − 173; II: 40 bp, − 165 to − 126). Finally, Bustos *et al.* (1989) identified two inverted AT-rich motifs in the far upstream region of the 7S gene from *Phaseolus vulgaris* that binds nuclear proteins and is important for activity. The glycinin genes contain similar motifs, and their elimination results in a two- to threefold decrease in promoter activity (Lelievre *et al.*, 1992a; Oliveira, 1994). Thus, while there are a few hints concerning how the legume storage proteins might be regulated developmentally, considerably more work needs to be done in this area before definitive statements can be made concerning tissue specific developmental control of their expression.

Protein engineering to improve nutritional quality

Ever since it was discovered that the legume seed proteins were nutritionally limiting in sulphur amino acids, a goal has been to improve the content of methionine in seeds. Advances in cloning, DNA manipulation, plant transformation, and an increased knowledge about seed protein genes, imply it will be feasible to use molecular genetic techniques to achieve this goal in the near term. Moreover, a number of strategies are being considered to approach this problem. From a historical perspective, the earliest approach considered was modification of existing proteins by site directed mutagenesis. Two other approaches that have received attention are either overexpression of existing seed proteins with desirable properties, or transfer and expression of genes from other species that encode proteins with elevated sulphur amino acid content. Because technology has developed which permits the rapid synthesis of DNA, the possibility exists to design novel proteins that will either satisfy nutritional inadequacies or provide new functional properties in food systems. Each of these approaches has merit, but all suffer from different technical problems. The factors to be considered when applying these approaches have been discussed (Sun and Larkins, 1993), and the interested reader is referred to that contribution for additional details.

Proteins abnormally rich in sulphur amino acids have been identified in a number of species. Higgins *et al.* (1986) described two low molecular weight albumins in pea that contain about 7.5% and 16% cysteine, respectively. A 2S albumin from Brazil nut contains 18% methionine and 8% cysteine (Ampe *et al.*, 1986; Altenbach *et al.*, 1987; Sun *et al.*, 1987), and a similar protein has been identified in sunflower (Lilley *et al.*, 1989). Finally, the 10 kDa and 15 kDa prolamines in maize contain 16–22.5% sulphur amino acids (Phillips and McClure, 1985; Pedersen *et al.*, 1986; Kirihara *et al.*, 1988). These proteins have the advantage that they originated from a seed system, and therefore may be stable in seeds of heterologous hosts (see below). Two of these proteins have already been used for heterologous expression. Altenbach *et al.* (1989) expressed the high methionine Brazil nut protein in tobacco under the control of the 7S phaseolin promoter, and recovered about 8% of the total seed protein as 2S albumin. A similar construction was expressed in transgenic *Brassica* (Guerche *et al.*, 1990) soybeans. Although high levels of expression were recorded in the case of soybean, this proved not to be the case in *Brassica*. Unfortunately, the 2S albumin from Brazil nut is a strong allergen (Nordlee *et al.*, 1994), so it is unlikely this technology will be useful in practical terms unless the antigenic epitopes can be identified and eliminated. Finally, the 2S sunflower has been expressed in alfalfa where it has caused the leaf methionine content to increase. These transgenic plants are being targeted to pastures for sheep because increased methionine can be demonstrated to increase wool quality.

Although the potential application of synthetic genes to improve seed nutritional quality has received attention from at least one company, there have been no reports about the successful application of this approach. However, Yang *et al.* (1989) synthesized a 292 bp DNA encoding a protein rich in the essential amino acids and expressed it in potato tubers under the control of a NOS promoter. Although expressed, only 0.02–0.35% of the total tuber protein could be attributed to this gene product, probably because of instability of the expressed protein.

At least three attempts have been made to modify legume seed protein genes to increase the methionine content. Hoffman *et al.* (1987) introduced a methionine-rich fragment of the 15 kDA β-zein from maize into the 7S phaseolin gene into tobacco, and Saalbach *et al.* (1988) performed a similar experiment with an 11S legumin from *Vicia faba*. Nielsen *et al.* (1990) introduced oligonucleotides encoding high methionine fragments into the hypervariable region of the *Gy4* glycinin gene, and expressed these constructions in tobacco. In each case, the engineered genes were transcribed, but translation products were either not present in the seed or accumulated at low levels. It is likely that the engineered proteins were transport competent but were degraded upon arrival in the storage vacuole. Nielsen *et al.* (1995) have shown that the high molecular weight asparaginyl endopeptidase required for assembly of the 11S trimers into hexamers degrade their engineered proteins, but that the unmodified protein is stable during extended incubations with the protease. One way to interpret these results is that the storage proteins evolved to be insensitive to the cadre of proteases found in the storage vacuoles and in the compartments through which they transit after they are synthesized in endoplasmic reticulum. Such a consideration would account for why heterologous seed proteins such as the 2S albumin from Brazil nut are stable while the engineered seed proteins do not accumulate. These proteins have evolved to be accumulated in the presence of the vacuolar proteases. More importantly, however, these results indicate that attention must be paid to the sensitivity of vacuolar proteases as attempts are made to engineer seed proteins.

Lipoxygenases

Lipoxygenases account for about 1–2% of the total soybean seed protein at physiological maturity (Vernooy-Gerritsen *et al.*, 1983). Axelrod *et al.* (1981) describe the preparation of four isozymes, denoted L-1, L-2, L-3a and L-3b. L-3a and L-3b are quite similar and for the purposes of this discussion will be denoted L3. These enzymes catalyze the addition of oxygen to the *cis,cis*-1,4-pentadienes of polyunsaturated fatty acids such that a conjugated hydroperoxide structure is formed. Each enzyme is 95–98 kDa, contains a single atom of tightly bound non-haem, non-Rieske iron per molecule (Vliegenthart and Veldink, 1982; Shibata *et al.*, 1988), and has been cloned (Shibata *et al.*, 1987, 1988; Yenofsky *et al.*, 1988). Upon germination, at least three new lipoxygenases appear in cotyledons (Kato *et al.*, 1992) and at least two lipoxygenases that are distinct from the cotyledon enzymes appear in the axis (Park *et al.*, 1994).

The L-1, L-2 and L-3 isozymes have been subdivided into two classes depending upon their substrate preferences, heat stability and pH optimum. L-1 is considered prototype for Class-1 lipoxygenases. It has a pH optimum around 9.0, is heat stable and prefers fatty acids as substrate. The Class-2 enzymes, L-2 and L-3, have pH optima near neutrality, are heat inactivated and prefer esterified substrates (Christopher *et al.*, 1970, 1972a,b). Additional differences in regiospecificity (Christopher, 1972), secondary and peroxidative reactions (Garssen *et al.*, 1971; Pistorius, 1974; Pistorius *et al.*, 1976; Bild *et al.*, 1977), anomalous dependence of reaction rate on substrate concentration (Christopher *et al.*, 1972a,b) and the generation of singlet oxygen (Kanofsky and Axelrod, 1986) also exist. The enzymes differ in the proportions of 9- and 13-hydroperoxides formed. L-1 forms exclusively the 13-hydroperoxide, whereas the other two enzymes form approximately equal amounts of both hydroperoxides (Christopher and Axelrod, 1971). Finally, the three-dimensional structure of L-1 has been deduced (Boyington *et al.*, 1993) and the structure of the active sites of several other isozymes published (Minor *et al.*, 1993).

The lipoxygenases have been the object of commercial interest because of their use in the food industry. Lipoxygenase is responsible for the generation of both objectionable (Mustakas *et al.*, 1969; Wolf, 1975; Hildebrand and Kito, 1984; Davies *et al.*, 1987) and of pleasant flavours (Axelrod, 1974; Eskin *et al.*, 1977; Galliard and Chan, 1980). The pleasant flavours and aromas arise from aromatic aldehydes and alcohols generated from the lipoxygenase-mediated hydroperoxidation of polyunsaturated fatty acids. Finally, defatted soy extract is an important component in bread-making, where it is responsible for bleaching carotenoid pigments to produce whiter bread. It also catalyses oxidations and subsequent cross-linking of wheat glutens to enhance bread texture (Eskin *et al.*, 1977).

Because of the undesirable off-flavours and aromas associated with full fat soy products, heat treatment has traditionally been used to suppress lipoxygenase activity. The amount of heat required to prevent the formation of lipoxygenase-induced flavours is not altogether satisfactory, however, because it often results in insolubilization of other soy proteins and created unpleasant 'cooked' odours (Wolf, 1975; MacLeod and Ames, 1988). During the past decade a number of mutant lines with recessive alleles (*lx1*, *lx2* and *lx3*) that cause each of the three lipoxygenase isozymes to be absent from the seed have been described (Hildebrand and Hymowitz, 1982; Kitamura *et al.*, 1983, 1985; Davies and Nielsen, 1986). From genetic segregation studies, the *Lx1* and *Lx2* gene loci were identified as linked, whereas the *Lx3* locus segregated independently from the other two genes. Mutagenesis of the *lx1lx1lx3lx3* double mutant was used to generate a triple null-line (Takamura *et al.*, 1991), and an apparent triple null was identified by searching progeny from a cross involving the *lx1lx1lx3lx3* and *lx2lx2* Century soybeans (Nielsen, unpublished results). These mutant alleles, when introduced into appropriate genetic backgrounds, have the potential to reduce the off-flavours associated with soybean products (Davies *et al.*, 1987) and could be useful in the development of new soybean products. Unfortunately, the recessive nature of the null alleles will create problems in trying to maintain pure lines that can be grown commercially, and statistical methodology has been developed that can be used in these instances (Evans *et al.*, 1994). To overcome the difficulty of commercially grown plants with the genetically recessive null-alleles, antisense genes could be used with dominant genetic characteristics.

Urease

Urease is an abundant protein in seeds of most Leguminosae, and carries out the breakdown of urea to NH_3 and CO_2. Two isozymes have been recognized in soybeans. One is localized exclusively in seeds and accounts for 0.2–0.5% of the total protein (Polacco and Havir, 1979; Winkler *et al.*, 1983). It is referred to as the embryo-specific urease. The other, which is less than one hundredth as prevalent as the embryo-specific isozyme (Polacco *et al.*, 1982; Polacco and Winkler, 1984), is distributed throughout the plant, and is designated the ubiquitous urease (Polacco and Holland, 1994).

Mutants that affect the structural genes for both of these enzymes have been identified. A recessive mutation (*eu4*) of a single gene was identified that caused the ubiquitous urease to be absent from leaf tissue (Polacco *et al.*, 1989). This recessive allele of *Eu4* was subsequently demonstrated to eliminate the enzyme from all vegetative tissues (Krueger *et al.*, 1987; Torisky and Polacco, 1990; Holland and Polacco, 1992) and embryonic tissue (Torisky *et al.*, 1994). Clones that encode the structural gene *Eu4* have been reported (Krueger *et al.*, 1987), and shown by RFLP genetic segregation analysis to map within the *Eu4* locus. A second recessive mutation, *eu1*, eliminates the presence of the embryo specific urease (Polacco *et al.*, 1982; Kloth and Hymowitz, 1985). Several alleles of *Eu1* have also been observed which affect the aggregation state of urease (Buttery and Buzzell, 1971; Polacco and Havir, 1979; Polacco and Sparks, 1982). Two other genetically unlinked genes, *Eu2* and *Eu3*, affect the maturation of urease, and result in the recovery of inactive embryo-specific and ubiquitous urease (Holland et al., 1987; Meyer-Bothling et al., 1987). Because in other organisms apourease is activated by the addition of nickel, it is likely that the gene products of Eu2 and Eu3 participate in similar functions in the plant (Polacco and Holland, 1994).

The function of urease in plants is still largely speculative. Arguments have been presented that it could perform a protective role against insects and other plant predators (Polacco and Holland, 1994). It is also likely that ubiquitous urease plays an important role in assimilating urea, because blocking its synthesis, either by genetic mutation or nickel deprivation, results in urea accumulation in all plant tissues and the appearance of necrotic leaf tips (reviewed, Polacco and Holland, 1994). Interestingly, the eu1 allele does not have the same effect (Polacco and Holland, 1994). Because nitrogen is fixed in purines in the nodules of soybeans and transported from there to the rest of the plant in the form of ureides, it was considered possible that the ureide allantoate would be degraded to two molecules of urea and glyoxylate (Polacco and Holland, 1994). According to this supposition, urease could participate in reassimilation of urea into other metabolites. However, allantoate is instead apparently broken down by the amidohydrolase enzyme. Nutritional studies with urease negative callus cultures (Polacco et al., 1989; Torisky et al., 1994), whose growth is blocked on urea nitrogen, exhibit normal growth on allantoin nitrogen.

Seed protease inhibitors

Soybean seeds contain two major classes of proteinase inhibitors specific for serine proteases, the Kunitz trypsin inhibitor and Bowman-Birk protease inhibitor family (for reviews, see Laskowski and Kato, 1980; Ryan, 1988). The soybean trypsin inhibitor (Kunitz) was isolated and crystallized from soybeans by Kunitz (1946). This inhibitor was the first seed compound proposed as an antinutritional compound in soybean. The Kunitz trypsin inhibitor consists of a single polypeptide chain of 181 amino acids, and has a molecular weight of 21,500 daltons (Koide and Ikenaka, 1973; Koide et al., 1973). It contains two methionine and four half cystine residues, two of which are critical for inhibitory activity. The three dimensional structure of Kunitz inhibitor is known (Sweet et al., 1974), and it contains a single reactive site at Arg63-lle64. There are three known functional alleles of the gene encoding Kunitz inhibitors (Ti^a, Ti^b, Ti^c) (Singh et al., 1969; Hymowitz and Hadley, 1972; Orf and Hymowitz, 1977), and a well characterized null allele exists (ti) (Orf and Hymowitz, 1979; Jofuku et al., 1989). It has been shown that chemical addition of glycine in the active site of the inhibitor inactivates the protein and leaves the active site susceptible to cleavage by trypsin. The altered inhibitor is still recognized by the inactivated inhibitor (Kowalski and Laskowski, 1976a,b), but now serves as a substrate rather than an inhibitor.

The second important group of protease inhibitors in soybeans are related to the Bowman-Birk protein first identified by Bowman (1946), but was purified and characterized by Birk *et al.* (1963). Subsequently, it became recognized as one of a family of related proteins (Hwang *et al.*, 1977). The Bowman-Birk family of inhibitors are all a single polypeptide chain 70 to 80 amino acids in length, and contain up to 14 cysteine residues (Ikenaka *et al.*, 1974). Because these cysteines are involved in disulphide bonds, the structure of these inhibitors is quite stable. The Bowman-Birk inhibitors can account for 4–6% of the total soybean seed protein (Hwang *et al.*, 1977). Despite their low content in the seed, it can be calculated that they account for nearly half of the cysteine. Hence, elimination of these molecules from the seed will aggravate an already low sulphur amino acid content.

The Bowman-Birk inhibitors contain two reactive sites per molecule. Because the N-terminal half of the molecule is homologous with the C-terminal half, the molecule is considered to be derived from gene duplication (Odani and Ikenaka, 1976). Because of this, these double-headed molecules can either inhibit two molecules of the same protease simultaneously, or each half can inhibit a different protease. In addition to trypsin, active sites in the Bowman-Birk group of inhibitors can also recognize chymotrypsin and elastase. Soybean cultivars are known that express up to 12 Bowman-Birk isoinhibitor forms at the same time (Tan-Wilson and Wilson, 1986). There are no null-alleles reported for the Bowman-Birk type inhibitors, but several genes encoding Bowman-Birk inhibitors have been cloned and characterized (Hammond *et al.*, 1984; Joudrier *et al.*, 1987). The three-dimensional structure (Werner and Wemmer, 1992), as well as inhibitory properties of synthetic Bowman-Birk loops, are known (Maeder *et al.*, 1992).

There is no homology between the Kunitz and Bowman-Birk inhibitors (Tan-Wilson and Wilson, 1986), but they do share a common mechanism of inhibition. The inhibitors act as a pseudosubstrate for serine proteases (for example, trypsin, chymotrypsin and elastase) which are important digestive enzymes in vertebrates. The enzymes bind the inhibitor and hydrolyse the peptide bond between two amino acids at the reactive site (see for example Laskowski and Kato, 1980). The reactive site is enclosed in a disulphide loop that prevents the two halves of the inhibitor from separating after the bond is hydrolysed, and the enzyme-inhibitor complex does not dissociate at physiological conditions. The proteases that are complexed in the inhibitor are excreted from the organism. Thus, sulphur is lost in two ways; first, through undigested inhibitor, and secondly, through the loss of serine proteases. Animal growth studies from feeding animals raw soybean meal show that secretion of proteases is stimulated beyond normal levels by the loss of the enzymes in complexes with the inhibitor. This results in 30–50% inhibition of growth and pancreatic hypertrophy (Rackis, 1965; Garlich and Nesheim, 1966). Similar results of increased secretory activity under inhibitor stress were obtained with the human pancreas (Liener *et al.*, 1988).

Soybean seed lectin

Seed extracts from soybean contain proteins with the capacity to agglutinate erythrocytes. These are commonly referred to as seed lectins, although the terms phytohemagglutinins and phytoagglutinins have also been used. As in many plants, soybean seeds have been shown to contain a group of closely related proteins called isolectins. Lis *et al.* (1966) recognized three minor components in addition to the prevalent seed lectin (SBA), all of which had similar amino acid and carbohydrate compositions. The origin of the minor seed lectins has not been resolved, although data reported by Goldberg *et al.* (1983) indicate that the soybean genome contains two homologous lectin genes. One of these genes encodes SBA, and while the other one is not expressed in seeds, it may be expressed at low levels in embryos and roots. As isolated from seeds, SBA is a glycoprotein that contains 4.5% mannose and 1.5% N-acetyl-glucosamine in the form of carbohydrate chains of the composition $(Man)_9$-$(GlcNAc)_2$. It has a molecular weight of about 122 kDa (Lotan *et al.*, 1974), and is a tetramer of 30 kDa subunits. The primary structure of the subunit was determined chemically by Lotan *et al.* (1974), and is rich in aspartate, leucine, serine, alanine and lysine. As in the case of glycinin and β-conglycinin, SBA has low levels of methionine and cysteine. Only about half of the glutamic acid/glutamine and aspartic acid/asparagine residues are the amide amino acids. Jaffé (1969) reported four weak binding sites for Mn^{2+} per 122,000 Da, but there are only two binding sites for DGalNAc (Lotan *et al.*, 1974). Upon storage, SBA can undergo aggregation into a high molecular weight complex with markedly higher hemagglutinating activity than freshly prepared proteins. The association constant of SBA for nGalNAc is about 3.0×10^4 litre. Pull *et al.* (1978) identified a lectin null allele during a survey of 107 soybean lines. The allele *Le*, which confers presence of SBA in the seed, is dominant to the null allele denoted *le*. The null allele is due to an 3.5 kilobase insertion in the gene for SBA, and no lectin mRNA is accumulated in seeds of the mutant (Goldberg *et al.*, 1983). Subsequent studies revealed that the insert contained inverted repeats that are typical of those associated with transposable elements (Vodkin *et al.*, 1983). No evidence exists, however, for transposition by this putative element.

Concluding Remarks

Over the course of the past half-century the soybean has evolved to be one of the most important crop plants grown in the United States, and much of this growth has been driven by the edible oil industry. At the present time the soybean yield potential permits more soybeans to be grown than frequently can be sold and, as a result, niche markets to which high-value soybeans with special traits can be directed are developing. With the advent of biotechnology, the agricultural industry has an increased potential to respond to the development of such markets and it is likely that soybean with oil compositions tailored to meet the various market niches will quickly develop. The spectacular progress made during the past decade in understanding the pathways leading to the storage triglycerides provides an opportunity to take advantage of these niche markets. It seems likely that specialty soybean crops with shorter chain fatty acids, less unsaturation and hence greater stability, or oil crops destined for the fuel and other industrial purposes will become available in the foreseeable future. With these opportunities, technical difficulties will also undoubtedly arise. One obvious problem concerns segregation of the specialty crop from the commodity seed storage and marketing mechanisms. Besides the expensive technical problems associated with segregated growth, harvest, storage and transport of the crops, such efforts are likely to promote vertical integration in the industry. Those companies who successfully develop plants with specialty traits that can be patented may seek to protect their investment by controlling the growth, harvest and marketing of these crops. In addition to affecting the normal course of business in the seed industry, such vertical integration could profoundly influence the structure and values of society.

Although changes in soybean oil composition can be expected in crops in the near term, their overall impact on the market place will, to some extent, be mitigated by competition from other oil crops that can be modified in similar fashion. An area where soybean enjoys an apparent advantage over many competitor crops is in protein content. Although the vast majority of soybean protein produced in the United States presently is used for animal feeds, important markets could potentially be generated if foods that depend on soy proteins could be developed. Were this to occur, it would increase the market value of soybean protein. For this to take place, however, a number of obstacles must be overcome. Outside the Orient, soy-based foods are not a traditional part of the human food chain, and people tend to discriminate against either the flavour or the texture of soy-based foods. While the use of low lipoxygenase soybeans may to some extent reduce the off-flavours associated with soybean products, and increase their acceptance, it seems likely that new uses of soybean protein in food systems will need to be developed before soy protein can exert a significant impact on the food industry. Indeed, efforts to increase the use of soybean protein as casein substitutes in processed cheese and other traditional dairy products is being tested and may prove to be the first place where it is accepted as a food on a commercially viable scale.

Although important advances in our knowledge about the soybean proteins have been gained during the past decade, substantially more fundamental knowledge about them will probably be necessary before significant inroads can be made to alter the texture and other functional properties of soy products. Although it seems likely that the nutritional imbalance of soybean proteins will be corrected in the near term by increasing seed sulphur amino acid content, it is not at all clear that opportunities exist to develop new soy based foods that will appeal to the average consumer. Rather, high-methionine seed proteins will probably impact the food industry via animal feed. Efforts to engineer the seed proteins to change digestibility, eliminate allergenic epitopes and change functional properties of the proteins such that the proteins will enter the food chain directly, must be considered longer term projects. Interestingly, it seems likely that the soybean plant can be used to produce rare and valuable products derived from genes introduced into them from other sources, and this type of approach may be viable on a commercial scale before changes to improve the food quality of the seed proteins are likely.

In reviewing the soybean literature, it became apparent that little research has been done on the carbohydrate portion of the seed. It is evident that considerably more research into this aspect of seed composition is warranted. The consequence of eliminating raffinose series sugars from the seed is one area of research that has merit, but opportunities probably also exist to manipulate the content of starch, sucrose and fructose. In this regard, it is worth pointing out that because soybeans have developed into such an important agronomic crop, and the cultural practices required for maximum yield are well studied, their seeds become an ideal vehicle from which to harvest valuable gene products from diverse sources.

References

Abe, Y., Shirane, K., Yokosawa, H., Matsushita, H., Mitta, M., Kato, I. and Ishii, S. (1993) Asparaginyl endopeptidase of jack bean seeds. *Journal of Biological Chemistry* 268, 3525–3529.

Adams, C.A., Broman, T.H., Norby, S.W. and Rinne, R.W. (1981a) Occurrence of multiple forms of α-amylase and absence of starch phosphorylase in soybean seeds. *Annals of Botany* 48, 895–903.

Adams, C.A., Broman, T.H. and Rinne, R.W. (1981b) Starch metabolism in developing and germinating soya beans is independent of β-amylase activity. *Annals of Botany* 48, 433–439.

Adams, C.A., Norby, S.W. and Rinne, R.W. (1983) Ontogeny of lipid bodies in developing soybean seeds. *Crop Science* 23, 757–759.

Allen, R.D., Bernier, F., Lessard, P.A. and Beachy, R.N. (1989) Nuclear factors interact with a soybean β-conglycinin enhancer. *Plant Cell* 1, 623–631.

Altenbach, S.B., Pearson, K.W., Leung, F.W. and Sun, S.S.M. (1987) Cloning and sequence analysis of a cDNA encoding a Brazil nut protein exceptionally rich in methionine. *Plant Molecular Biology* 8, 239–250.

Altenbach, S.B., Pearson, K.W., Meeker, G., Staraci, L.C. and Sun, S.S.M. (1989) Enhancement of the methionine content of seed proteins by the expression of a chimeric gene encoding a methionine-rich protein in transgenic plants. *Plant Molecular Biology* 13, 513–522.

Ampe, C., Van Damme, J., Castro, L.A.B., Sampaio, M.J.A.M, Van Montagu, M. and Vandekerckhove, J. (1986) The amino-acid sequence of the 2S sulphur-rich proteins from seeds of Brazil nut (*Bertholletia excelsa* H.B.K.). *European Journal of Biochemistry* 159, 597–604.

Argos, P., Narayana, S.V.L. and Nielsen, N.C. (1985) Structural similarity between legumin and vicilin storage proteins from legumes. *EMBO Journal* 4, 1111–1117.

Axelrod, B. (1965) Mono and oligosaccharides. In: Bonner, J. and Varner, J.E. (eds) *Plant Biochemistry*. Academic Press, New York, pp. 231–257.

Axelrod, B. (1974) Lipoxygenases. In Whitaker, J.R. (ed.) *Advances in Chemistry Series, No. 136*. American Chemistry Society, Washington, DC, pp. 324–348.

Axelrod, B., Cheesbrough, T.M. and Laasko, S. (1981) Lipoxygenase from soybeans. In: Lowenstein, J.M. (ed.) *Methods in Enzymology, Vol. 71*. Academic Press, New York, pp. 441–451.

Badley, R.A. Atkinson, D., Hauser, H., Oldani, D., Green, J.P. and Stubbs, J.M. (1975) The structure, physical and chemical properties of the soybean protein glycinin. *Biochimica et Biophysica Acta* 412, 214–228.

Bäumlein, H., Wobus, U., Pustell, J. and Kafatos, F.C. (1986) The legumin gene family: Structure of a B type gene of *Vicia faba* and a possible legumin gene specific regulatory element. *Nucleic Acids Research* 14, 2707–2720.

Bäumlein, H., Nagy I., Villarroel, R., Inzé, D. and Wobus, U. (1992) Cis-analysis of a seed protein gene promoter: The conservative RY repeat CATGCATG within the legumin box is essential for tissue-specific expression of a legumin gene. *Plant Journal* 2, 233–240.

Beaver, J.S. and Cooper, R.L. (1982) Dry matter accumulation patterns and seed yield components of two indeterminate soybean cultivars. *Agronomy Journal* 74, 380–383.

Bild, G.S., Ramadoss, C.S. and Axelrod, B. (1977) Multiple dioxygenation by lipoxygenase of lipids containing all- *cis*-1,4,7-octatriene moieties. *Archives of Biochemistry and Biophysics* 184, 36–41.

Birk, Y., Gertler, A. and Khalef, S. (1963) A pure trypsin inhibitor from soya beans. *Biochemical Journal* 87, 281–284.

Bowman, D.E. (1946) Differentiation of soybean antitryptic factors. *Proceedings of the Society for Experimental Biology* 63, 547–550.

Boyington, J.C., Gaffney, B.J. and Amzel, L.M. (1993) The three-dimensional structure of an arachidonic acid 5-lipoxygenase. *Science* 260, 1482–1486.

Brim, C.A. and Burton, J.W. (1979) Recurrent selection in soybeans. II. Selection for increased percent protein in seeds. *Crop Science* 19, 494–498.

Brim, C.A., Schutz, W.M. and Collins, F.I. (1968) Maternal effect on fatty acid composition and oil content of soybeans, *Glycine max* (L.) Merrill. *Crop Science* 8, 517–518.

Burks, W.W. Jr., Brooks, J.R. and Sampson, H.A. (1988) Allergenicity of major component proteins of soybean determined by enzyme-linked immunosorbent assay (ELISA) and immunoblotting in children with atopic dermatitis and positive soy challenges. *Journal of Allergy and Clinical Immunology* 81, 1135–1142.

Burton, J.W. and Brim, C.A. (1981) Recurrent selection in soybeans. III. Selection for increased percent oil in seeds. *Crop Science* 21, 31–34.

Burton, J.W. and Brim, C.A. (1983) Registration of a soybean germplasm population. *Crop Science* 23, 191.

Burton, J.W., Wilson, R.F. and Brim, C.A. (1983) Recurrent selection in soybeans. IV. Selection for increased oleic acid percentage in seed oil. *Crop Science* 23, 744–747.

Bustos, M.M., Guiltinan, M.J., Jorfano, J., Begum, D., Kalkan, F.A. and Hall, T.C. (1989) Regulation of β-glucuronidase expression in transgenic tobacco plants by an A/T-rich, *cis*-acting sequence found upstream of a French bean *β*-phaseolin gene. *Plant Cell* 1, 839–853.

Buttery, B.R. and Buzzell, R.I. (1971) Properties and inheritance of urease isozymes in soybean seeds. *Canadian Journal of Botany* 49, 1101–1105.

Carlson, J.B. and Lersten, N.R. (1987) Reproductive morphology. In: Wilcox, J.R. (ed.) *Soybeans: Improvement, Production and Uses.* American Society of Agronomy, Madison, Wisconsin, pp. 95–134.

Casey, R., Domoney, C. and Ellis, N. (1986) Legume storage proteins and their genes. *Oxford Surveys of Plant Molecular and Cell Biology* 3, 1–95.

Chen, Z.L., Pan, N.S. and Beachy, R.N. (1988) A DNA sequence element that confers seed-specific enhancement to a constitutive promoter. *EMBO Journal* 7, 297–302.

Cho, T.J., Davies, C.S. and Nielsen, N.C. (1989) Inheritance and organization of glycinin genes in soybean. *Plant Cell* 1, 329–337.

Chrispeels, M.J., Higgins, T.J.V., Craig, S. and Spencer, D. (1982a) Role of the endoplasmic reticulum in the synthesis of reserve proteins and the kinetics of their transport to protein bodies in developing pea cotyledons. *Journal of Cell Biology* 93, 5–14.

Chrispeels, M.J., Higgins, T.J.V. and Spencer, D. (1982b) Assembly of storage protein oligomers in the endoplasmic reticulum and processing of the polypeptides in the protein bodies of developing pea cotyledons. *Journal of Cell Biology* 93, 306–313.

Christopher, J.P. (1972) Isoenzymes of soybean lipoxygenase: Isolation and partial characterization. PhD thesis. Purdue University, West Lafayette, Indiana.

Christopher, J. and Axelrod, A. (1971) On the different positioned specificities of peroxidation of linoleate shown by two isozymes of soybean lipoxygenase. *Biochemical and Biophysical Research Communications* 44, 731–736.

Christopher, J., Pistorius, E. and Axelrod, B. (1970) Isolation of an isozyme of soybean lipoxygenase. *Biochimica et Biophysica Acta* 198, 12–19.

Christopher, J.P., Pistorius, E.K. and Axelrod, B. (1972a) Isolation of a third isoenzyme of soybean lipoxygenase. *Biochimica et Biophysica Acta* 284, 54–62.

Christopher, J.P., Pistorius, E.K., Regnier, F.E. and Axelrod, B. (1972b) Factors influencing the positional specificity of soybean lipoxygenase. *Biochimica et Biophysica Acta* 289, 82–87.

Circle, S.J. and Smith, A.K. (1972) *Soybeans: Chemistry and Technology. Vol. 1. Proteins.* Avi Publishing Co. Inc., Westport, Connecticut, p. 294.

Coates, J.B., Medeiros, J.S., Thanh, V.H. and Nielsen, N.C. (1985) Characterization of the subunits of *β*-conglycinin. *Archives of Biochemistry and Biophysics* 243, 184–194.

Collins, F.I. and Howell, R.W. (1957) Variability of linolenic and linoleic acids in soybean oil. *Journal of the American Oil Chemists Society* 34, 491–493.

Craig, S., Goodchild, D.J. and Miller, C. (1980a) Structural aspects of protein accumulation in developing pea (*Pisum sativum*) cotyledons: II. 3-dimensional reconstructions of vacuoles and protein bodies from serial sections. *Australian Journal of Plant Physiology* 7, 329–338.

Craig, S., Millerd, A. and Goodchild, D.J. (1980b) Structural aspects of protein accumulation in developing pea cotyledons. III. Immunocytochemical localization of legumin and vicilin using antibodies shown to be specific by the enzyme linked immunosorbent assay. *Australian Journal of Plant Physiology* 7, 339.

Danielsson, C.E. (1949) Seed globulins of the Gramineae and Leguminosae. *Biochemical Journal* 44, 387–400.

Davies, C.S. and Nielsen, N.C. (1986) Genetic analysis of a null-allele for lipoxygenase-2 in soybean. *Crop Science* 26, 460–462.

Davies, C.S., Coates, J.B. and Nielsen, N.C. (1985) Inheritance and biochemical analysis of four electrophoretic variants of β-conglycinin from soybean. *Theoretical and Applied Genetics* 71, 351–358.

Davies, C.S., Nielsen, S.S. and Nielsen, N.C. (1987) Flavor improvement of soybean preparations by genetic removal of lipoxygenase. *Journal of the American Oil Chemists Society* 64, 1428–1433.

Dickinson C.D., Hussein, E.H.A. and Nielsen, N.C. (1989) Role of post-transitional cleavage in the assembly of glycinin hexamers. *Plant Cell* 1, 459–469.

Diers, B.W., Beilinson, V., Nielsen, N.C. and Shoemaker, R.C. (1994) Genetic mapping of the *Gy4* and *Gy5* glycinin genes in soybean and analysis of a variant *Gy4*. *Theoretical and Applied Genetics* 89, 297–304.

Doyle, J.J., Schuler, M.A., Godette, W.D., Zenger, V., Beachy, R.N. and Slightom, J.L. (1986) The glycosylated seed storage proteins of *Glycine max* and *Phaseolus vulgaris*. *Journal of Biological Chemistry* 261, 9228–9238.

Egin-Buhler, B. and Ebel, J. (1983) Improved purification and further characterization of acetyl-CoA carboxylase from cultured cells of parsley (*Petroselinum hortense*). *European Journal of Biochemistry* 133, 335–339.

Egli, D.B. and Leggett, J.E. (1973) Dry matter accumulation patterns in determinate and indeterminate soybeans. *Crop Science* 13, 220–222.

Egli, D.B. and Leggett, J.E. (1976) Rate of dry matter accumulation in soybean seeds with varying source-sink ratios. *Agronomy Journal* 68, 371–374.

Egli, D.B., Leggett, J.E. and Wood, J.M. (1978) Influence of soybean seed size and position on the rate and duration of filling. *Agronomy Journal* 70, 127–130.

Egli, D.B., Fraser, J., Leggett, J.E. and Poneleit, C.G. (1981) Control of seed growth in soyabeans. (*Glycine max* (L.) Merr.) *Annals of Botany* 48, 171–176.

Eskin, N.A.M., Grossman, S. and Pinsky, A. (1977) Biochemistry of lipoxygenase in relation to food quality. *Critical Reviews in Food Science and Nutrition* 9, 1–40.

Evans, D.E., Nyquist, W.E., Santini, J.B., Bretting, P. and Nielsen, N.C. (1994) The immunological identification of seed lipoxygenase genotypes in soybean. *Crop Science* 34, 1529–1537.

Fukazawa, C., Momma, T., Hirano, H., Harada, K. and Udaka, K. (1985) Glycinin A3B4mRNA: Cloning and sequencing of double-stranded cDNA complementary to a soybean storage protein. *Journal of Biological Chemistry* 260, 6234–6239.

Galliard, T. and H.W.-S. Chan, (1980) Lipoxygenases. In: Stumpf, P.K. and Conn, E.E. (eds) *The Biochemistry of Plants*, Vol. 4. Academic Press, New York, pp. 132–162.

Garlich, J.D. and Nesheim, M.C. (1966) Relationship of fractions of soybeans and a crystalline soybean trypsin inhibitor to the effects of feeding unheated soybean meal to chicks. *Journal of Nutrition* 88, 100–110.

Garssen, G.J., Vleigenthart, J.F.G. and Boldingh J. (1971) An anaerobic reaction between lipoxygenase, linoleic acid and its hydroperoxides. *Biochemical Journal* 122, 327–332.

Gatehouse, J.A., Evans, I.M., Croy, R.R.D. and Boulter, D. (1986) Differential expression of genes during legume seed development. *Philosophical Transactions of the Royal Society* B314, 367–384.

Gay, S., Egli, D.B. and Reicosky, D.A. (1980) Physiological aspects of yield improvement in soybeans. *Agronomy Journal* 72, 387–391.

Gent, M.P.N. (1983) Rate of increase in size and dry weight of individual pods of field grown soya bean plants. *Annals of Botany* 51, 317–329.

Gibbs, P.E.M., Strongin, K.B. and McPherson, A. (1989) Evolution of legume seed storage proteins – A domain common to legumins and vicilins is duplicated in vicilins. *Molecular Biology and Evolution* 6, 614–623.

Goldberg, R.B. (1986) Regulation of plant gene expression. *Philosophical Transactions of the Royal Society* B 314, 343–353.

Goldberg, R.B., Hoschek, G. and Vodkin, L.O. (1983) An insertion sequence blocks the expression of a soybean lectin gene. *Cell* 33, 465–475.

Gomyo, T. and Nakamura, M. (1966) Biosynthesis of raffinose from uridine diphosphate galactose and sucrose by an enzyme preparation of immature soybeans. *Agricultural and Biological Chemistry* 30, 425-427.

Graham, D. and Patterson, B.D. (1982) Response of plants to low, non-freezing temperatures: Proteins, metabolism, and acclimation. *Annual Review of Plant Physiology* 33, 347–372.

Guerche, P., De Almeida, E.R.P., Schwarztein, M.A., Gander, E., Krebbers, E. and Pelletier, G. (1990) Expression of the 2S albumin from *Bertholletia excelsa* in *Brassica napus*. *Molecular and General Genetics* 221, 306–314.

Gupta, K. and Dhindsa, K.S. (1982) Fatty acid composition of oil of different varieties of soybean. *Journal of Food Science and Technology (Tokyo)* 19, 248–250.

Handley, L.W., Parr, D.M. and McFeeters, R.F. (1983) Relationship between galactinol synthease activity and sugar composition of leaves and seeds of several crop species. *Journal of the American Society of Horticultural Science* 108, 600–605.

Hammond, R.W., Foard, D.E. and Larkins, B.A. (1984) Molecular cloning and analysis of a gene coding for the Bowman-Birk protease inhibitor in soybean. *Journal of Biological Chemistry* 259, 9883–9890.

Hara-Nishimura, I., Inoue, K. and Nishimura, M. (1991) A unique vacuolar processing enzyme responsible for conversion of several proprotein precursors into the mature forms. *FEBS Letters* 294, 89–93.

Hara-Nishimura, I., Shimada, T., Hiraiwa, N. and Nishimura, M. (1995) Vacuolar processing enzyme responsible for maturation of seed proteins. *Journal of Plant Physiology* 145, 632–640.

Harada, J.J., Barker, S.J. and Goldberg, R.B. (1989) Soybean β-conglycinin genes are clustered in several DNA regions and are regulated by transcriptional and post-transcriptional processes. *Plant Cell* 1, 415–425.

Hartwig, E.E. and Hinson, K. (1972) Association between chemical composition of seed and seed yield of soybeans. *Crop Science* 12, 829–830.

Herman, E.M., Leland, M.S. and Chrispeels, M.J. (1986) The Golgi apparatus mediates the transport and post-translational modification of protein bodies. In: Shannon, L.M. and Chrispeels, M.J. (eds) *Molecular Biology of Seed Storage Proteins and Lectins*. American Society of Plant Physiologists, Rockville, Maryland, pp. 163–173.

Higgins, T.J.V., Chandler, P.M., Randall, P.J., Spencer, D., Beach, L.R., Blagrove, R.J., Kortt, A.A. and Inglis, A.S. (1986) Gene structure, protein structure, and regulation of the synthesis of a sulfur-rich protein in pea seeds. *Journal of Biological Chemistry* 261, 11124–11130.

Hildebrand, D.F. and Hymowitz, T. (1982) Inheritance of lipoxygenase-1 activity in soybean seeds, *Crop Science* 22, 851–853.

Hildebrand, D.F. and Hymowitz, T. (1980a) Inheritance of β-amylase nulls in soybean seed, *Crop Science* 20, 727–730.

Hildebrand, D.F. and Hymowitz, T. (1980b) Rapid test for Kunitz trypsin inhibitor activity in soybean seeds. *Crop Science* 20, 818–819.

Hildebrand, D.F. and Kito, M. (1984) Role of lipoxygenases in soybean seed protein quality. *Journal of Agricultural and Food Chemistry* 32, 815–819.

Hoffman, L.M., Donaldson, D.D., Bookland, R., Rashka, K. and Herman, E.M. (1987) Synthesis and protein body deposition of maize 15-kd zein in transgenic tobacco seeds. *EMBO Journal* 6, 3213–3221.

Holland, M.A. and Polacco, J.C. (1992) Urease-null and hydrogenase-null phenotypes of a phylloplane bacterium reveal altered nickel metabolism in two soybean mutants. *Plant Physiology* 98, 942–948.

Holland, M.A., Griffin, J.D., Meyer-Bothling, L.E. and Polacco, J.C. (1987) Developmental genetics of the soybean urease isozymes. *Developmental Genetics* 8, 375–387.

Howell, R.W., Brim, C.A. and Rinne, R.W. (1972) The plant geneticist's contribution toward changing lipid and amino acid composition of soybeans. *Journal of the American Oil Chemists Society* 49, 30–32.

Hu, B. and Esen, A. (1982) Heterogeneity of soybean proteins: Two-dimensional electrophoretic maps of three solubility fractions. *Journal of Agricultural and Food Chemistry* 30, 21–25.

Hwang, D.L.-R., Lin, K.-T.D., Yang, W.-K. and Foard, D.E. (1977) Purification, partial characterization, and immunological relationships of multiple low molecular weight protease inhibitors of soybean. *Biochimica et Biophysica Acta* 495, 369–382.

Hymowitz, T. and Hadley, H.H. (1972) Inheritance of a trypsin inhibitor variant in seed protein of soybeans. *Crop Science* 12, 197–200.

Hymowitz, T., Walker, W.M., Collins, F.I. and Panczner, J. (1972) Stability of sugar content in soybean strains. *Communications in Soil Science and Plant Analysis* 3, 367–373.

Ikenaka, T., Odani, S. and Koide, T. (1974) Chemical structure and inhibitory activities of soybean proteinase inhibitors. In: Fritz, H. *et al.* (eds) *Bayer Symposium V. Proteinase inhibitors*. Springer Verlag, Berlin, pp. 325–343.

Jaffé, W.G. (1969) Hemagglutinins. In: Liener, I.E. (ed.) *Toxic Constituents of Plant Foodstuff*. Academic Press, New York, pp. 69–101.

Jofuku, K.D., Okamuro, J. and Goldberg, R.B. (1987) Interaction of an embryo DNA binding protein with a soybean lectin gene upstream region. *Nature* 328, 734–737.

Jofuku, K.D., Schipper, R.D. and Goldberg, R.B. (1989) A frameshift mutation prevents Kunitz trypsin inhibitor mRNA accumulation in soybean embryos. *Plant Cell* 1, 427–435.

Joudrier, P.E., Foard, D.E., Floener, L.A. and Larkins, B.A. (1987) Isolation and sequence of cDNA encoding the soybean protease inhibitors PI IV and C-II. *Plant Molecular Biology* 10, 35–42.

Kandler, O. and Hopf, H. (1980) Occurrence, metabolism, and function of oligosaccharides. In: Stumpf, P.K. and Conn, E.E. (eds) *The Biochemistry of Plants*, Vol. 3. Academic Press, New York, pp. 221–270.

Kannangara, C.G. and Stumpf, P.K. (1972) Fat metabolism in higher plants: LIV. A pro-caryotic type acetyl CoA carboxylase in spinach chloroplasts. *Archives of Biochemistry and Biophysics* 152, 83–91.

Kanofsky, J.R. and Axelrod, B. (1986) Singlet oxygen production by soybean lipoxygenase isozymes. *Journal of Biological Chemistry* 21, 1099–1104.

Kato, T., Ohta, H., Tanaka, K. and Shibata, D. (1992) Appearance of new lipoxygenases in soybean cotyledons after germination and evidence for expression of a major new lipoxygenase gene. *Plant Physiology* 98, 324–330.

Kim, C.H. and Choi, Y.D. (1989) Molecular cloning of a cDNA encoding the precursor to the glycinin A2B1a subunit of soybean. *Korean Biochemical Journal* 22, 233–241.

Kirihara, J.A., Petri, J.B. and Messing, J. (1988) Isolation and sequence of a gene encoding a methionine-rich 10-kDa zein protein from maize. *Gene* 71, 359–370.

Kitamura, K., Davies, C.S., Kaizuma, N. and Nielsen, N.C. (1983) Genetic analysis of a nullallele for lipoxygenase-3 in soybean seeds. *Crop Science* 23, 924–927.

Kitamura, K., Kumagai, T. and Kikuchi, A. (1985) Inheritance of lipoxygenase-2 and genetic relationships among genes for lipoxygenase-1, -2, and -3 isozymes in soybean seeds. *Japanese Journal of Breeding* 35, 413–420.

Kitamura, Y., Arahira, M., Itoh, Y. and Fukazawa, C. (1990) The complete nucleotide sequence of soybean glycinin A2B1a gene spanning to another glycinin gene A1aB1b. *Nucleic Acids Research* 18, 4245.

Kloth, R.H. and Hymowitz, T. (1985) Re-evaluation of the inheritance of urease in soybean seed. *Crop Science* 2, 352–354.

Ko, T.-P., Ng, J.D. and McPherson, A. (1992) The three-dimensional structure of canavalin from jack bean (*Canavalia ensiformis*). *Plant Physiology* 101, 729–774.

Koide, T. and Ikenaka, T. (1973) Studies on soybean trypsin inhibitors. III. Amino acid sequence of the carboxyl-terminal region and the complete amino acid sequence of soybean trypsin inhibitor (Kunitz). *European Journal of Biochemistry* 32, 417–431.

Koide, T., Tsunasawa, S. and Ikenaka, T. (1973) Studies on soybean trypsin inhibitors. II. Amino acid sequence around the reactive site of soybean trypsin inhibitor (Kunitz). *European Journal of Biochemistry* 32, 408–416.

Kollman, G.E., Streeter, J.G., Jeffers, D.L. and Curry, R.B. (1974) Accumulation and distribution of mineral nutrients, carbohydrate, and dry matter in soybean plants as influenced by reproductive sink size. *Agronomy Journal* 66, 549–554.

Konno, S. (1979) Changes in chemical composition of soybean seeds during ripening. *Japan Agricultural Research Quarterly* 13, 186–194.

Kowalski, D. and Laskowski, M., Jr. (1976a) Chemical-enzymatic replacement of Ile64 in the reactive site of soybean trypsin inhibitor (Kunitz). *Biochemistry* 5, 1300–1309.

Kowalski, D. and Laskowski, M., Jr. (1976b) Chemical enzymatic insertion of an amino acid residue in the reactive site of soybean trypsin inhibitor (Kunitz). *Biochemistry* 15, 1309.

Krueger, R.W., Holland, M.A., Chisholm, D. and Polacco, J.C. (1987) Recovery of a soybean urease genomic clone by sequential library screening with two synthetic oligonucleotides. *Gene* 54, 41–50.

Kunitz, M. (1946) Crystalline soybean trypsin inhibitor. *Journal of General Physiology* 29, 149–154.

Laskowski, M., Jr. and Kato, I. (1980) Protein inhibitors of proteinases. *Annual Review of Biochemistry* 49, 593–626.

Lawrence, M.C., Suzuki, E., Varghese, J.N., Davis, P.C., VanDonkelaar, A., Tulloch, P.A. and Colman, P.M. (1990) The three-dimensional structure of the seed storage protein phaseolin at 3 Å resolution. *EMBO Journal* 9, 9–15.

Lawrence, M.C., Izard, T., Beuchat, M., Blagrove, R.J. and Colman, P.M. (1994) Structure of phaseolin at 2.2 Å resolution. Implications for a common vicilin/legumin structure and the genetic engineering of seed storage proteins. *Journal of Molecular Biology* 238, 748–776.

Lelievre, J.M., Oliveira, L.O. and Nielsen, N.C (1992a) 5'-CATGCAT-3' elements modulate the expression of glycinin genes. *Plant Physiology* 98, 387–391.

Lelievre, J.M. Dickinson, C.D., Dickinson, L.F. and Nielsen, N.C. (1992b) Assembly of soybean β-conglycinin in vitro. *Plant Molecular Biology* 18, 259–274.

Leske, K.L., Jevne, C.J. and Coon, C.N. (1993) Effect of oligosaccharide additions on nitrogen corrected true metabolizable energy of soy protein concentrate. *Poultry Science* 72, 664–668.

Liener, I.E., Goodale, R.L., Deshmukh, A., Satterberg, T.L., Ward, G., DiPeitro, C.M., Bankey, P.E. and Borner, J.W. (1988) Effect of a trypsin inhibitor from soybeans (Bowman-Birk) on the secretory activity of the human pancreas. *Gastroenterology* 94, 419–427.

Lilley, G.G., Caldwell, J.B., Kortt, A.A., Higgins, T.J. and Spencer, D. (1989) In: Applewhite, T.H. (ed.) *Proceedings of the World Congress on Vegetable Protein Utilization in Human Foods and Animal Feedstuffs*. Champaign, Illinois, America Oil Chemists' Society, pp. 497–502.

Lis, H., Fridman, C., Sharon, N. and Katchalski, E. (1966) Multiple hemagglutins in soybean. *Archives of Biochemistry and Biophysics* 117, 301–309.

Lotan, R., Siegelman, H.W., Lis, H. and Sharon, N. (1974) Subunit structure of soybean agglutinin. *Journal of Biological Chemistry* 249, 1219–1224.

MacLeod, G. and Ames, J. (1988) Soy flavor and its improvement. *CRC Critical Reviews of Food Science and Nutrition* 27, 219–400.

Maeder, D.L., Sunde, M. and Botes, D.P. (1992) Design and inhibitory properties of synthetic Bowman-Birk loops. *International Journal of Peptide Protein Research* 40, 97–102.

Marco, Y.A., Thanh, V.H., Turner, N.E., Scallon, B.J. and Nielsen, N.C. (1984) Cloning and structural analysis of DNA encoding an A2B1a subunit of glycinin. *Journal of Biological Chemistry* 259, 13436–13441.

Meyer-Bothling, L.E., Polacco, J.C. and Cianzio, S.R. (1987) Pleiotropic soybean mutants defective in both urease isozymes. *Molecular and General Genetics* 209, 432–438.

Minor, W., Steczko, J., Bolin, J.T., Otwinowski, Z. and Axelrod, B. (1993) Crystallographic determination of the active site iron and its ligands in soybean lipoxygenase L-1. *Biochemistry* 32, 6320–6323.

Miquel, M., James, D., Dooner, H. and Browse, J. (1993) *Arabidopsis* requires polyunsaturated lipids for low-temperature survival. *Proceedings of the National Academy of Sciences, USA* 90, 6208–6212.

Momma, T., Negoro, T., Udaka, K. and Fukazawa, C. (1985a) A complete cDNA coding for the sequence of glycinin A_2B_{1a} subunit precursor. *FEBS Letters* 188, 117–122.

Momma, T., Negoro, T., Hirano, H., Matsumoto, A., Udaka, K. and Fukazawa, C. (1985b) Glycinin $A_5A_4B_3$ mRNA: cDNA cloning and nucleotide sequencing of a splitting storage protein subunit of soybean. *European Journal of Biochemistry* 149, 491–496.

Moreira, M.A., Hermodson, M.A., Larkins, B.A. and Nielsen, N.C. (1979) Partial characterization of the acidic and basic polypeptides of glycinin. *Journal of Biological Chemistry* 254, 9921–9926.

Muramatsu, M. and Fukazawa, C. (1993) A high-order structure of plant storage proprotein allows its second conversion by an asparagine-specific cystein proteinase, a novel proteolytic enzyme. *European Journal of Biochemistry* 215, 123–132.

Mustakas, L.C., Albrecht, W.J., McGlee, J.E., Black, L.T., Bookwalter, G.N. and Griffin, J.J. (1969) Lipoxidase deactivation to improve stability, odor and flavor of full-fat soy flours. *Journal of the American Oil Chemists Society* 46, 623–626.

Ng, J.D., Ko, T.-P. and McPherson, A. (1992) Cloning, expression, and crystallization of jack bean (*Canavalia ensiformis*) canavalin. *Plant Physiology* 101, 713–728.

Nielsen, N.C. (1984) The chemistry of legume storage proteins. *Philosophical Transactions of the Royal Society Series* B304, 287–296.

Nielsen, N.C. (1989) Soybean Gy1 gene for glycinin subunit G1. Accession X15121. NIH gene database.

Nielsen, N.C., Dickinson, C.D., Cho, T.J., Thanh, V.H., Scallon, B.J., Fischer, R.L., Sims, T.L., Drews, G.N. and Goldberg, R.B. (1989) Characterization of the glycinin gene family. *Plant Cell* 1, 313–328.

Nielsen, N.C., Scott, M.P. and Lago, W.J.P. (1990) Assembly properties of modified subunits in the gycinin subunit family. In: Hermann, R. and Larkins, B. (eds) *NATO Advanced Study Institute on Plant Molecular Biology, Schloss Elmau, Germany, May 14–23, 1990*, Plenum Press, New York, pp. 635–640.

Nielsen N.C., Jung, R., Nam, Y.-W., Beaman, T.W., Oliveira, L.O. and Bassüner, R. (1995) Synthesis and assembly of 11S globulins. *Journal of Plant Physiology* 145, 641–647.

Nordlee, J.A., Taylor, S.L., Townsend, J.A. and Thomas, L.A. (1994) High methionine Brazil nut protein binds human IGE. *Journal of Allergy and Clinical Immunology* In press.

Odani, S. and Ikenaka, T. (1976) The amino acid sequences of two soybean double headed proteinase inhibitors and evolutionary considerations on the legume proteinase inhibitors. *Journal of Biochemistry (Tokyo)* 80, 641.

Oliveira, L.O. (1994) Studies on modulation of the expression of *Gy2* glycinin gene and on assembly properties of modified beta-conglycinin subunits. PhD thesis, Purdue University, West Lafayette, Indiana.

Openshaw, S.J. and Hadley, H.H. (1978) Maternal effects on sugar content in soybean seeds. *Crop Science* 18, 581–584.

Openshaw, S.J. and Hadley, H.H. (1981) Selection to modify sugar content of soybean seeds. *Crop Science* 21, 805–808.

Openshaw, S.J., Hadley, H.H. and Brokoski, C.E. (1979) Effects of pod removal upon seeds of nodulating and nonnodulating soybean lines. *Crop Science* 19, 289–290.

Orf, J.H. and Hymowitz, T. (1977) Inheritance of a second trypsin inhibitor variant in seed protein of soybeans. *Crop Science* 17, 811–813.

Orf, J.H. and Hymowitz, T. (1979) Inheritance of the absence of the Kunitz trypsin inhibitor in seed protein of soybeans. *Crop Science* 19, 107–109.

Osborne, T.B and Campbell, G.F. (1898) Proteids of the soy bean (*Glycine hispida*). *Journal of the American Chemical Society* 20, 419–428.

Park, T.K., Holland, M.A., Laskey, J.G. and Polacco, J.C. (1994) Germination-associated lipoxygenase transcripts persist in maturing soybean plants and are induced by jasmonate. *Plant Science* 95, 109–117.

Parker, W.B., Marshall, L.C., Burton, J.D., Somers, D.A., Wyse, D.L., Gronwald, J.W. and Gengenbach, B.G. (1990a) Dominant mutations causing alterations in acetyl-coenzyme A carboxylase confer tolerance to cyclohexanedione and aryloxyphenoxypropionate herbicides in maize. *Proceedings of the National Academy of Sciences, USA* 87, 7175–7179.

Parker, W.B., Somers, D.A., Wyse, D.L., Keith, R.A., Gronwald, J.W. and Gengenbach, B.G. (1990b) Selection and characterisation of sethoxydim-tolerant maize tissue cultures. *Plant Physiology* 92, 1220–1225.

Pavlinova, O.A. and Turkina, M.V. (1978) Biosynthesis and the physiological role of sucrose in the plant. *Soviet Plant Physiology* 25, 815–828.

Pedersen, K., Argos, P., Naravana, S.V.L. and Larkins, B.A. (1986) Sequence analysis and characterization of a maize gene encoding a high-sulfur zein protein of M_r 15,000. *Journal of Biological Chemistry* 261, 6279–6284.

Pedersen, H.S. and Djurtoft, R (1989) Antigenic and allergenic properties of acidic and basic peptide chains from glycinin. *Food and Agricultural Immunology* 1, 101–109.

Phillips, R.L. and McClure, B.A. (1985) Elevated protein-bound methionine in seeds of a maize line resistant to lysine plus threonine. *Cereal Chemistry* 62, 213–218.

Pistorius, E.K. (1974) Studies on isoenzymes of soybean lipoxygenase. PhD thesis. Purdue University, West Lafayette, Indiana.

Pistorius, E.K., Axelrod, B. and Palmer, G. (1976) Evidence for participation of iron in lipoxygenase reaction from optical and electron spin resonance studies. *Journal of Biological Chemistry* 251, 7144–7148.

Plietz, P. and Damaschun, G. (1986) The structure of the 11S seed globulins from various plant species: Comparative investigations by physical methods. *Studies in Biophysics* 3, 153–173.

Plietz, P., Drescher, B. and Damaschun, G. (1988) Structure and evolution of the 11S globulins: Conclusions from comparative evaluation of amino acid sequences and x-ray scattering data. *Biochemie und Physiologie Pflanzen* 183, 199–203.

Polacco, J.C. and Havir, E.A. (1979) Comparison of soybean urease isolated from seed and tissue culture. *Journal of Biological Chemistry* 254, 1707–1715.

Polacco, J.C. and Holland, M.A. (1994) Genetic control of plant ureases. In: J.K. Setlow (ed.) *Genetic Engineering, Vol. 16*, Plenum Press, New York, pp. 33–48.

Polacco, J.C. and Sparks, Jr., R.B. (1982) Patterns of urease synthesis in developing soybeans. *Plant Physiology* 70, 189–194.

Polacco, J.C. and Winkler, R.G. (1984) Soybean leaf urease: A seed enzyme? *Plant Physiology* 74, 800–803.

Polacco, J.C., Thomas, A.L. and Bledsoe, P.J. (1982) A soybean seed urease-null produces urease in cell culture. *Plant Physiology* 69, 1233–1240.

Polacco, J.C., Judd, A.K., Dybing, J.K. and Cianzio, S.R. (1989) A new mutant class of soybean lacks urease in leaves but not in leaf-derived callus or in roots. *Molecular and General Genetics* 217, 257–262.

Post-Beittenmiller, M.A., Jaworski, J. and Ohlrogge, J.B. (1991) In vivo pools of free and acylated acyl carrier protein in spinach: evidence for sites of regulation of fatty acid biosynthesis. *Journal of Biological Chemistry* 266, 1858–1865.

Preiss, J. and Levi, C. (1980) Starch biosynthesis and degradation. In: Stumpf, P.K. and Conn, E.E. (eds) *The Biochemistry of Plants*, vol. 3. Academic Press, New York, pp. 371–423.

Probst, A.H., Laviolette, F.A., Athow, K.L. and Wilcox, J.R. (1971) Registration of Protana soybean. *Crop Science* 11, 312.

Pull, S.P., Pueppke, S.G., Hymowitz, T. and Orf, J.H. (1978) Soybean lines lacking the 120,000-Dalton seed lectin. *Science* 200, 1277–1279.

Rackis, J.J. (1965) Physiological properties of soybean trypsin inhibitors and their relationship to pancreatic hypertrophy and growth inhibition of rats. *Federation Proceedings. Federation of American Societies for Experimental Biology* 24, 1488–1493.

Reichelt, R., Schwenke, K.-D., König, T., Pähtz, W. and Wangermann, G. (1980) Electron microscopic studies for estimation of the quaternary structure of the 11S globulin (helianthinin) from sunflower seed (*Helianthus annuus* L.). *Biochemie und Physiologie Pflanzen* 175, 653–663.

Reicosky, D.A., Orf, J.H. and Poneleit, C. (1982) Soybean germplasm evaluation for length of the seed filling period. *Crop Science* 22, 319–322.

Robinson, D.G., Hinz, G., Hoh, B. and Jeong, B.K. (1994) One vacuole or two vacuoles: Do protein storage vacuoles arise *de novo* during pea cotyledon development? *Journal of Plant Physiology* 145, 654–664.

Roesler, K.R., Shorrosh, B.S. and Ohlrogge, J.B. (1994) Structure and expression of an *Arabidopsis* acetyl-coenzyme A carboxylase gene. *Plant Physiology* 105, 611–617.

Ryan, C.A. (1988) Proteinase inhibitor gene families: Tissue specificity and regulation. In: Verma, D.P.S. and Goldberg, R.B. (eds) *Temporal and Spacial Control of Plant Genes*. Springer-Verlag, Vienna, pp. 223–233.

Saalbach, G., Jung, R., Saalbach, I. and Müntz, K. (1988) Construction of storage protein genes with increased number of methionine codons and their use in transformation experiments. *Biochemie und Physiologie Pflanzen* 183, 211–218.

Sasaki, Y., Hakamada, K., Suama, Y., Nagano, Y., Furusava, I. and Matsuno, R. (1993) Chloroplast-encoded protein as a subunit of acetyl-CoA carboxylase in pea plant. *Journal of Biological Chemistry* 268, 25118–25123.

Sato, K. and Ikeda, T. (1979) The growth responses of soybean to photoperiod and temperature. IV. The effect of temperature during the ripening period on the yield and characters of seeds. *Japanese Journal of Crop Science* 48, 283–290.

Scallon, B.J., Thanh, V.H., Floener, L.A. and Nielsen, N.C. (1985) Identification and characterization of DNA clones encoding Group II glycinin subunits. *Theoretical and Applied Genetics* 70, 510–519.

Schou, J.B., Jeffers, D.L. and Streeter, J.G. (1978) Effects of reflectors, black boards, or shades applied at different stages of plant development on yield of soybeans. *Crop Science* 18, 29–34.

Schuler, M.A., Schmitt, E.S. and Beachy, R.N. (1982) Closely related families of genes code for the α and α' subunits of the soybean 7S storage protein complex. *Nucleic Acids Research* 10, 8225–8243.

Schweizer, T.F., Horman, I. and Wursch, P. (1978) Low molecular weight carbohydrates from leguminous seeds; a new disaccharide: galactopinitol. *Journal of the Science of Food and Agriculture* 29, 148–154.

Scott, M.P., Jung, R., Müntz, K. and Nielsen, N.C. (1992) A protease responsible for post-translational cleavage of a conserved Asn-Gly linkage in glycinin, the major seed storage protein of soybean. *Proceedings of the National Academy of Sciences, USA* 89, 658–662.

Sebastiani, F.L., Farrell, L.B., Schuler, M.A. and Beachy, R.N. (1990) Complete sequence of a cDNA of the α subunit of soybean β-conglycinin. *Plant Molecular Biology* 15, 197–201.

Sengupta, C., Deluca, V., Bailey, D.S. and Verma, D.P.S. (1981) Post-translational processing of 7S and 11S components of soybean storage proteins. *Plant Molecular Biology* 1, 19–34.

Shibasaki, M., Suzuki, S., Tajima, S., Nemoto, H. and Kuroume, T. (1980) Allergenicity of major component proteins of soybean. *International Archives of Allergy and Applied Immunology* 61, 441–448.

Shibata, D., Steczko. J., Dixon, J.E., Hermodson, M., Yazdanparast, R. and Axelrod, B. (1987) Primary structure of soybean lipoxygenase-1. *Journal of Biological Chemistry* 262, 10080–10085.

Shibata, D., Steczko, J., Dixon, J.E., Andrews, P.C., Hermodson, M. and Axelrod, B. (1988) Primary structure of soybean lipoxygenase L-2. *Journal of Biological Chemistry* 263, 6816–6821.

Shotwell, M.A. and Larkins, B.A. (1989) The biochemistry and molecular biology of seed storage proteins. In: Marcus, A. (ed.) *The Biochemistry of Plants*, vol 15. Academic Press, San Diego, California, pp. 297–345.

Singh, L., Wilson, C.M. and Hadley, M.H. (1969) Genetic differences in soybean trypsin inhibitors separated by disc electrophoresis. *Crop Science* 9, 489–491.

Staswick, P.E., Hermodson, M.A. and Nielsen, N.C. (1981) Identification of the acidic and basic subunit complexes of glycinin. *Journal of Biological Chemistry* 256, 8752–8755.

Staswick, P.E., Hermodson, M.A. and Nielsen, N.C. (1984a) The amino acid sequence of the A_2B_{1a} subunit of glycinin. *Journal of Biological Chemistry* 259, 13424–13430.

Staswick, P.E., Hermodson, M.A. and Nielsen, N.C. (1984b) Identification of the cystines which link the acidic and basic components of the glycinin subunits. *Journal of Biological Chemistry* 259, 13431–13435.

Streeter, J.G. and Jeffers, D.L. (1979) Distribution of total nonstructural carbohydrates in soybean plants having increased reproductive load. *Crop Science* 19, 729–734.

Sturm, A., van Kuik, J.A., Vliegenhart, J.F.G. and Chrispeels, M.J. (1987) Structure, position and biosynthesis of the high mannose and the complex oligosaccharide chains of the bean storage protein phaseolin. *Journal of Biological Chemistry* 262, 13392–13403.

Sun, S.S.M. and Larkins, B.A. (1993) Transgenic plants for improving seed storage proteins. In: *Transgenic Plants, Vol. 1. Engineering and Utilization*. Academic Press, pp. 339–372.

Sun, S.S.M., Leung, F.W. and Tomic, J.C. (1987) Brazil nut (*Bertholletia excelsa* H.B.K.) proteins: Fractionation, composition, and identification of a sulfur-rich protein. *Journal of Agricultural and Food Chemistry* 35, 232–235.

Sweet, R.M., Wright, M.T., Janin, T, Chothia, C.H. and Blow, D.M. (1974) Crystal structure of the complex of porcine trypsin with soybean trypsin inhibitor (Kunitz) at 2.6 Å resolution. *Biochemistry* 13, 4212–4228.

Takamura, H., Kitamura, K. and Kito, M. (1991) Inhibition by lipoxygenase -3 of *n*-hexanal generation in soybeans. *FEBS Letters* 292, 42–44.

Tan-Wilson, A.L. and Wilson, K.A. (1986) Relevance of multiple soybean trypsin inhibitor forms to nutritional quality. In: M. Friedman (ed.) *Nutritional and Toxicological Significance of Enzyme Inhibitors in Food*. Plenum Press, New York, pp. 392–411.

Thanh, V.H. and Shibasaki, K. (1976) Heterogeneity of beta-conglycinin. *Biochimica et Biophysica Acta* 439, 326–338.

Thanh, V.H. and Shibasaki, K. (1977) Beta-conglycinin from soybean proteins. Isolation and immunological and physicochemical properties of the monomeric forms. *Biochimica et Biophysica Acta* 490, 370–384.

Thanh, V.H. and Shibasaki, K. (1978a) Major proteins of soybean seeds. Reconstitution of *β*-conglycinin from its subunits. *Journal of Agricultural and Food Chemistry* 26, 695–698.

Thanh, V.H. and Shibasaki, K. (1978b) Major proteins of soybean seeds. Subunit structure of *β*-conglycinin. *Journal of Agricultural and Food Chemistry* 26, 692–695.

Thanh, V.H. and Shibasaki, K. (1978c) Major proteins of soybean seeds. Reconstitution of *β*-conglycinin from its subunits. *Journal of Agricultural and Food Chemistry* 26, 695–698.

Thanh, V.H. and Shibasaki, K. (1979) Major proteins of soybean seeds. Reversible and irreversible dissociation of *β*-conglycinin. *Journal of Agricultural and Food Chemistry* 27, 805–809.

Torisky, R.S. and Polacco, J.C. (1990) Soybean roots retain the seed urease isozyme synthesized during embryo development. *Plant Physiology* 94, 681–689.

Torisky, R.S., Griffin, J.D., Yenofsky, R.L. and Polacco, J.C. (1994) A single gene (Eu4) encodes the tissue-ubiquitous urease of soybean. *Molecular and General Genetics* 242, 404–414.

Utsumi, S. (1992) Plant food protein engineering. *Advances in Food and Nutrition Research* 36, 89–208.

Utsumi, S., Inaba, H. and Mori, T. (1981) Heterogeneity of soybean glycinin. *Phytochemistry* 20, 585–589.

Utsumi, S., Gidamis, A.B., Mikami, B. and Kito, M. (1993) Crystallization and preliminary x-ray crystallographic analysis of the soybean proglycinin expressed in *Escherichia coli*. *Journal of Molecular Biology* 233, 177–178.

Vernooy-Gerritsen, M., Bos, A.L.M., Veldink, G.A. and Vliegenthart, J.F.G. (1983) Localization of lipoxygenase -1 and -2 in germinating soybean seeds by an indirect immunofluorescence technique. *Plant Physiology* 73, 262–267.

Vliegenthart, J.F.G. and Veldink, G.A. (1982) In: Pryor, W.A. (ed.) *Free Radicals in Biology*, Vol. V. Academic Press, New York, pp. 29–64.

Vodkin, L.O., Rhodes, P.R. and Goldberg, R.B. (1983) cA lectin gene insertion has the structural features of a transposable element. *Cell* 34, 1023–1031.

Weber, C.R. and Fehr, W.R. (1970) Registration of Provar soybeans. *Crop Science* 10, 728.

Wehrman, V.K., Fehr, W.R., Cianzio, S.R. and Cavins, J.F. (1987) Transfer of high seed protein to high-yielding soybean cultivars. *Crop Science* 27, 927–931.

Werner, M.H. and Wemmer, D.E. (1992) Three-dimensional structure of soybean trypsin/chymotrypsin Bowman-Birk inhibitor in solution. *Biochemistry* 31, 999–1010.

Wilcox, J.R. and Cavins, J.F. (1995) Backcrossing high seed protein to a soybean cultivar. *Crop Science* 35, 1036–1041.

Wilson, R.F. (1987) Seed Metabolism. In: Wilcox, J.R. (ed.) *Soybeans: Improvement, Production and Uses.* American Society of Agronomy, Inc., Madison, Wisconsin, pp. 643–686.

Wilson, R.F., Burton, J.W., Buck, J.A. and Brim, C.A. (1978) Studies on genetic male-sterile soybeans. I. Distribution of plant carbohydrate and nitrogen during development. *Plant Physiology* 61, 838–841.

Winkler, R.L., Polacco, J.C., Eskew, D.L. and Welch, R.M. (1983) Nickel is not required for apourease synthesis in soybean seeds. *Plant Physiology* 72, 262–263.

Wolf, J.W. (1975) Lipoxygenase and flavor of soybean protein products. *Journal of Agricultural and Food Chemistry* 23, 136–141.

Wolf, W.J. and Briggs, D.R. (1958) Studies on the cold-insoluble fraction of the water-extractable soybean proteins. II. Factors influencing conformation changes in the 11S component. *Archives of Biochemistry and Biophysics* 76, 377–393.

Wright, D.J. (1988) The seed globulins – part II. In: Hudson, B.J.F. (ed.) *Developments in Food Proteins – 6.* Elsevier, London, pp. 119–178.

Yamauchi, F. and Yamagishi, T. (1979) Carbohydrate sequence of a soybean 7S protein. *Agricultural and Biochemical Chemistry* 43, 505–510.

Yamauchi, F., Saio, K. and Yamagishi, T. (1984) Isolation and partial characterization of a salt-extractable globulin from soybean seeds. *Agricultural and Biochemical Chemistry* 48, 645–650.

Yang, M.S., Espinoza, N.O., Nagpala, P.G. Dodds, J.H., White, F.F., Schnorr, K.L. and Jaynes, J.M. (1989) Expression of a synthetic gene for improved protein quality in transformed potato plants. *Plant Science* 64, 99–111.

Yazdi-Samadi, B., Rinne, R.W. and Seif, R.D (1977) Components of developing soybean seeds: Oil, protein, sugars, starch, organic acids, and amino acids. *Agronomy Journal* 69, 481–486.

Yenofsky, R.L., Fine, M. and Liu, C. (1988) Isolation and characterization of a soybean (*Glycine max*) lipoxygenase-3 gene. *Molecular and General Genetics* 211, 215–222.

Zeiher, C., Egli, D.B., Leggett, J.E. and Reicosky, D.A. (1982) Cultivar differences in N redistribution in soybeans. *Agronomy Journal* 74, 375–379.

Genetic Modification of Soybean Oil Quality

8

N.S. Yadav

Agricultural Products, DuPont Co., PO Box 80402, Wilmington, Delaware 19880-0402, USA.

Introduction

Plant breeding goals have recently been shifting from crop yield to quality traits, fuelled in part by agricultural overabundance in the developed world and in part by biotechnology. Vegetable oils are a major agricultural commodity accounting for *ca.* 60 billion tonnes worldwide (Battey *et al.*, 1989). Soybean accounts for *ca.* one-third of the world oil production and *ca.* three-quarters of the 15 billion pounds of edible oil consumed in the United States (USDA, 1994). In the US, soybean is a major agricultural commodity: *ca.* 60 million acres of soybean were planted and 17 million metric tonnes of soybeans exported in 1992/93 (USDA, 1994). Soybean is the lowest-cost producer of vegetable oil, since the oil is a co-product of its protein-rich meal. Its oil is used in a multitude of food applications, including baking, frying, salad dressing, margarine, and processed foods. The low cost and ready availability of soybean oil provides an excellent opportunity to genetically upgrade this commodity oil into higher-value specialty oils.

Oil is composed of triacylglycerols with three same or different fatty acids esterified to the glycerol backbone. The fatty acid composition and distribution on the triglyceride molecule largely determine oil quality – nutritional value, flavour, and physical properties, such as oxidative stability and melting point. Soybean oil, like most edible oils, is composed of five common fatty acids: palmitate (16:0), stearate (18:0), oleate (18:1), linoleate (18:2) and linolenate (18:3). 16:0 and 18:0 are, respectively, 16- and 18-carbon-long saturated fatty acids. 18:1, 18:2 and 18:3 are 18-carbon-long unsaturated fatty acids containing one, two and three methylene-interrupted *cis* double bonds, respectively. The positions of the double bonds on the acyl chain from the carboxyl end are δ-9 in 18:1, δ-9 and -12 (or ω-6, counting from the methyl end) in 18:2, and δ-9, -12 (or ω-6), and -15 (or ω-3) in 18:3. 18:1 is also referred to as a monounsaturated fatty acid, while 18:2 and 18:3 are also referred to as diene and triene polyunsaturated fatty acids. The same fatty acids found in the oil are also major constituents of cell membranes, where they have an important function in the physiology and development of plants. Thus, although oil is normally a seed storage product, its constituent fatty acids are synthesized constitutively in all cells.

This chapter briefly reviews some of the key steps in oil biosynthesis, the recent progress in cloning lipid biosynthesis genes, desired fatty acid compositions, and the progress of both traditional breeding and transgenic approaches to achieve the desired compositions in soybean oil.

Desired Oil Qualities and Compositions

Increased oxidative stability

A key goal for improving soybean oil quality has been to increase its oxidative stability by reducing its linolenate (18:3) content. 18:3 oxidizes readily to result in off-flavours, rancidity, and reduced performance (Frankel, 1980). In fact, because of its poor quality, soybean oil was used as much for food as for industrial use early in this century. The rise in its edible consumption since the early 1950s is tied to the use of hydrogenation – one of chemistry's success stories of this century. Hydrogenation, which chemically reduces the double bonds, results in greatly improved oil stability and flavour. However, the need for hydrogenation reduces the economic attractiveness of soybean oil. Furthermore, even hydrogenated soybean oil has some flavour 'reversion' – off-odours and flavours that have been attributed to the presence of isolinoleate, a hydrogenation product (Frankel, 1980). Thus, a naturally stable soybean oil will result in lower processing cost and improved flavour, shelf life, and frying performance. 18:2 also contributes to oxidative instability and poor flavour of soybean oil (Liu and White, 1992). Removal of the last few traces of 18:2 (below 10%) contributed most to oxidative stability in high 18:1 sunflower oils (Purdy, 1985). Thus, soybean oil without polyunsaturates is highly desirable for similar stability.

Reduced polyunsaturate content is also important in making almost all other specialty oils, due in part to its oxidative stability and in part because biosynthesis of polyunsaturated fatty acids competes with that of most specialty fatty acids. Thus, a 'zero' polyunsaturate content is a key desired trait in oils biotechnology.

Improved functionality

Higher saturates, 16:0 or 18:0, have improved functional properties for certain specific food applications, such as in margarines and shortenings. For example, 15% or more of 16:0 would favour β' crystal with improved texture (Hammond, 1992). An example of a specialty confectionery fat is a substitute for cocoa butter, the most expensive edible oil (*ca.* six times more expensive than soybean oil). Cocoa butter's annual worldwide production is about 2 billion pounds and several hundred million dollars worth of it are imported in the US. Its high and volatile prices and uncertain supply have encouraged the development of substitutes. Its fatty acid composition is 26% 16:0, 34% 18:0, 35% 18:1 and 3% 18:2 acids and *ca.* 72% of its triglycerides have the structure in which saturated fatty acids occupy the *sn*-1 and -3 positions and 18:1 occupies the *sn*-2 position on the glycerol backbone. These characteristics confer on it properties eminently suitable for confectionery end-uses: it is hard and non-greasy at ordinary temperatures and melts very sharply in the mouth. It is also extremely resistant to rancidity. For these reasons, producing soybean oil with increased levels of 18:0 and 16:0 and highly reduced levels of polyunsaturated fatty acids may lead to an oil which can function as a cocoa butter substitute.

Improved nutritional quality

Since vegetable oil is a major component of our diet, its nutritional value is important. Unfortunately, medical opinion of an ideal oil composition is not established and appears to be constantly changing. Nevertheless, there is consensus that reducing the saturate content, especially 16:0, is desirable. When compared to canola oil, soybean oil has a higher percent of saturated fatty acids (15% vs 6%). There is great interest in reducing the saturate contents of soybean oil, at least as low as in canola.

Reducing polyunsaturates is apparently also nutritionally beneficial. Although 18:2 and 18:3 are essential fatty acids, vegetable oils, including soybean oil, contain polyunsaturated fatty acids at levels far in excess of human essential dietary requirement. Recent clinical studies suggest that diets high in 18:1 may be beneficial (see Gurr, 1989): fats rich in 18:1 reduce both total cholesterol and the 'bad' (low-density lipoprotein) cholesterol while maintaining the 'good' (high-density lipoprotein) cholesterol and total triglycerides. These results corroborate previous epidemiological studies of people living in Mediterranean countries where a relatively high intake of monounsaturated fat and low consumption of saturated fat correspond with low coronary heart disease mortality (Keys, 1980).

Hydrogenated oil contains unnatural *trans* fatty acids, by-products of hydrogenation, believed to be unhealthy. A recent report has renewed serious concerns that *trans* fatty acid intake is causally related to risk of coronary disease in humans (Willet and Ascherio, 1994). Thus, an oil high in 18:1, which does not require hydrogenation, may be healthy.

Industrial oils

Linseed oil is an inedible drying oil due to its high 18:3 content (*ca.* 55%). If the 18:3 content of soybean oil is similarly increased it could be used as replacement for the higher value linseed oil, most of which is imported. In addition, although the five fatty acids found in soybean oil are common in most edible vegetable oils, there are more than 200 chemically diverse uncommon fatty acids found in different rare plant species (van de Loo *et al.*, 1993). These include hydroxy-, epoxy-, acetylenic, keto fatty acids and fatty acids with unusual positions of conjugated and non-conjugated double bonds. Some of these are of known industrial value, including ricinoleate in castor oil, erucate in high erucic rapeseed, medium-chain (C_{12} or C_{14}) fatty acids in coconut oil, α-eleostearate in tung oil, and wax esters in jojoba wax. Other chemically interesting fatty acids could also be valuable, if sufficient quantities of them can be obtained for test applications.

It is likely that the biosynthesis of some of these uncommon fatty acids requires one or a few specific genes. Thus, the prospect of using transgenic approaches to tailor soybean oil for specialty fatty acids appears attractive.

Thus, besides opportunities with oils with specialty fatty acids, the key targets for modifying the five common fatty acids of soybean oil are:

- an oil low in saturates and polyunsaturates and high in monounsaturates would provide significant health benefits, improved stability, and an economic benefit to oil processors;
- an oil with increased 18:0 and/or 16:0 is desirable in solid fats and confectionery uses;
- an oil with 60% or more 18:3 as a substitute for linseed oil.

Oil Biosynthesis

Oil biosynthesis consists of two overlapping stages: fatty acid biosynthesis and triacylglycerol assembly (Fig. 8.1). Details of the pathway can be obtained from the several excellent reviews on the subject published recently (Browse and Somerville, 1991; Frentzen, 1993; Heinz, 1993; Ohlrogge *et al.*, 1993; Kinney, 1994).

Fatty acid biosynthesis in plants involves different lipids and different cellular compartments. The biosynthesis of 16:0, 18:0, and 18:1 occurs on acyl carrier protein (ACP) in the plastid and the subsequent biosynthesis of 18:2

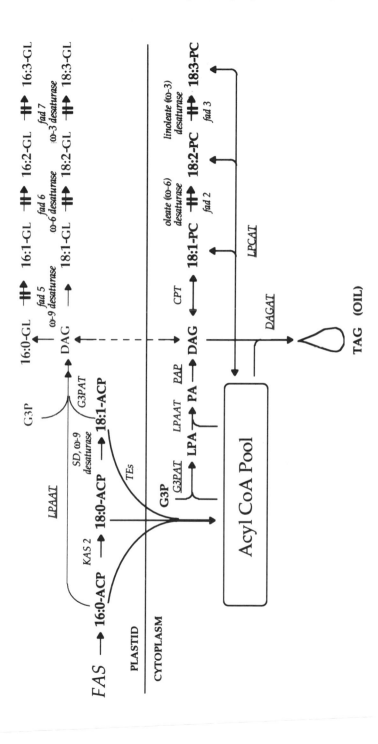

Fig. 8.1. Schematic representation of plant oil biosynthetic pathway. The enzymes are shown in italics; the genes for enzymes underlined remain to be cloned. The plastid glycerolipid biosynthetic pathway is unlikely to be significant in developing oilseeds. The dotted line between the plastid and extraplastid DAG denotes glycerolipid exchange between the two cellular compartment by unknown mechanism(s). The abbreviations are described in the text. Desaturation (*fad*) mutants in *Arabidopsis* are indicated.

and 18:3 occurs on glycerolipids following acylation of 18:1 on glycerol-3-phosphate backbone in the endoplasmic reticulum (ER) or plastid. The discussion below highlights some of the oil biosynthetic steps that can be targeted for improving oil quality.

Biosynthesis of 16:0, 18:0, and 18:1 fatty acids

De novo fatty acid biosynthesis has been reviewed recently in plants (Ohlrogge et al., 1993) and in E. coli (Magnuson et al., 1993); the bacterial review also provides useful insights into plant fatty acid biosynthesis. Fatty acid synthase (FAS) catalyses cycles of two-carbon unit elongation of acyl chains esterified to acyl carrier protein (ACP). It consists of four soluble enzymes: β-ketoacyl-ACP synthase (KAS), β-ketoacyl-ACP reductase, β-hydroxyacyl-ACP dehydratase, and enoyl-ACP reductase. Of these only the KAS isoforms have been shown to have specificity for substrate length. The primary products of FAS are 16:0-ACP and 18:0-ACP. KAS 2 isoform is largely responsible for the elongation of 16:0-ACP to 18:0-ACP. Stearoyl-ACP (δ-9) desaturase (SD) introduces the first double bond in 18:0-ACP to form 18:1-ACP. It is a soluble desaturase requiring O_2 and reduced ferredoxin as the electron donor. SD is an important determinant of oil quality, since it determines the degree of overall unsaturation. The 16:0-ACP, 18:0-ACP, and 18:1-ACP can serve as substrates for acyl-ACP thioesterase(s) (TEs) and, in the leaf, also for the plastid acyl-ACP acyltransferases. 'Oleoyl-specific' isoforms of TE have been reported from safflower (Knutzon et al., 1992a) and soybean (Hitz, W.D., personal communication). The soybean enzyme had activities against 16:0-ACP and 18:0-ACP of between 5 and 20% of the activity on 18:1-ACP. Other isoforms with different specificities may be present.

During oil biosynthesis 16:0-ACP and 18:0-ACP are branch points for two competing reactions: the ratio of KAS 2 and TE activities on 16:0-ACP and the ratio of SD and TE activities on 18:0-ACP determine the amounts of 16:0 and 18:0, respectively, synthesized. Thus, the interplay of at least three key enzymes, viz., KAS 2, SD, TE, determines the relative amounts 16:0, 18:0 and 18-carbon unsaturated fatty acids synthesized. The fatty acids released by the thioesterases in the plastid are exported as acyl-CoA to the cytoplasm.

Biosynthesis of 18:2 and 18:3 fatty acids

In plants there are two pathways for polyunsaturated fatty acid biosynthesis, one localized in the ER and the other in the plastid (Heinz, 1993). In both, biosynthesis of 18:2 and 18:3 is via O_2-dependent and glycerolipid-linked desaturations catalysed by oleate desaturase and linoleate desaturase, respectively. The immediate electron donors for the ER and plastid desaturases are cytochrome b_5 and ferredoxin, respectively. Phosphatidylcholine (PC) is the primary glycerolipid substrate for the biosynthesis of polyunsaturated fatty acids in the ER and galactolipids (GL), especially monogalactosyl diacylglycerol and digalactosyl diacylglycerol, are the primary substrates in the plastid membrane. In *Arabidopsis thaliana* ER oleate and linoleate desaturation are controlled by *fad 2* and *fad 3* loci, respectively (Lemieux *et al.*, 1990) and plastid oleate and linoleate desaturations are controlled by *fad 6* and *fad 7* loci, respectively (Browse and Somerville, 1991). While both pathways operate in the leaf tissue with significant exchange between the intermediates of the two pathways, the ER pathway predominates during oil biosynthesis in developing seeds and in other non-green tissues. Nevertheless, the relative contribution of the plastid pathway might be significant in a low polyunsaturate oil.

Triacylglycerol assembly

Acyl-CoAs from the cytoplasmic pool are assembled into triacylglycerol (TAG) by the action of three microsomal acyltransferases (Frentzen, 1993). The sequential addition of acyl moieties to the *sn*-1 and *sn*2- positions of glycerol-3-P (G3P) catalysed by glycerol-3-P acyltransferase (GPAT) and lysophosphatidic acyltransferase (LPAAT), respectively, results in the synthesis of phosphatidic acid (PA). PA is converted to diacylglycerol (DAG) by phosphatidic acid phosphatase (PAP). DAG is a branch point for membrane and oil biosynthesis. For example, for membrane biosynthesis it can be converted by a reversible reaction catalysed by cholinephosphotransferase (CPT) into phosphatidylcholine (PC), which, as mentioned above, serves as a substrate for ER polyunsaturated fatty acid biosynthesis. The other fate of DAG is the final acylation of *sn*-3 position to form triacylglycerol (TAG), which is catalysed by diacylglycerol acyltransferase (DAGAT) – the only step unique in oil biosynthesis (Frentzen, 1993).

GPAT has a strong preference for saturated fatty acids over 18:1, while LPAAT strongly prefers C18 unsaturated fatty acids over saturated fatty acids. The combined acyl group specificities of the three acyltransferases generally results in TAG molecules with saturated fatty acids at the *sn*-1 position, unsaturated fatty acids at the *sn*-2 position, and either type at *sn*-3 position. These specificities are not absolute and the fatty acid composition of the acyl pool can, to some extent, drive by mass action a corresponding change in the fatty acid composition of the oil. For example, given the correct composition of acyl-CoA pool plants can produce cocoa butter substitutes (Bafor *et al.*, 1990). However, these specificities could limit the range of other fatty acid compositions. For example, the specificity of LPAAT has been proposed to be responsible for the selective exclusion of uncommon fatty acids, such as erucic and medium chain fatty acids from the *sn*-2 position (Frentzen, 1993).

The levels of 18:2 and 18:3 in oil depend on their rates of biosynthesis as well as on their availability for oil biosynthesis. There are two mechanisms by which the latter can occur: first, the reverse reaction of CPT converts PC back to DAG, which can then be used for oil biosynthesis via DAGAT (see above). Second, an 'exchange' lysophosphatidylcholine acyltransferase (LPCAT) cata- lyses the reversible acyl exchange between 18:1-CoA and the acyl group at the sn-2 position of PC (Frentzen, 1993). This results in enrichment of the acyl- CoA pool with polyunsaturated fatty acids for use in oil biosynthesis. Thus, CPT and LPCAT can play a significant role in determining the degree of poly- unsaturation in vegetable oils. Their relative contributions, however, are not known and may vary in different tissues. For example, LPCAT is not detectable in avocado mesocarp and cocoa seeds, which lack polyunsaturated fatty acids in their oil. Finally, PC may have a central role in oil biosynthesis even when it is not involved in a PC-dependent modification of fatty acids, such as in the case of the assembly of petroselinic acid in coriander oil (Cahoon and Ohlrogge, 1994).

Cloning of Lipid Biosynthesis Genes

The genetics of plant lipid biosynthesis and cloning of lipid biosynthesis genes were recently reviewed (Ohlrogge *et al.*, 1991, 1993). The list of plant lipid genes reported cloned is shown in Table 8.1 and is growing rapidly. Some of the first genes cloned were for the soluble plastid enzymes, which were puri- fied and the protein sequence used to clone the gene. These include KAS 2, SD, and TEs. Recently, the cDNA of plastid membrane oleate desaturase was cloned following enzyme purification (Schmidt *et al.*, 1993).

However, the extraplastidic fatty acid biosynthetic enzymes are membrane proteins and have been, for the most part, recalcitrant to purifica- tion and study. Fortunately, genetic approaches are proving useful in cloning the genes for some of them. For example, microsomal linoleate desaturase gene was isolated by map-based cloning that is becoming increasingly more refined and powerful (Arondel *et al.*, 1992), microsomal oleate and linoleate desaturase genes were isolated by T-DNA tagging (Yadav *et al.*, 1993; Okuley *et al.*, 1994), and cDNAs encoding LPAAT (Brown *et al.*, 1994) and CPT (Dewey *et al.*, 1994) were isolated by complementation in *E. coli* and yeast, respectively. Finally, the random sequencing of cDNAs from different plants, especially *Arabidopsis* and rice, is expected to result in 'isolating' genes by homology to stretches of conserved amino acids in known enzymes. The cDNAs for plastid linoleate (ω-3) desaturases were isolated by low stringency hybridization using the microsomal linoleate (ω-3) desaturase cDNA (Iba *et al.*, 1993; Yadav *et al.*, 1993). The cDNA for plastid oleate (ω-6) desaturase was cloned by using oligomers made to a stretch of amino acids conserved between plant linolenate (ω-3) desaturases and cyanobacterial ω-6 desaturase (Hitz *et al.*, 1994a).

The soybean homologues of most of these genes have been isolated and used for determining the gene copy number, tissue-specificity of expression

Table 8.1. List of selected cloned plant lipid genes.

Enzyme	Reference
Plastid acetyl CoA carboxylase	See Ohlrogge *et al.*, 1993
Acyl carrier protein (ACP)	See Ohlrogge *et al.*, 1993
Enoyl-ACP reductase	See Ohlrogge *et al.*, 1993
3-keto acyl-ACP reductase	See Ohlrogge *et al.*, 1993
β-keto acyl-ACP synthase (KAS I isoform)	See Ohlrogge *et al.*, 1993
β-keto acyl-ACP synthase (KAS II isoform)	Thompson, 1994
β-keto acyl-ACP synthase (KAS III isoform)	Tai and Jaworski, 1993
Palmitoyl-ACP (δ-4) desaturase (coriander)	Cahoon *et al.*, 1992
Stearoyl-ACP (δ-9) desaturase	Ohlrogge *et al.*, 1993
Oleoyl-ACP thioesterase (type I)	Knutzon *et al.*, 1992a
Lauryl-ACP thioesterase (type II)	Voelker *et al.*, 1992
Plastid G3PAT	Ishikazi *et al.*, 1988
Microsomal 1-acyl-G-3-P acyltransferase (LPAAT)	Brown *et al.*, 1994
Choline phosphotransferase	Dewey *et al.*, 1994
Microsomal oleate desaturase (*Fad 2*)	Okuley *et al.*, 1994
Microsomal linoleate desaturase*	Arondel *et al.*, 1992
Plastid oleate desaturase*	Hitz *et al.*, 1994a
Plastid linoleate desaturase*	Yadav *et al.*, 1993
Cytochrome b_5	Napier *et al.*, 1994
Ferredoxin	Vorst *et al.*, 1990
NADPH-ferredoxin reductase	Jansen *et al.*, 1988
Oleosin (oil-body protein)*	Kalinski *et al.*, 1991

*Denotes soybean gene.

levels, and for RFLP mapping to soybean linkage groups (Yadav, Miao and Kinney, unpublished data).

Breeding for Soybean Oil Quality

The plasticity in the fatty acid composition of vegetable oils is illustrated by the dramatically altered composition in some agronomically useful oilseed mutants. Notable examples include 'zero' erucate (22:1) rapeseed (compared to *ca.* 40% erucate in parent) (Stefansson *et al.*, 1994), 90% 18:1 sunflower (compared to *ca.* 36% 18:1 in parent) (Miller *et al.*, 1987), and *ca.* 1.6% 18:3 flax (compared to 40% 18:3 in parent) (Green, 1986).

Good progress has been made in the last decade in discovering soybean variants with a relatively wide range of seed fatty acid compositions (Table 8.2). The mutants are useful as breeding material as well as research tools for studying regulation of oil biosynthesis. The fatty acid composition of selected mutants is compared to that of their parents grown under the same condi-

Table 8.2. Seed fatty acids (%) in normal and mutant soybeans.

Fatty acid	Normal soybean	Range in mutant soybeans
16:0	11	4–27
18:0	4	3–28
18:1	24	20–60
18:2	54	20–60
18:3	7	2–14

Data from Hammond, 1992.

tions (Table 8.3). Environment can have a significant impact on fatty acid compositions (Cherry *et al.*, 1985), as illustrated by the fatty acid compositions of mutant A5 in two different experimental conditions (Table 8.3). Generally, warmer temperatures during seed fill are positively correlated with increased oil content, increased rate of oil biosynthesis, increased 18:1, and decreased

Table 8.3. Seed fatty acid composition of selected soybean mutants.

	Fatty acids (%)					
	16:0	18:0	18:1	18:2	18:3	locus
Fehr *et al.*, 1992						
Century (parent of C1640)	9.8	3.8	24.3	54.6	7.4	
C1640, low 18:3	10.3	3.9	24.9	57.2	3.7	*fan*
A5, low 18:3	9.4	4.3	46.6	36.3	3.4	*fan, ?*
A23, high 16:0	13.0	4.6	38.5	38.3	5.6	*fan 2*
A16(A5 × A23) double mutant	10.1	5.3	41.1	41.2	2.2	*fan, fan 2*
Hammond and Fehr, 1983a						
FA9525 (parent of A5)	9.3	3.1	39.1	42.2	6.2	
A5, low 18:3	9.3	3.9	39.8	42.9	4.1	*fan, ?*
Hammond and Fehr, 1983b						
FA8077 (parent of A6)	8.4	4.4	42.8	36.7	7.6	
A6, high 18:0	8.0	28.1	19.8	35.5	6.6	*fas*
Erickson *et al.*, 1988a						
Century (parent)	11.1	3.0	21.3	57.1	7.5	
C1727, high 16:0	16.6	3.0	19.7	52.4	8.0	*fap 2*
C1726, low 16:0	8.4	3.0	24.6	56.7	7.3	*fap 1*

18:2 and 18:3 levels (Howell and Collins, 1957; Wolf *et al.*, 1982; Rennie and Tanner, 1989a; Wilcox and Cavins, 1992). Selected soybean mutations affecting oil composition are described below.

Reduced levels of polyunsaturated fatty acids (18:2 and 18:3)

One of the earliest goals of oil quality breeding in soybean has been to reduce its 18:3 content for oxidative stability and has involved recurrent selection, mutagenesis, and germplasm screening.

Recurrent selection for high 18:1 resulted in line N78-2245 with high 18:1 (51%) and low 18:3 (4.2%) (Wilson *et al.*, 1981). Both traits showed maternal influence (Carver *et al.*, 1987), involved several loci (Rennie and Tanner, 1991) and were temperature-sensitive: up to 70% 18:1 and 1.9% 18:3 at 40/30°C day/night temperatures from 35% 18:1 and 4.5% 18:3 at 28/22°C (Rennie and Tanner, 1989a). The temperature-sensitivity of 18:1 levels is much greater in the mutant than in normal soybeans (Martin *et al.*, 1986). This is the highest level of 18:1 reported in soybean seed and mutagenesis has had no significant success in further increasing the 18:1 content.

However, breeding approaches have been more successful in reducing 18:3. Low 18:3 was discovered in two germplasms and three chemically-induced mutants. All have *ca.* 50% lower 18:3 than normal lines and map to *Fan* locus in linkage groups 17 (Rennie and Tanner, 1991). Mutants C1640 and 9509 with 18:3 contents of 3.4% and 4.1%, respectively, were derived from Century cultivar (Wilcox *et al.*, 1984; Wilcox and Cavins, 1992). In Century × C1640 cross, the low 18:3 trait showed no maternal influence, was inversely correlated with 18:1 rather than 18:2, and was controlled by two alleles at the *Fan* locus with additive effects – each *fan* allele contributing to 2% reduction in 18:3 (Wilcox and Cavins, 1985, 1986, 1987). The low 18:3 alleles in soybean germplasms PI 361088B and PI 1223440 also showed additive effect at the *Fan* locus (Rennie *et al.*, 1988; Rennie and Tanner, 1989b). The low 18:3 phenotype in C1640 is partially expressed in root but not shoot tissues (Wang *et al.*, 1989) and showed less temperature dependence than the 18:3 in Century (Wilcox and Cavins, 1992).

Mutant A5 with 4.1% 18:3 was derived from line FA9525, which had previously been selected from germplasms PI 80476 and PI 85671 for its moderately low (6.3%) 18:3 content (Hammond and Fehr, 1983a). The low 18:3 trait in A5 is conditioned by two or more genes, one of which is the *Fan* locus; no transgressive segregants were found in the A5 × C1640 cross and the low 18:3 trait was largely conditioned by a single allele equidistant from the *Idh* 1 locus in both N78-2245 × C1640 and N78-2245 × A5 crosses (Rennie and Tanner, 1991). The low 18:3 trait exhibited both maternal influence and quantitative inheritance in the A5 × Pella and A5 × Weber crosses and segregated in a continuous distribution in the A5 × C1640 cross (Rennie and Tanner, 1991; Fehr *et al.*, 1992). The authors proposed that since both N78-

2245 and FA9525, the parent of A5, were selected for high 18:1/low 18:3, they shared common genes that allowed detection of the *fan* locus in A5.

Some low 18:3 mutations have been combined to obtain 18:3 levels lower than either parent. For example, an individual with 2.3% 18:3 content was identified among F_4 progeny of the A5 × N78-2245 cross (Rennie and Tanner, 1991). In combining the low 18:3 trait in A5 and the high 16:0 trait in A23 (see below), segregant A16 with < 2.5% 18:3 and normal 16:0 was unexpectedly discovered (Fehr *et al.*, 1992). When A16 was crossed to Century, C1640, and its parental lines, a maternal effect was observed in crosses with Century and C1640 but not in crosses with its parents. In the A16 × Century cross, the low 18:3 trait was inherited quantitatively, although its segregation suggested control at two independent loci. The new locus was designated *fan 2*(A23). Based on these crosses between the mutants and between the mutants and their parents, the *Fan* locus appears to contribute *ca.* 2–4% 18:3 and the *Fan 2* locus to *ca.* 0.5–1.5% 18:3. Thus, the lowest 18:3 (2.2%) reported involves at least two major loci. However, the low 18:3 in A16 showed evidence of the involvement of other minor genes and behaved as a quantitative character (Fehr *et al.*, 1992).

C1640 was 24% lower-yielding than its parent (Wilcox and Cavins, 1986) but further study showed that the lower yield was not linked to the *fan* allele (Wilcox *et al.*, 1993). The molecular basis of low 18:3 mutations are unknown. However, since the *fan* locus in C1640 appears to control microsomal linoleate desaturation (Wang *et al.*, 1989) and since A5, which also has a mutation in the *fan* locus, is deleted for an ER linoleate desaturase gene (J. Byrum, personal communication), it is likely that the *Fan* locus encodes a microsomal linoleate desaturase. The modifier genes for low 18:3 content may represent other desaturase activities, such as the plastid ones, or differences in the flux of fatty acids through polyunsaturated fatty acid biosynthesis. The flux could be decreased by an overall reduction of fatty acid biosynthesis, for example, by low temperature or by reduced activity of enzymes such as CPT and LPCAT (Fig. 8.1). Maternal and graft-transmissible influences on levels of polyunsaturation have been reported (Brim *et al.*, 1968; Carver *et al.*, 1987) and may also reflect altered flux in fatty acid biosynthesis due to different amounts of photosynthate translocating to the developing soybean seed.

Increased levels of 18:0

Chemical mutagenesis has resulted in several mutants with three- to seven-fold increases in 18:0 content (Graef *et al.*, 1985a; Erickson *et al.*, 1988a; Bubeck *et al.*, 1989). Of these, six (A6, A9, ST1, ST2, ST3, and ST4) were derived from different low 18:3 parents, one (A9) from cultivar Coles, and two (C1728 and C1730) from Century. Since modifying genes can affect the expression of the high 18:0 trait (see below), it is possible that some of the differences in 18:0 levels in these mutants may be attributed to their different genetic backgrounds.

The high 18:0 trait in all mutants is controlled by a recessive allele at a single locus, with no maternal influence. Most, if not all, mutations are at the *Fas* locus. In crosses with their parents, the low 18:0 phenotype is partially dominant to the high one. Crosses between some of these mutants revealed interesting dominance relationships: the *fas*[a] allele in mutant A6 (with 30% 18:0) and the *fas*[b] allele in mutant A10 (with 16% 18:0) were completely dominant over the *fas* allele in mutant A9 (with 19% 18:0); the *fas*[a] allele was codominant with the *fas* alleles of mutants A10, ST1, ST3 and ST4 (Bubeck *et al.*, 1989). These relationships have to be accounted for in providing a biochemical basis for these mutations.

On average, *ca.* three-quarters of the increased 18:0 in the high 18:0 mutants was at the expense of 18:1, with *ca.* one-fifth at the expense of 18:2, and only slight (but consistent) decrease in 16:0; no significant changes in 18:3 were found (Graef *et al.*, 1985b). The level of arachidate (20:0) was reported only for mutant A6, where it was found to be 2% – significantly higher than normal soybeans (Hammond and Fehr, 1983b). The increase in 20:0 is related to the increase in 18:0, since both are reduced when the A6 mutant is grown at cool temperatures (see below). A similar increase in 20:0 was reported in high 18:0 transgenic canola (Knutzon *et al.*, 1992b).

The biochemical basis of the high 18:0 mutations has not been determined, though it is consistent with a mutation in SD. The high 18:0 trait in A6 is cold temperature-sensitive: when the temperature regime during seed development and maturation was 28°C day/22°C night the 18:0 level was 22.3% but when it was 15°C/12°C the 18:0 level was only 2.3%, which is near normal (Rennie and Tanner, 1989a). The 20% drop in 18:0 is accompanied by a 6% drop in 18:1 and a 14% and 12% increase in 18:2 and 18:3 content, respectively. The biochemical basis of this temperature sensitivity is also unknown. One explanation may be that the flux of fatty acid biosynthesis through FAS at the low temperature is reduced such that 18:0 desaturation is not rate limiting. However, other possible explanations cannot be ruled out: a high temperature-sensitive protein encoded by the *fas*[a] allele, inability of developing soybean seeds to incorporate high levels of 18:0 into triacylglycerol at low temperature, and assuming the mutation is in SD, low temperature induction of another stearoyl-ACP desaturase (SD) isozyme. It would be interesting to determine whether this temperature-sensitivity is also expressed in other high 18:0 mutants.

Changes in fatty acid composition during seed development was compared in mutant A6 and its parent line (Graef *et al.*, 1985a). The sharp decrease in 18:0 observed in the parent line between 15 and 20 DAF was reversed in the mutant. The increase in 18:0 level in the mutant started at 19 DAF and was maximum at 25 DAF, when the mutant had 45.4% 18:0. The seed weight, oil and protein contents of the mutant and parent were not significantly altered.

The effect of the high 18:0 content on agronomic yield was tested by crossing mutants A6, A9 and A10 to high-yielding genotypes and analysing the pools of high 18:0 and low 18:0 segregants (Lundeen *et al.*, 1987). In each of the three crosses, the mean 18:0 in the high 18:0 segregants was lower than the high 18:0 mutant parents: 16% compared to 28.2% in the A6 cross, 15.5% compared to 16.3% in the A9 cross, and 12.4% compared to 14.6% in the A10 cross. This suggested that modifying genes contributed to the expression of the high 18:0 trait. The ranges of 18:0 in the high 18:0-segregants were: 13–19% for the A6 cross, 12–19% for the A9 cross, and 11–14% for the A10 cross. Comparison of the fatty acid compositions of the high and low 18:0 populations in each cross showed that while the high 18:0 trait was associated largely with reduced 18:1, it was also associated with *ca.* 1.5% reduced 16:0 and *ca.* 1% increased 18:3. There was no significant difference in seed weight or oil content associated with the high 18:0 traits. While the high and low 18:0 segregants of the A9 and A10 crosses did not differ significantly in yield, those of the A6 cross did; the A6 mutation was associated with an average of 7.7%, reduction in yield. Since the 18:0 levels in A6 and A9 crosses were comparable, this yield reduction was not attributed to the *fas*^a^ locus, but to another locus linked to it. However, the 18:0 levels in non-seed tissues could still be different enough in the A6 and A9 crosses to affect yield. The polar lipid fraction in roots of the A6 mutant also had a higher 18:0 level (Martin and Rinne, 1985). In any case, it remains to be demonstrated that the 28–30% 18:0 levels in seeds do not affect germination or yield. The availability of other mutants with *ca.* 28% 18:0, such as ST1, should make it possible to address this important issue.

Increased levels of 16:0

Chemical mutagenesis of soybeans has resulted in six mutants with increased 16:0 (Erickson *et al.*, 1988a,b; Bubeck *et al.*, 1989; Wilcox and Cavins, 1990). Mutant C1727, derived from cultivar Century, has *ca.* 6% more 16:0 than the parent. The mutant phenotype is controlled by single recessive gene at *Fap 2* locus with no maternal influence. The *fap 2* alleles in C1727 show an additive effect, that is, each increases 16:0 by 3%. The *Fap 2* locus maps to linkage group D of the SDA-ARS/ISU soybean molecular genetic map (Nickell *et al.*, 1994).

Mutant A21 (formerly A1937NMU-85) with 20% 16:0 and mutant A24 (formerly ElginEMS-421) with 18% 16:0 were derived from cultivars A1937 and Elgin, respectively (Fehr *et al.*, 1991b; Schnebly *et al.*, 1994). F$_2$ progeny from the A21 × A24 cross resulted in transgressive segregants, suggesting the presence of two independent loci for high 16:0; the homozygous double mutant showed a mean 16:0 content of 27.3% (Fehr *et al.*, 1991b). Since the high 16:0 mutations were tightly linked in the A21 × C1727 cross, the mutant allele in A21 was designated *fab-2-b* (Fehr *et al.*, 1991b); while that in A24 was designated *fap 4* (Schnebly *et al.*, 1994).

High 16:0 (13–16% 16:0) mutants PA1, PA2, PA3, and PA4 were derived from different parent lines (Bubeck *et al.* 1989). In crosses of all four with cultivar Coles, the mutant trait segregated as a single locus and the normal 16:0 trait was additive or showed slight dominance over the high 16:0 trait. Their genetics has not been characterized further.

Mutant C1727 has a seed oil and protein content per seed comparable to that in the parent, however its yield is reduced *ca.* 11% (Wilcox and Cavins, 1990). The effect of *fap 2* and *fap 4* alleles, alone or together, on agronomic yield and the tissue-specificity of the high 16:0 traits remains to be determined. The biochemical basis of the high 16:0 mutations are unknown. *Fap 2* and *Fap 4* loci could encode β-ketoacyl-ACP synthase 2 (KAS 2) enzymes (Fig. 8.1).

Reduced levels of 16:0

Chemical mutagenesis resulted in two low palmitate soybean lines, C1726 (Erickson *et al.*, 1988a,b; Wilcox and Cavins, 1990) and A22 (Fehr *et al.*, 1991a). C1726, derived from Century, has *ca.* 3% less 16:0 than its parent. The mutant phenotype is controlled by a recessive gene at *Fap 1* locus, with no maternal influence. *fap 1* alleles in C1726 show an additive effect, that is, each lowers 16:0 by 1.5%. Mutant A22 (formerly A937NMU-173), derived from 'A1937', has 6.8% 16:0 compared to 12.1% in its parent and its alleles also exhibit an additive effect. F_2 progeny of a cross between A22 × C1726 show transgressive segregants suggesting that the two low 16:0 mutations are at different loci (Schnebly *et al.*, 1994); the 16:0 content for the double homozygous mutant, designated A18, was *ca.* 4.0% compared with 7.1% and 8.0% for A22 and C1726, respectively (Fehr *et al.*, 1991a). The mutant locus in A22 was designated *fap 3*. Loci *fap 1* and *fap 3* are different from the two loci for increased 16:0 (see above).

Recurrent selection for high 18:1 resulted in the discovery of a low 16:0 line N79-2077-12 with a mean 16:0 content of *ca.* 6% (Burton *et al.*, 1994). This line was crossed to other lines resulting in a line, N87-2122-4 with 5.3% 16:0 (Wilcox *et al.*, 1994). In a cross of N87-2122-4 × C1726, transgressive segregants were observed suggesting that independent loci exist for reduced 16:0; the homozygous double mutant has a 16:0 content of 4.3% (Wilcox *et al.*, 1994). It remains to be determined whether the low 16:0 mutation in N87-2122-4 is at *fap 3* locus.

The low 16:0 trait in *fap 1* and *fap 3* double mutant are influenced by modifying genes and are not detrimental to yield; however, the presence of these alleles is correlated with a slightly reduced oil content (Horejsi *et al.*, 1994). The tissue-specificity of the low 16:0 traits and the biochemical basis of the low 16:0 mutations is unknown. The low 16:0 mutations could be in genes encoding an acyl-ACP TE with selectivity towards 16:0-ACP and/or an acyltransferase (Fig. 8.1).

In summary, breeding approaches have identified several loci affecting oil quality, including two that control the low 18:3 trait (fan and fan 2), two that control high 16:0 (fap 1 and fap 3), two that control low 16:0 (fap 3 and fap 4), and one that controls high 18:0 (fas). Almost all of these loci have some modifying genes associated with the expression of the trait. The lines homozygous for the double mutant (fap 3 and fap 4) have ca. 4% 16:0 making oil from this line quite similar in its saturate content to that of canola oil. Combination of high 16:0 and 18:0 traits have resulted in almost 50% total saturates (Hammond, 1992). The agronomic penalty of some of these traits have not been determined. In contrast, the success in reducing total polyunsaturates has not been as significant.

Limited Success of Mutagenesis in Modifying Soybean Oil

Mutagenesis for oil variants is effective because a simple assay, gas chromatography, can screen for changes in any one of the numerous biochemical steps that affect the relative amounts of all the fatty acids. However, it has the obvious limitation of not discovering mutations in genes whose loss is lethal or does not result in a phenotype. An example for the latter occurs when two genes encode the same enzyme (isozyme); mutation in one will have no phenotype if the other produces 100% of the required activity. In soybean, there are at least four oleate desaturase genes that can be grouped into two pairs of highly homologous genes; one or both members of each pair are expressed strongly in developing seeds, although only one pair is seed-specific (G.H. Miao, personal communication). This may explain why a soybean mutant higher than ca. 50% 18:1 under normal growth conditions has not been discovered yet. Another drawback of mutation breeding for altered oil composition is that both membrane and oil lipid compositions may be affected, since both share a common biosynthetic pathway. That is, in as much as the genes are also common for oil and membrane biosynthesis, mutations in them for altered oil composition will also affect the membrane composition of seed. In the case of constitutively expressed genes, vegetative tissues will also be affected. Consequently, the degree of phenotypic change in oil composition is limited and even then may carry a penalty in terms of agronomic performance. This is unlikely to be acceptable in the market place.

    The agronomic penalty in some oil mutants is illustrated by the altered physiology of some high 18:1 mutants of Arabidopsis (Miquel and Browse, 1994) and canola (Hitz et al., 1994b). Mutation in an oleate desaturase gene of Westar canola resulted in a 78% seed 18:1 line, IMC129, with normal root and leaf lipid profiles as well as good agronomy. Its remutagenesis produced a double mutant with mutations in two separate oleate desaturase genes resulting in higher seed 18:1 (87%). In this case, however, the 18:1 content in roots and non-plastid leaf lipids were also increased. The resulting line exhibited poor agronomic properties, which might be attributable to the lipid changes in non-seed tissues (Chen, DeBonte, Hitz, and Miao, unpublished data).

Thus, notwithstanding the commercially adapted sunflower mutant with 90% 18:1 (Fernandez-Martinez *et al.*, 1993), mutagenesis has had limited success in reducing polyunsaturation in oil crops, such as soybean, canola and corn. In the case of low 18:3 soybean mutants with normal agronomy, the breeding timelines are long because the trait is multigenic and environmentally-sensitive. This is especially true when combining different traits, such as low polyunsaturate and altered saturates.

Transgenic Approaches to Modifying Oil Quality

The three key enzymes, KAS 2, SD, and TE, involved in determining the ratio of 16:0, 18:0 and 18-carbon unsaturated fatty acids in seed oil and oleate and linoleate desaturases controlling biosynthesis of the polyunsaturated fatty acids, 18:2 and 18:3, have been manipulated by transgenic approaches (see Fig. 8.1). Because of poor efficiency of obtaining transgenic soybean plants, most of the recent results have come from rapeseed or *Arabidopsis*.

Altering 16:0 and 18:0 contents in transgenic seeds

Overexpression of KAS 2 in transgenic rapeseed were reported to result in reduced 16:0 (Bleibaum *et al.*, 1993). Seeds of transgenic *Brassica rapa* and *B. napus* seeds containing a chimeric gene comprising of a *B. rapa* stearoyl-ACP desaturase cDNA in an antisense orientation under the control of a seed specific promoter showed increased 18:0, some up to 40% 18:0 (Knutzon *et al.*, 1992b). The increased 18:0 was associated with some increases in both 18:3 and 20:0. Similarly, in soybean somatic embryos, introduction of soybean SD cDNA in either sense or antisense orientation under the control of a seed-specific promoter resulted in increased levels of 18:0 (up to *ca.* 30% 18:0) (Yadav, unpublished).

Seed-specific overexpression of soybean oleoyl-ACP thioesterase in transgenic canola resulted in increased 16:0 and 18:0 to *ca.* 10% each (Hitz, W.D., personal communication). Inhibition of the soybean oleoyl-ACP thioesterase by cosuppression in transgenic somatic soybean embryos resulted in decreased saturates (Hitz, W.D., personal communication). The seed-specific expression of California Bay lauryl-ACP thioesterase cDNA in *Arabidopsis* resulted in 25% laurate in seeds (Voelker *et al.*, 1992).

In addition, the saturated fatty acid content of transgenic tobacco leaf was reduced by the expression of the structured gene for δ-9 desaturase from yeast or rat under the control of a 35S promoter (Grayburn *et al.*, 1992; Polashock *et al.*, 1992). This enzyme converts palmitoyl-CoA and stearoyl-CoA into palmitoleoyl-CoA and oleoyl-CoA, respectively.

Increasing 18:3 contents in transgenic plants

Overexpression of rapeseed linoleate desaturase in wild type *Arabidopsis* roots resulted in a 1.6 fold increase in 18:3 content (Arondel *et al.*, 1992). Overexpression of the *Arabidopsis* enzyme under the control of 35S promoter in carrot hairy roots resulted in a more than a sevenfold increase in the 18:3 content and almost all endogenous 18:2 was converted to 18:3 (Yadav *et al.*, 1993). Expression of the *Arabidopsis* enzyme under the control of a seed-specific promoter resulted in a 12-fold increase in 18:3 content in *Arabidopsis* seed when compared with an untransformed *fad 3* mutant and a twofold increase when compared with untransformed wild type (Yadav *et al.*, 1993). Thus, the ω-3 desaturase appears to be a rate-limiting step in the biosynthesis of 18:3 in *Arabidopsis* seeds as well as in *Arabidopsis* roots and carrot hairy roots. This observation is supported by genetic studies with *fad 3* mutants that indicate gene dosage dependence of the *fad 3* phenotype (Lemieux *et al.*, 1990). If, as seems likely, the ω-3 desaturase is also rate-limiting in soybean, then the alteration of the 18:3 content in the triacylglycerols of these plants by transgenic approaches should prove practicable to make a soybean oil comparable to linseed oil.

Reducing polyunsaturates in transgenic soybean

Kinney and co-workers have used a transgenic approach with soybean ER oleate and linoleate desaturase cDNAs to obtain reduced levels of polyunsaturation in soybean seeds (Hitz *et al.*, 1994b). Fertile soybean plants were regenerated from somatic embryos transformed with chimeric genes consisting of an antisense soybean oleate or linoleate desaturase cDNA under the control of seed-specific conglycinin promoter. A single copy of an antisense linoleate desaturase DNA reduced the seed 18:3 level from 9% to less than 2% (Table 8.4). This level is lower than any currently available soybean mutant and behaved as a dominant single locus trait. An antisense oleate desaturase cDNA increased the 18:1 level from 20% to *ca.* 80% and decreased the saturate (16:0 + 18:0) level from 15% to less than 11% in seeds (Table 8.4). A transgene with soybean oleate desaturase cDNA in sense orientation was, at least, as effective in reducing 18:1 desaturation in somatic embryos apparently due to cosuppression of the native gene.

Transgenic traits can be further enhanced by combining them with appropriate mutants. This is illustrated by the work of Hitz and co-workers (Hitz *et al.*, 1994b). Seeds of transgenic canola lines homozygous for a single chimeric transgene comprising a canola oleate desaturase cDNA in an antisense orientation under the control of a seed-specific promoter had an 18:1 content above 80% and the high 18:1 trait segregates as a single, almost dominant locus. Further increase in seed 18:1 content was achieved by crossing it with high 18:1 mutant IMC129. F_2 progeny homozygous for both the mutant oleate

Table 8.4. Fatty acid composition in transgenic seeds.

	Fatty acids (%)				
	16:0	18:0	18:1	18:2	18:3
Soybean&#15 Control line	11.3	3.8	21.5	55.0	8.9
Antisense 18:2 desaturase, line #1	10.1	3.4	20.1	65.0	1.4
Antisense 18:1 desaturase, line #2	7.9	3.0	78.9	3.0	5.8
Canola					
Control	4.2	2.2	63.0	19.9	10.6
High 18:1 mutant line, IMC129	4.0	2.0	77.8	10.4	5.9
Transgene line #8, napin: OD antisense	3.7	2.2	83.3	5.0	5.8
Transgene line #8 × IMC129	3.6	1.7	86.2	3.6	2.9

The data for transgenic soybean lines is an average of ten T2 seeds homozygous for the transgene. The data for canola lines is of bulk seeds homozygous for the mutant genes (IMC129), for the transgene (line #8), and for the transgene and the mutant gene (line #8 × IMC129). OD, oleate desaturase.

desaturase gene of IMC129 and the transgene showed 18:1 levels of *ca.* 86% (Table 8.4), which is close to that in the double mutant but apparently without the agronomic penalty of the double mutant (see discussion above).

Prospects of producing specialty oils in soybean by transgenic approaches

Transgenic approaches can offer several advantages over the traditional breeding approaches. The ability to obtain seed-specific expression of foreign genes, which, in combination with antisense inhibition and cosuppression (transwitch), can result in seed-specific gene silencing is important in overcoming potential pleiotrophic affects of mutational breeding. Further, since these methods of gene silencing rely on nucleotide homology, they may be effective in a dominant manner, not only affecting both alleles at the same locus but also alleles at different loci encoding homologous genes. In addition, transgenes for different phenotypes may be introduced together for breeding as a single locus.

Application of transgenic approaches, however, requires cloning of genes for identified target enzymes, an efficient transformation and regeneration system, and the expression of the transgene at the appropriate level and time during seed development. The track record of the stability of transgenic traits in the field is not as well established as that for mutations and commercialization of transgenic traits is expected to face public perception issues and regulatory hurdles that mutants do not have to. Nevertheless, the recent results discussed above bolster the confidence that transgenic approaches alone or in combination with known soybean mutants should result in soybean oils with a wide range of fatty acid compositions suited for specific food and non-food applications without agronomic penalty and with easier breeding.

Acknowledgements

I thank my colleagues W.D. Hitz, A.J. Kinney, G.H. Miao and Dr J. Burton for making available their unpublished data, and W.D. Hitz and R. Broglie for their critical review of the manuscript.

References

Arondel, V., Lemieux, B., Hwang, I., Gibson, S., Goodman, H.M. and Somerville, C.R. (1992) Map-based cloning of a gene controlling omega-3 fatty acid desaturation in *Arabidopsis. Science* 258, 1353–1355.

Bafor, M., Stobart, A.K. and Stymne, S. (1990) Properties of the glycerol acylating enzymes in microsomal preparations from the developing seeds of safflower (*Carthaamus tinctorius*) and turnip rape (*Brassica campestris*) and their ability to assemble cocoa-butter type fats. *Journal of the American Oil Chemists Society* 67, 217–225.

Battey, J.F., Schmid, K.M. and Ohlrogge, J.B. (1989) Genetic engineering for plant oils: potential and limitations. *Trends in Biotechnology* 7, 122–126.

Bleibaum, J.L., Genez, A., Fayet-Faber, J., McCarter, D.W. and Thompson, G.A. (1993) Modifications in palmitic acid levels via genetic engineering of *b*-ketoacyl-ACP synthases. In: *Abstracts, National Plant Lipid Symposium, Minneapolis, Minnesota.*

Brim, C.A., Schutz, W.M. and Collins, F.I. (1968) Maternal effects on fatty acid composition and oil content of soybean, *Glycine max* (L.) Merrill. *Crop Science* 8, 517–518.

Brown, A.P., Tommey, A.M., Watson, M.D. and Slabas, A.R. (1994) Complementation of an *E. coli* 1-acyl-*sn*-glycerol-3-phosphate acyltransferase mutant with a *Zea mays* cDNA clone. *4th International Congress of Plant Molecular Biology* Abstract 1465.

Browse, J. and Somerville, C. (1991) Glycerolipid synthesis: biochemistry and regulation. *Annual Review of Plant Physiology and Plant Molecular Biology* 42, 467–506.

Bubeck, D.M., Fehr, W.R. and Hammond, E.G. (1989) Inheritance of palmitic and stearic acid mutants of soybean. *Crop Science* 29, 652–656.

Burton, J.W., Wilson, R.F. and Brim, C.A. (1994) Registration of N79-2077-12 and N87-2122-4, two soybean germplasm lines with reduced palmitic acid in seed oil. *Crop Science* 34, 313.

Cahoon, E.B. and Ohlrogge, J.B. (1994) Apparent role of phosphatidylcholine in the metabolism of petroselinic acid in developing Umbelliferae endosperm. *Plant Physiology* 104, 845–855.

Cahoon, E.B., Shanklin, J. and Ohlrogge, J.B. (1992) Expression of a novel coriander desaturase results in petroselinic acid production in transgenic tobacco. *Proceedings of the National Academy of Sciences, USA* 89, 11184.

Carver, B.F., Burton, J.W. and Wilson, R.F. (1987) Graft-transmissible influence on fatty acid composition of soybean seed. *Crop Science* 27, 53–57.

Cherry, J.H., Bishop, L., Hasegawa, P.M. and Leffler, H.R. (1985) Differences in the fatty acid composition of soybean seed produced in northern and southern areas of the USA. *Phytochemistry* 24, 237–241.

Dewey, R.E., Wilson, R.F., Novitzky, W.P. and Goode, J.H. (1994) The *AAPT1* gene of soybean complements a choline phosphotransferase-deficient mutant of yeast. *Plant Cell* 6, 1495–1507.

Erickson, E.A., Wilcox, J.R. and Cavins, J.F. (1988a) Fatty acid composition of the oil in reciprocal crosses among soybean mutants. *Crop Science* 28, 644–646.

Erickson, E.A., Wilcox, J.R. and Cavins, J.F. (1988b) Inheritance of altered palmitic acid percentage in two soybean mutants. *Journal of Heredity* 79, 465–468.

Fehr, W.R., Welke, G.A., Hammond, E.G., Duvick, D.N. and Cianzo, S.R. (1991a) Inheritance of reduced palmitic acid content in seed oil of soybean. *Crop Science* 31, 88–89.

Fehr, W.R., Welke, G.A., Hammond, E.G., Duvick, D.N. and Cianzo, S.R. (1991b) Inheritance of elevated palmitic acid content in soybean seed oil. *Crop Science* 31, 1522–1524.

Fehr, W.R., Welke, G.A., Hammond, E.G., Duvick, D.N. and Cianzo, S.R. (1992) Inheritance of reduced linolenic acid content in soybean genotypes A16 and A17. *Crop Science* 32, 903–906.

Fernandez-Martinez, J., Munoz, J. and Gomez-Arnau, J. (1993) Performance of near-isogenic high and low oleic acid hybrids of sunflower. *Crop Science* 33, 1158–1163.

Frankel, E.N. (1980) Soybean oil flavor stability. *Handbook of Soy Oil Processing and Utilization.* St. Louis, American Soybean Association and American Oil Chemists' Society, pp. 229–244.

Frentzen, M. (1993) Acyltransferase and triacylglycerols. In: Moore, T.S. (ed.) *Lipid Metabolism in Plants.* CRC Press, Boca Raton, pp. 195–230.

Graef, G.L., Miller, L.A., Fehr, W.R. and Hammond, E.G. (1985a) Fatty acid development in a soybean mutant with high stearic acid. *Journal of the American Oil Chemists Society* 62, 773–775.

Graef, G.L., Fehr, W.R. and Hammond, E.G. (1985b) Inheritance of three stearic acid mutants of soybean. *Crop Science* 25, 1076–1079.

Grayburn, W.S., Collins, C.G. and Hildebrand, D.F. (1992) Fatty acid alteration by a Δ9 desaturase in transgenic tobacco tissue. *Biotechnology* 10, 675–678.

Green, A.G. (1986) A mutant genotype of flax (*Linum usitatissimum* L.) containing very low levels of linolenic acid in its seed oil. *Canadian Journal of Plant Science* 66, 499–503.

Gurr, M.I. (1989) Dairy fats: nutritional nasties or dietary delights. In: Cambie, R.C. (ed.) *Fats for the Future.* Ellis Harwood Ltd, Chichester, pp. 41–61.

Hammond, E.G. (1992) The seeds of new technology. *INFORM* 3, 1288–1292.

Hammond, E.G. and Fehr, W.R. (1983a) Registration of A5 germplasm line of soybean. *Crop Science* 23, 192.

Hammond, E.G. and Fehr, W.R. (1983b) Registration of A6 germplasm line of soybean. *Crop Science* 23, 192–193.

Heinz, E. (1993) Biosynthesis of polyunsaturated fatty acids. In: Moore, T.S. (ed.) *Lipid Metabolism in Plants.* CRC Press, Boca Raton, pp. 33–89.

Hitz, W.D., Carlson, T.J., J.R., B., Kinney, A.J., Stecca. K.L. and Yadav, N.S. (1994a) Cloning of a higher plant plastid ω-6 fatty acid desaturase cDNA and its expression in a cyanobacterium. *Plant Physiology* 105, 635–641.

Hitz, W.D., Yadav, N.S., Reiter, R.S., Mauvais, C.J. and Kinney, A.J. (1994b) *Reducing poly-unsaturation in oils of transgenic canola and soybean.* In: Kader, J.-C. and Mazliak, P. (eds) *Plant Lipid Metabolism.* Kluwer Academic Publishers, Dordrecht, The Netherlands, pp. 506–508.

Horejsi, T.F., Fehr, W.R., Welke, G.A., Duvick, D.N., Hammond, E.G. and Cianzio, S.R. (1994) Genetic control of reduced palmitate content in soybean. *Crop Science* 34, 331–334.

Howell, R.W. and Collins, F.I. (1957) Factors affecting linolenic and linoleic acid content in soybean oil. *Agronomy Journal* 49, 593–597.

Iba, K., Gibson, S., Nishiuchi, T., Fuse, T., Nishimura, M., Arondel, V., Hugly, S. and Somerville, C. (1993) A gene encoding a chloroplast omega-3 fatty acid desaturase complements alterations in fatty acid desaturation and chloroplast copy number of the *fad7* mutant of *Arabidopsis thaliana. Journal of Biological Chemistry* 268, 24099–24105.

Ishikazi, O., Nishida, I., Agata, K., Eguchi, G. and Murata, N. (1988) Cloning and nucleotide sequence of cDNA for the plastid glycerol-3-phosphate acyltransferase from squash. *FEBS Letters* 238, 424.

Jansen, T., Reilaender, H., Steppuhn, J. and Herrmann, R.G. (1988) Analysis of cDNA clones encoding the entire precursor for ferredoxin: $NADP^+$ oxireductase from spinach. *Current Genetics* 13, 517–522.

Kalinksi, A., Loer, D.S., Weisemann, J.M., Natthews, B.F. and Herman, E.M. (1991) Isoforms of soybean seed oil body membrane protein 24 kDa oleosin are encode by closely related cDNAs. *Plant Molecular Biology* 17, 1095–1098.

Keys, A. (1980) *Seven Countries: A Multivariate Analysis of Death and Coronary Heart Disease.* Harvard University Press, Cambridge, USA.

Kinney, A.J. (1994) Genetic modification of the storage lipids of plants. *Current Opinions in Biotechnology* 5, 144–151.

Knutzon, D.S., Bleibaum, J.L., Nelsen, J., Kridl, J.C. and Thompson, G.A. (1992a) Isolation and characterization of two safflower oleoyl-acyl carrier protein thioesterase cDNA clones. *Plant Physiology* 100, 1751.

Knutzon, D.S., Thompson, G.A., Radke, S.E., Johnson, W.B., Knauf, V.C. and J.C., K. (1992b) Modification of *Brassica* seed oil by antisense expression of a stearoyl-ACP carrier protein desaturase gene. *Proceedings of the National Academy of Sciences, USA* 89, 2624–2628.

Lemieux, B., M., M., Somerville, C. and Browse, J. (1990) Mutants of *Arabidopsis* with alterations in seed lipid fatty acid composition. *Theoretical and Applied Genetics* 80, 234–240.

Liu, H.R. and White, P.J. (1992) Oxidative stability of soybean oils with altered fatty acid composition. *Journal of the American Oil Chemists Society* 69, 528–532.

Lundeen, P.O., Fehr, W.R. Hammond, E.G. and Cianzio, S.R. (1987) Association of alleles for high stearic acid with agronomic characters of soybean. *Crop Science* 27, 1102–1105.

Magnuson K., Jackowski, S., Rock, C.O. and Cronan, J.E. (1993) Regulation of fatty acid biosynthesis in *Escherichia coli. Microbiology Reviews* 57, 522–542.

Martin, B.A. and Rinne, R.W. (1985) Relationships between fatty acid composition of vegetative and reproductive structures of six soybean genotypes. *Crop Science* 25, 1055–1058.

Martin, B.A., Wilson, R.F. and Rinne, R.W. (1986) Temperature effects upon the expression of a high oleic acid trait in soybean. *Journal of the American Oil Chemists Society* 63, 346–352.

Miller, J.F., Zimmerman, D.C. and Vick, B.A. (1987) Genetic control of high oleic content in sunflower. *Crop Science* 27, 923.

Miquel, M.F. and Browse, J.A. (1994) High oleate oilseeds fail to develop at low temperature. *Plant Physiology* 106, 421–427.

Napier, J.A., Smith, M.A., Browne, R., Shewry, P.R. and Stobart, K. (1994) Isolation of cDNAs encoding tobacco cytochrome *b*5: expression analysis, targetting and overexpression in *E. coli*. *4th International Congress of Plant Molecular Biology*, Abstract 1470.

Nickell, A.D., Wilcox, J.R., Lorenzen, L.L., Cavins, J.F., Guffy, R.G. and Shoemaker, R.C. (1994) The *Fap 2* locus in soybean maps to linkage group D. *Journal of Heredity* 85, 160–162.

Ohlrogge, J.B., Browse, J. and Somerville, C.R. (1991) The genetics of plant lipids. *Biochimica et Biophysica Acta* 1082, 1–26.

Ohlrogge, J.B., Jaworski, J.G. and Post-Beittenmiller, D. (1993) De novo fatty acid biosynthesis. In: Moore, T.S. (ed.) *Lipid Metabolism in Plants*. CRC Press, Boca Raton, pp. 3–32.

Okuley, J., Lightner, J., Feldmann, K., Yadav, N., Lark, E. and Browse, J. (1994) *Arabidopsis* FAD2 gene encodes the enzyme that is essential for polyunsaturated lipid synthesis. *Plant Cell* 6, 147–158.

Polashock, J.J., Chin, C.-K. and Martin, C.F. (1992) Expression of the yeast Δ-9 desaturase in *Nicotiana tobacum*. *Plant Physiology* 100, 894–901.

Purdy, R.H. (1985) Oxidative stability of high oleic sunflower and safflower. *Journal of the American Oil Chemists Society* 62, 523–525.

Rennie, B.D. and Tanner, J.W. (1989a) Fatty acid composition of oil from soybean seeds grown at extreme temperatures. *Journal of the American Oil Chemists Society* 66, 1622–1624.

Rennie, B.D. and Tanner, J.W. (1989b) Genetic analysis of low linolenic acid levels in the line PI123440. *Soybean Genetics Newsletter* 16, 25–26.

Rennie, B.D. and Tanner, J.W. (1991) New allele at the *Fan* locus in the soybean line A5. *Crop Science* 31, 297–301.

Rennie, B.D., Zilka, J., Cramer, M.M. and Beversdorf, W.D. (1988) Genetic analysis of low linolenic acid levels in the soybean line P361088B. *Crop Science* 28, 655–657.

Schmidt, H., Sperling, P. and Heinz, E. (1993). New *in vivo* and *in vitro* evidence for lipid-linked desaturation in plants. In: *Biochemistry and Molecular Biology of Membrane and Storage Lipids of Plants*. American Society of Plant Physiologists, Rockville, Maryland.

Schnebly, S.R., Fehr, W.R., Welke, G.A., Hammond, E.G. and Duvick, D.N. (1994) Inheritance of reduced and elevated palmitate in mutant lines of soybean. *Crop Science* 34, 829–833.

Stefansson, B.R., Hougen, F.W. and Downey, R.K. (1994) The isolation of rape plant with seed oil free from erucic acid. *Canadian Journal of Plant Science* 41, 218–219.

Tai, C. and Jaworski, J.G. (1993) 3-ketoacyl-ACP carrier protein synthase III from spinach (*Spinacia oleracea*) is not similar to other condensing enzymes of fatty acid synthase. *Plant Physiology* 103, 1361–1367.

Thompson, G.A. (1994) Genetic engineering of saturated fatty acid content in seed oils. In: *Abstracts of the 11th International Meeting on Plant Lipids, Paris, June 26–July 1, 1994.* Abstract no. L54.

USDA (1994) *Oilcrops Situation and Outlook Report.*

van de Loo, F.J., Fox, B.G. and Somerville, C. (1993) Unusual fatty acids. In: Moore, T.S. (ed.) *Lipid Metabolism in Plants.* CRC Press, Boca Raton, pp. 91–126.

Voelker, T.A., Worrell, A.C., Anderson, L., Bleibaum, J., Fan, C., Hawkins, D.J., Rdke, S.E. and Davies, H.M. (1992) Fatty acid biosynthesis redirected to medium chains in transgenic oilseed plants. *Science* 257, 72–74.

Vorst, O., van Dam, F., Oosterhoff-Teerstra, R., Smeekens, S. and Weisbeek, P. (1990) Tissue-specific expression directed by an *Arabidopsis thaliana* pre-ferredoxin promoter in transgenic tobacco plants. *Plant Molecular Biology* 14, 491–499.

Wang, X.-M., Norman, H.A., St. John, J.B., Yin, T. and Hildebrand, D.F. (1989) Comparison of fatty acid composition in tissues of low linolenate mutants of soybean. *Phytochemistry* 28, 411–414.

Wilcox, J.R. and Cavins, J.F. (1985) Inheritance of low linolenic acid content of the seed oil of a mutant in *Glycine max. Theoretical and Applied Genetics* 71, 74–78.

Wilcox, J.R. and Cavins, J.F. (1986) Registration of C1640 soybean germplasm. *Crop Science* 26, 209–210.

Wilcox, J.R. and Cavins, J.F. (1987) Gene symbol assigned for low linolenic acid mutant in the soybean. *Journal of Heredity* 78, 410.

Wilcox, J.R. and Cavins, J.F. (1990) Registration of C1726 and C1727 soybean germplasm with altered levels of palmitic acid. *Crop Science* 30, 240.

Wilcox, J.R. and Cavins, J.F. (1992) Normal and low linolenic acid soybean strains: Response to planting date. *Crop Science* 32, 1248–1251.

Wilcox, J.R., Cavins, J.F. and Nielsen, N.C. (1984) Genetic alteration of soybean oil composition by a chemical mutagen. *Journal of the American Oil Chemists Society* 61, 97–100.

Wilcox, J.R. Nickell, A.D. and Cavins, J.F. (1993) Relationships between the *fan* allele and agronomic traits in soybean. *Crop Science* 33, 87–89.

Wilcox, J.R., Burton, J.W., Rebetzke, G.J. and Wilson, R.F. (1994) Transgressive segregation for palmitic acid in seed oil of soybean. *Crop Science* 34, 1248–1250.

Willett, W.C. and Ascherio, A. (1994) *Trans* fatty acids: Are the effects only marginal? *American Journal of Public Health* 84, 722–724.

Wilson, R.F., Burton, J.W. and Brim, C.A. (1981) Progress in the selection for altered fatty acid composition in soybeans. *Crop Science* 21, 788–791.

Wolf, R.B., Cavins, J.F., Kleiman, R. and Black, L.T. (1982) Effect of temperature on soybean seed constituents: Oil, protein, moisture, fatty acids, amino acids and sugars. *Journal of the American Oil Chemists Society* 59, 230–232.

Yadav, N.S., Wierzbicki, A., Aegerter, M., Caster, C.S., Perez-Grau, L., Kinney, A.J., Hitz, W.D., Booth, J.R., Schweiger, B., Stecca, K.L., Allen, S.M., Blackwell, M., Reiter, R.S., Carlson, T.J., Russell, S.H., Feldmann, K.A., Pierce, J. and Browse, J. (1993) Cloning of higher plant ω-3 fatty acid desaturases. *Plant Physiology* 103, 467–476.

Molecular Genetic Analysis of Soybean Nodulation Mutants

P.M. Gresshoff

Plant Molecular Genetics and Center for Legume Research, University of Tennessee, Knoxville, Tennessee 37901-1071, USA.

Introduction

This chapter deals with the genetic and molecular analysis of plant symbiotic mutants. Emphasis will be given to new insights into some components of the soybean genome, and methods that are used to analyse these. Using chemical mutagenesis, our laboratory has isolated several soybean mutants which are altered in their ability to form nodules (Carroll *et al.*, 1986; Mathews *et al.*, 1992). Additionally, we isolated soybean mutants which have an increased nodule number and nodule mass per plant. These plants (dubbed supernodulating mutants) also were able to nodulate in the presence of nitrate (Carroll *et al.*, 1984, 1985a,b). Our recent focus has been the molecular analysis of these genetic loci (Gresshoff, 1993a; Kolchinsky *et al.*, 1994). To do so required additional methods of soybean genome analysis. Following is presented a summary of our current research status.

Molecular Approaches to Plant Developmental Genes

Plant developmental processes such as flowering, pattern development, tracheary element development, and nodulation are controlled by specific sets of genes. Mutant phenotypes and their inheritance in some of these pathways have been documented repeatedly. Some sites of mutation are understood at the molecular level, especially, if the gene product is detectable either in the form of a differentially expressed cDNA clone or a protein product. For those genes the DNA sequence and regulatory elements are more or less understood. In nodulation biology, the leghaemoglobin gene is an excellent example for this level of investigation (see Marcker *et al.*, 1984). For seed development of cereals the waxy and shrunken loci are similar examples.

Molecular biology employs multiple strategies to select cDNA clones corresponding to developmental changes or mutational blockages. Such clones allow the determination of the DNA sequence as well as the isolation of the genomic copy, together with its promoter and regulatory elements. The isolation of cDNA clones may utilize methods like synthetic oligonucleotide synthesis derived from N-terminal sequence analysis, expression libraries, PCR amplification from conserved sequence motifs, and cloning by homology.

However, often a gene product is not detectable, although phenotype and inheritance point towards a clear genetic basis. In these situations a positional or map-based cloning strategy (Wicking and Williamson, 1991) has been used. This method utilizes the genetics of a mutation and couples the phenotype of a mutation with a molecular marker. This marker acts as a 'beacon' in a molecular search. In its earliest version, chromosome walking utilized overlapping pieces of high molecular weight DNA (Gibson and Somerville, 1993). These pieces of DNA, at times as large as 1000 kilobases (kb), are cloned into vectors which replicate in yeast. They are called yeast artificial chromosomes (YACs). The complexity of most plant genomes and the general distribution of repeated DNA make this approach difficult, as the terminal DNA regions of a cloned piece of DNA, needed to find an overlap, frequently target the search onto several putative neighbouring DNA pieces, only one of which is the 'true' molecular contiguous neighbour (see Funke and Kolchinsky, 1994).

A second approach for positional cloning requires the detection of many molecular markers close to, and flanking, the gene of interest. Proximity is judged by recombination frequency. The aim then is to obtain a YAC, which contains two flanking markers. By logic the desired gene must exist between these markers (Leyser *et al.*, 1993). This candidate YAC clone can then serve as a probe for cDNA clones specifically expressed in the affected tissue. This approach was successfully demonstrated for a locus (*pto*) in tomato conferring resistance to *Pseudomonas syringae* (Martin *et al.*, 1993). Sequence analysis suggested that the *pto* gene codes for a protein kinase, consistent with its function in a signal transmission pathway.

There is no a priori way to know which approach works better. Even in a plant with a small genome like *Arabidopsis thaliana* (about 120 megabases), known to have a low amount of highly repeated DNA, one finds that YACs and the generation of contiguous physical maps by overlapping clones (contigs) are difficult. In the chromosome walking approach, effort is needed to sort out the overlaps and generate the contig. In the high density mapping effort, one may be spared the 'walk', but needs to find many markers close to the gene of interest and to map and order these.

Nodulation: A Plant Developmental Process

Nodulation and subsequent nitrogen fixation represent many normal plant developmental processes (see Verma, 1992; Gresshoff, 1993a) Following recognition of an outside organism, the plant defence response needs to be adjusted, cell divisions are induced and subsequently regulated to prevent neoplastic outgrowth. Systemic regulation and source-sink relations of various nitrogen and carbon metabolites are established. Nodules differentiate vascular tissue, transfer cells, infected and uninfected cells. Cellular patterns and hormonal control mimic the root developmental patterns.

The study of nodules has a genetic advantage, as it is possible to isolate conditional mutants. In most other plant developmental steps, recessive mutations lead to genetic or physiological death. For example, most mutations in flower initiation render the plant genetically dead; likewise a rootless mutant is a physiological cripple, needing special cell culture for propagation. Mutations in the nodulation and nitrogen fixation pathway are critical for the plant and give a pronounced phenotype, but they are able to be rescued by the addition of an alternative nitrogen source, such as nitrate. This conditionality reminds one of the situation with auxotrophs, which contributed significantly to the development of microbial and biochemical genetics.

Nodulation is the process by which specialized root structures develop on the roots (and sometimes stems) of legume plants. These organs are induced by bacteria, collectively labelled rhizobia. The recent discovery that empty, bacteria-free nodules form in the absence of *Rhizobium* demonstrated that the entire developmental information for nodule morphogenesis is plant-encoded (Caetano-Anollés *et al.*, 1991c; Joshi *et al.*, 1991). Such spontaneous nodules developed if the plant carried the appropriate dominant gene (Caetano-Anollés and Gresshoff, 1992). They also formed when lipo-oligosaccharides (nod-factor; see review by Verma, 1992) or phytohormone antagonists were applied to the root in the absence of bacteria (Hirsch *et al.*, 1989).

Nodules are complex organs consisting of several cell types commonly found in other plant structures (for example, trachids, sieve elements, transfer cells, endodermis, parenchyma) as well as specialized cells needed for the nitrogen-fixing symbiosis. The predominant symbiotic cell type is the infected cell, which in soybean may harbour as many as 30,000 bacteroids (Gresshoff and Rolfe, 1978). These bacteroids complete the reduction of nitrogen gas to ammonia, which in turn is assimilated by plant enzymes. Uninfected cells in the nodule interior function to assimilate the fixed nitrogen to form the nitrogenous compounds (e.g. ureides in soybean).

Molecular Genetics of Soybean Nodulation

The molecular genetic analysis of the bacterial symbiotic partner has made significant advances in the last decade (for review, see Stacey et al., 1992). Likewise, plant research defined many facets of the symbiosis. For example, it was recognized that the nodule contains specifically or preferentially expressed plant proteins. These are termed 'nodulins' (Legocki and Verma, 1979; Marcker et al., 1984; Verma, 1992), which are divided into two broad classes, i.e. early and late nodulins (Sanchez et al., 1991). Late nodulins occur in the nodule around the onset of nitrogen fixation; by definition early nodulins appear before that point. Some nodulins, like leghaemoglobin, glutamine synthetase and uricase have defined biochemical functions (oxygen transport, ammonia assimilation, and ureide synthesis, respectively). Others are only known through their cDNA sequences, deduced protein sequence and structure, and in a few cases, nodular location as determined by in situ hybridization or cell separation. Whether nodulins are required for the symbiosis is not known and requires a more complete genetic analysis. Unfortunately, none of the symbiotic mutants of legumes lacks a nodulin gene, and plants lacking certain nodulins are still symbiotically active.

Reverse genetics using antisense mutagenesis (Miao and Verma, 1993) may provide a method for further elucidation of the function of nodulins. It is one of the goals of research in the future to find biochemical and physiological roles for 'orphan nodulin genes'. A further task is to find those early nodulins which are functional in the recognition and interpretation of the extracellular Rhizobium-derived signal, which causes cell division as well as root hair curling, and infection. One may assume that these nodulins are rare, thus difficult to detect. Furthermore, it is likely, contrary to expectations, that these essential steps of early nodule initiation are not nodule-specific. Accordingly, it is of value to study genes of general metabolism, which are expressed in roots to understand nodule ontogeny. One example of this type of new focus involves the cell cycle genes (Hata et al., 1991; Miao et al., 1993; Deckert et al., 1994).

Molecular studies of nodulins were pre-dated by a classical genetics approach (Williams and Lynch, 1954; Nutman, 1968; Carroll et al., 1986). In several legumes, nodulation and nitrogen-fixation mutants (see Caetano-Anollés and Gresshoff, 1991; Gresshoff, 1993a for reviews) were isolated. As yet none of these mutations has been correlated with a molecular change. Pleiotropy, the extent of molecular change, and lack of relative abundance of the mutant gene product have made biochemical analysis a difficult task.

Significant efforts were focused on the supernodulation (nts) gene of soybean, because of its scientific importance as well as the potential implication towards nitrate-tolerant symbioses and increased nitrogen fixation. Supernodulation in soybean is the absence of autoregulation of nodulation, caused by a Mendelian recessive mutation at a single locus on RFLP linkage group H

(Landau-Ellis *et al.*, 1991) of the USDA-ARS RFLP map (Keim *et al.*, 1990). The first supernodulation and nitrate-tolerant symbiotic mutants were described by Carroll *et al.* (1984, 1985a,b) for soybean. Concurrently, one nitrate-tolerant mutant of pea (*Pisum sativum*) was isolated by Jacobsen and Feenstra (1984). Further details on the genetic and physiological properties of the soybean mutants have been reviewed (Gresshoff, 1993a). Additional supernodulation mutants were isolated in soybean by others (Gremaud and Harper, 1989; Buzzell *et al.*, 1990; Akao and Kouchi, 1992) as well as in other legumes such as French bean, pea, and *Medicago truncatula* (Park and Buttery, 1988; Duc and Messager, 1989; Duc, personal communication), confirming the phenomenon as well as several of the physiological characteristics. So far only the *nts* mutant of Akao and Kouchi (1992) has been tested for allelism with the Canberra isolated *nts* mutants and was found to be allelic to *nts382* and *nts1007* described by Carroll *et al.* (1985a,b).

Supernodulation is an alteration in the development of a plant organ. Soybean regulates the number of developing nodule foci through a developmental feedback loop (Pierce and Bauer, 1983; Calvert *et al.*, 1984; Mathews *et al.*, 1989a). Once a sufficient number of nodule foci has been established, the plant suppresses or slows the development of ontogenetically younger stages. This autoregulation occurs prior to the onset of nitrogen fixation, does not involve the nutritional status of the plant, and limits the mass of symbiotic tissue. We demonstrated that younger nodule foci were arrested and not aborted (Caetano-Anollés *et al.*, 1991a). We also found interaction of lateral root foci and nodule formation at a regulatory (not anatomical) level. It can be hypothesized that autoregulation evolved from an ancestral developmental programme limiting the extent of lateral root proliferation. Accordingly, a study of autoregulation, and the mutational state causing supernodulation, may reveal plant developmental phenomena beyond nodulation biology.

The plant-controlled regulation of nodulation, altered in supernodulation mutants, may exist in all legumes. Grafting determined that all supernodulation mutants [except for the NOD3 mutant of pea isolated by Jacobsen and Feenstra (1984)] are controlled through the shoot and not the root of the plant (Delves *et al.*, 1986). This implies systemic regulation by which the root signals the top of the plant about the initial stages of the nodulation process, so that the top of the plant signals the root terminating further nodulation in ontogenetically younger root tissue, or slowing nodulation sufficiently to permit the cessation of further nodulation by other developmental processes. As yet there is no clear evidence concerning the chemical nature of this regulatory circuit. Gresshoff (1993b) advanced an auxin burst control hypothesis which proposes that the plant regulates nodulation initiation through a shoot-derived increase in auxin translocation. Meyer and Mellor (1993) proposed that nod-off substances, being defined as compounds that lower the expression of bacterial nodulation genes, may be altered in supernodulation mutants. One such factor may be riboflavin, which was shown to inhibit the induction of *nodA-lacZ* fusions of *Bradyrhizobium japonicum* (Mellor and Rosendahl, 1994).

Multicellular organisms have a general need to regulate cell division. A plant with multiple and persisting growth centres (or apexes) has a special requirement as some meristems must grow (e.g. root tips), while others are arrested or slowed (e.g. the lateral root primordia, nodule initials, flower buds, tendrils, etc.) during certain periods of development. Such regulation requires more than cellular control over cell division. The process must involve a systemic regulation in which different plant parts perceive growth activity of others.

Without doubt the understanding of plant growth *per se* is an imperative for future agricultural research. Why do some plants of the same species grow better than others? Why does one plant have more flowers, fruits, or seeds than another? Why do root systems possess different patterns of lateral root branching?

Autoregulation, and its genetic alteration leading to supernodulation, may reflect a plant developmental programme, which, for example, may control lateral root development and pattern formation (Caetano-Anollés *et al.*, 1991a). Thus, the study of supernodulation provides information on the nodulation process itself, but also contributes to the understanding of plant development and differentiation.

RFLP Mapping of Nodulation Mutant Genes

Genetic analysis showed that the supernodulation mutations were all recessive to wild type, and were in the same complementation group (Delves *et al.*, 1988). The two non-nodulation mutants nod49 and nod139 were separate loci, also showing recessive inheritance (Mathews *et al.*, 1989b). Supernodulation was epistatically suppressed by non-nodulation (Mathews *et al.*, 1990). The *nts* locus segregated independently from the two non-nodulation loci (now termed rj_1 and rj_6; Pracht *et al.*, 1993). Mutant nod49 was shown to be allelic with the naturally occurring mutation rj_1 (Williams and Lynch, 1954). This genetic knowledge formed the basis for a more detailed molecular genetic study.

Glycine max and *Glycine soja* are two subspecies which form totally fertile hybrids. Their karyotype as determined by pachytene chromosome analysis is identical (Singh and Hymowitz, 1988), and suggests that they should be considered one species.

Soybean has 21 to 23 genetic linkage groups reflecting its 20 chromosomes, although as yet only one of these has been assigned. The analysis of trisomic lines of soybean showed that RFLP linkage group L is chromosome 13 (Kollipara *et al.*, 1994). At present as many as 16 chromosomes exist in a trisomic condition. This represents about 90% of the soybean genome. The

further coupling of RFLP groups and the parallel placement of mutant phenotypes to chromosomes are exciting developments of soybean molecular
genetics.

Shoemaker and associates (at Iowa State University, Ames, Iowa) used the
genetic divergence at the DNA level to detect restriction fragment length polymorphisms (RFLP) between *G. soja* and *G. max* (Keim *et al.*, 1990). The segregation and linkage of the RFLP markers was determined in an F_2 population
to develop a molecular map of soybean. The present status of this map and the
soybean genome are listed in Table 9.1. Other maps exist for a *G. soja* × *G. max*
cross (from DuPont Company) and from *G. max* × *G. max* crosses (including
valuable recombinant inbred lines (RILs); Lark *et al.*, 1993).

To map the supernodulation and nitrate tolerant symbiosis locus identified in our laboratory, a new F_2 population derived from a nts382 (*G. max*) ×
wild-type (*G. soja*) cross was analysed for linkage of the supernodulation phenotype to a particular RFLP probe. RFLP marker pA-132 from the Iowa State
University map was tightly linked (Landau-Ellis *et al.*, 1991; Landau-Ellis and
Gresshoff, 1992). However, the probe gave a complex hybridization pattern
(Fig. 9.1). This was simplified after the discovery that pA-132, although being
a random genomic *Pst*I clone, contained three unlinked inserts (Landau-Ellis
and Gresshoff, 1994; Fig. 9.1). The three inserts detected polymorphisms, when
DNA was digested with different restriction nucleases. All polymorphisms
segregated independently of each other (Table 9.2), confirming the conclusion
reached from the Southern blot analysis (Fig. 9.1) that the inserts are not contiguous. The largest insert clone, called pUTG-132a ('UT' to define the geographic origin, 'G' to signify *Glycine*), was linked to *nts*, by about 0.3
centimorgans. This placed *nts* onto linkage group H of the USDA/ARS map.

Table 9.1. Currently known features of the soybean
genome and genetic map.

---

1. One billion base pairs genome
2. 20 chromosomes (2n = 40)
3. 23 RFLP linkage groups
4. 500 linked RFLP, morphological and enzyme markers
5. About 3000 centimorgans
6. 35% highly repeated DNA
7. One nucleolus organizer region
8. Duplicated genomic regions
9. True diploid segregation of meiosis
10. Recessive mutations available
11. Trisomics (20) available
12. 92 bp satellite (0.7% of genome)
13. YACs and endclones available
14. Linkage group L is chromosome 13
15. RILs and NILs available

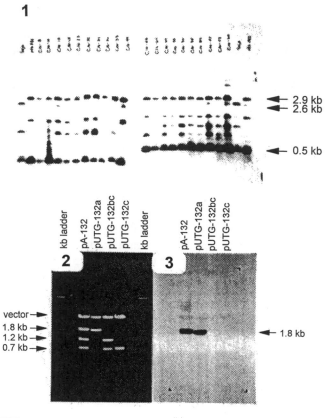

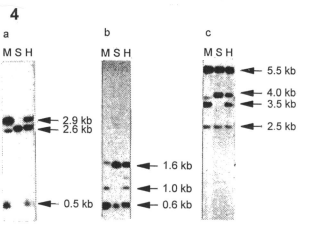

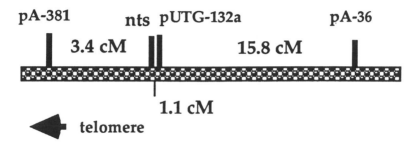

Fig. 9.2. Regional map of RFLP markers around the *nts* locus. Determined from 113 F₂ plants using MAPMAKER program. Inclusion of other mapping data from different crosses gives a nts–pUTG-132a distance of about 0.3 cM.

Another allele at the *nts* locus (*nts1007*), previously demonstrated by complementation to be allelic to *nts382* (Delves *et al.*, 1988), was also mapped to the same marker. Flanking marker pA-381 (distal) and pA-36 (proximal) were placed 3.4 and 15.8 cM from *nts*, respectively (see Fig. 9.2).

The pUTG-132a probe detected three genomic fragments in the *nts* parent, but only one in *G. soja*. To verify the utility of the probe for further analysis it was necessary to determine the nature of the polymorphism. Both the 1.7 kb *G. max* and the 0.9 kb *G. soja* genomic fragments were sequenced (Kolchinsky *et al.*, 1995b). The internal *Dra*I sites responsible for the polymorphism were determined. A deletion of 877 bp removes one of the *Dra*I sites in *G. soja*, causing the superpositioning of two genomic fragments at about 2.6 kb. The deletion removes the small 500 bp *Dra*I fragment from *G. max*. It is perhaps significant for the explanation of the deletion that two 40 bp direct repeats were found exactly at the site of the deletion, suggesting that an internal recombination event either caused the insertion or deletion. Sequence analysis of pUTG-132a failed to demonstrate any significant similarity to other known sequences, or large open reading frames, suggesting that the pUTG-132a clone represents non-coding DNA. Surprisingly, the DNA sequence over

---

Fig. 9.1. (opposite) RFLP pattern of *Glycine soja* and *Glycine max* nts382 probed with pA-132. Panel 1: parental and F₂ plants restricted with *Dra*I and probed with the original pA-132. Panel 2: ethidium bromide-stained fragments of pA-132 digested with *Pst*I. Panel 3: Southern blot of panel 2 with a lambda clone λ132a (14 kb size) homologous to pUTG-132a, the largest of the inserts. The absence of hybridization to the middle and smaller fragment demonstrates that they are not contained on λ132a and therefore are unlikely to be contiguous. Panel 4: RFLPs of the three pA-132 inserts on *G. max*, *G. soja* and F₁ hybrid. (a) *Dra*I digest with pUTG-132a; (b) *Dra*I digest with pUTG-132b; (c) *Taq*I digest with pUTG-132c. The *Dra*I digest did not reveal a RFLP with pUTG-132c. (D. Landau-Ellis, University of Tennessee, unpublished data).

Table 9.2. Segregation of RFLP markers among fifteen supernodulating F_2 plants from the cross *G. max* (nts 1007) × *G. soja* (PI468.397).

	Markers/banding patterns				
F_2 plant no.	nts	pUTG-132a	pUTG-132b	pUTG-132c	pA-36
A3-104	M	M	H	S	M
A3-105	M	M	M	H	M
A3-108	M	M	S	S	M
A3-114	M	M	H	S	M
A3-116	M	M	H	H	M
A3-118	M	M	H	H	M
A3-120	M	M	H	M	H
A3-123	M	M	S	H	M
A3-127	M	M	H	S	M
A3-134	M	M	H	M	M
A3-152	M	M	H	H	M
A3-159	M	M	H	S	S
A3-171	M	M	H	H	M
A3-175	M	M	H	H	M
A3-197	M	M	S	H	M
Ratio M:H:S	15:0:0	15:0:0	1:11:3	2:8:5	13:1:1

M = *G. max* banding pattern or supernodulating phenotype/genotype, H = heterozygous banding pattern, S = *G. soja* banding pattern.

about 1700 bp is conserved among soybean cultivars (i.e. Noir I, Minsoy, Enrei, DPS3546, Bragg). Only cv. Peking showed nine changes from the pUTG-132a clone.

The RFLP linked to the supernodulation gene was caused by an internal molecular alteration. This allowed the conversion of the probe to a PCR set of primers. Three PCR primers, which flanked the deletion (two forward, one reverse) were synthesized. One forward primer site was placed inside the deleted, polymorphic sequence and gave an amplification product only with *G. max*, but not *G. soja*. These PCR primers generated polymorphic products (caused by the 877 bp deletion), which allowed the investigation of large F_2 populations by the less laborious PCR method (Kolchinsky *et al.*, 1995). Accordingly, the linked region of the pA-132 marker is defined by the pUTG-132a probe as well as the PCR primers. This makes this region one of the first SCARs (sequence characterized amplified regions; Paran and Michelmore, 1993) linked to a plant developmental locus. These primers also eased the search of genomic libraries such as ordered YAC arrays (see below) for homologous clones.

Probe pUTG-132a and the PCR primers were tested for their ability to detect polymorphisms between *G. max* cultivars. None were detected, confirming the sequencing data. However the RFLP probe allowed the isolation of a lambda genomic clone (14 kb), which contained neighbouring sequences that may detect RFLPs between commercial cultivars. The PCR primers were used on related legumes. Alfalfa DNA failed to produce an amplification product, but cowpea DNA was amplified giving an identically sized product as soybean. This further underscores the contention that cowpea and soybean share a lot of DNA sequences and genome arrangement (synteny; see Moore *et al.*, 1993).

These studies demonstrate a general strategy. The knowledge of close linkage of an RFLP and a phenotype marker was converted to find either PCR primers or additional markers with greater resolving power.

Physical Mapping Close to the Supernodulation Region

The total genetic distance between the known RFLP markers on the USDA-ARS map is about 3000 cM (Shoemaker *et al.*, 1994; Gresshoff, 1993a). The genome size of soybean is estimated at about $1-1.26 \times 10^9$ base pairs (Blackhall *et al.*, 1991; Gurley *et al.*, 1979). This gives an average distance of 340 to 400 kb per centimorgan. However, eukaryotic genomes are characterized by regions of high and low recombination. It is therefore important to get a measure for the relationship between genetic and physical distances.

A pilot study in soybean focused on a cluster of RFLP markers on linkage group H within 16 cM of the *nts* gene (see Fig. 9.2). The cluster places four markers (pA36, pA69, pA-89d and pK-9a) within 2.8 cM of each other (Fig. 9.3). Two RFLP markers (pK-9a and pA-89d) were not able to be separated in the USDA mapping population of 66 plants. The two probes do not share hybridization homology.

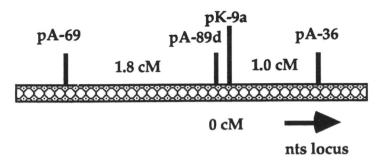

Fig. 9.3. RFLP marker arrangement used for physical mapping close to the *nts* (pA-132) marker. The direction of the *nts* locus is assumed based on presumptive genome duplication to the left of the pA-36 marker, while the arm of linkage group H to the right of pA-36 is diploid.

The strategy involved the isolation of leaf protoplasts, which were lysed *in situ* to release high molecular weight DNA. This DNA was cut with restriction nucleases that recognize eight base pair recognition sites (so-called rare cutters such as *Not*I and *Sfi*I). Restricted DNA was separated by pulsed field gel electrophoresis (Schwartz and Cantor, 1984), blotted onto Nylon membranes, and probed with relevant RFLP clones. Hybridization of separate probes to the same band indicated putative physical linkage, which allowed calculation for physical distance based on the fragments molecular size and recombination between the two markers (Funke *et al.*, 1993).

Major problems arose, because of the ancestral and partial tetraploid nature of the soybean genome. Most RFLP probes mapped are characterized by detecting two or more genomic regions (Landau-Ellis *et al.*, 1991; Shoemaker *et al.*, 1992; Funke *et al.*, 1993). The markers of the RFLP cluster detected at least two high molecular weight fragments. It was, therefore, unclear which of these was the polymorphic and mapped copy. To resolve this problem, two-dimensional electrophoresis was used. In this approach the fragments separated by PFGE were digested *in situ* with the restriction nuclease that yielded the original mapped RFLP. This liberated the RFLP and monomorphic fragments from the larger PFGE fragments. Separation of the excised PFGE lane occurred in a perpendicular direction by agarose electrophoresis. Probing relative to a total genomic DNA sample revealed which PFGE band contained the polymorphic or monomorphic copies. The application of this approach allowed the construction of a partial physical map in the region of the pA-36 marker. In this region of the soybean genome one centimorgan was equivalent to less than 500 kb. Other regions of the genome need to be tested to verify this estimate, but are expected to differ as in rice (Ashikawa *et al.*, 1992) and tomato (Ganal *et al.*, 1989) nearly 50-fold differences were found, showing variable recombination rates in plant genomes.

From the physical mapping, the nature of the RFLP clones and the recessive condition of the *nts* mutations, we presume that the arm of linkage group H is present as a single copy in the soybean genome. Why the pUTG-132a region is so strongly conserved is presently not understood. However, since other regions of the genome have not diploidized (see Shoemaker *et al.*, 1994), one can presume that other developmental genes governing nodulation and nitrogen fixation exist, but these are not able to be defined either by chemical or insertion mutagenesis.

Soybean Yeast Artificial Chromosomes

Yeast artificial chromosomes (YACs) serve as a cloning vehicle for large pieces of DNA. They take advantage of the yeast's eukaryotic ability to condense and segregate DNA with high fidelity at mitosis. There are some potential disadvantages, which require further study. Yeast possesses an efficient homologous recombination system, which leads to the excision of insert sequences between homologous regions. Since soybean DNA contains frequent stretches of satellite DNA it is likely that soybean DNA is lost in some cloned fragments. Such instabilities lower the value of a clone. Accordingly, YAC clones need to be verified for physical sizes of insert DNA relative to pulse field gel separated genomic DNA. While minor deletions cannot be detected in this way, one would see major changes. Because of the apparent efficiency of sub-cloning a large genome as smaller, more easily assessable, units inside YACs, this cloning system has become popular for library construction. Workable YAC libraries exist for *Arabidopsis*, barley, rice, sugarbeet, and tomato.

The general construction of a YAC carrying soybean DNA is depicted in Fig. 9.4. In general, about three genomes' worth of insert DNA are required to achieve a 95% chance to cover the entire genome. For soybean this means about 12 to 15,000 YAC clones with an average size of 200 kb are required. At present our laboratory can produce 100 YACs per week per person.

The construction of soybean DNA containing YACs has been a challenge to several laboratories. The major barriers were the difficulty of isolating high molecular weight DNA from soybean, caused by the presence of polysacchar-

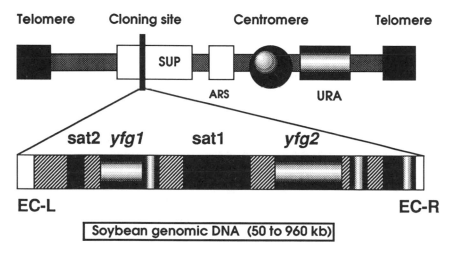

Fig. 9.4. General components of a yeast artificial chromosome (YAC). The telomeres are derived from *Tetrahymena*, the centromere is of yeast origin as are the suppressor (SUP) and uracil biosynthetic (URA) genes. ARS is an autonomously replication sequence which serves as an origin of replication. The suppressor gene inactivates a non-sense mutation in an adenine metabolism gene, causing the yeast colony carrying this gene to appear white. Insertional inactivation of the suppressor gene results in the accumulation of a red adenine intermediate, providing a visual marker for a YAC with a soybean DNA insert. The insert DNA may contain not only genes *yfg1* and *yfg2* but also satellite DNAs (sat1 and sat2) and other repeated units. The terminal sequences are labelled as EC-L and EC-R and if cloned, can be used as probes for neighbouring and overlapping YACs.

ides, difficulty of leaf protoplast handling, and variable degrees of nuclease restriction (Funke *et al.*, 1993). Our research with pulse field gel separated soybean DNA pointed towards the possibility of cloning some of the detected fragments in yeast artificial chromosomes. YACs were produced by two related methods. In the first, soybean protoplasts were lysed and digested to give partial *Eco*RI digests. The high molecular weight DNA, embedded in low melting point agarose plugs, was permitted to soak in buffer for a short period to allow diffusion of small DNA molecules out of the plug (Kolchinsky *et al.*, 1994). The remaining DNA was released from the agarose plug by melting, and was ligated into the *Eco*RI site of vector pYAC4. Ligated DNA was transformed into yeast spheroplasts, which were selected by prototrophy, then screened by colony colour. Resultant yeast colonies were tested for the presence of YACs by PFGE separation of chromosomes. The initial set of characterized colonies showed extra chromosomes on ethidium bromide stained PFGE gels. Hybridization to both vector sequences and total soybean genomic DNA was positive. The YAC inserts, however, were relatively small, averaging about 60–75 kb. The largest YAC produced by this method was 175 kb. This direct cloning approach yielded thousands of clones. While such small inserts may have some applications, for example as linking clones, it was decided to attempt the construction of larger YACs.

The second approach involves a size selection of partially digested DNA. This involved a short PFGE run of the partially digested DNA to remove smaller molecules from the compression zone. The zone containing DNA of greater than 250 kb was transformed as above into pYAC4. Transformed colonies showed YACs at a lower frequency. However, an average YAC insert size of 200 kb, with a maximum of 960 kb, was obtained (Funke and Kolchinsky, 1994; Funke *et al.*, 1994; Kolchinsky, 1994). At present two YAC libraries are being constructed using both size-selected and non-selected methods. A total of one thousand clones have been characterized and stored. About 7% of the clones represent chloroplast DNA. Initial studies on YAC stability using their size on pulsed field gels have removed concerns as YACs were stable through repeated subculture (Funke *et al.*, 1994). However, small rearrangements or deletions caused by homologous recombination cannot be detected. Moreover, any insert that is incompatible with yeast, for example causing a direct or indirect effect on yeast viability, will not be detected as the colony is not represented in the culture. For this reason it seems wise to construct parallel libraries in bacterial artificial chromosomes (BACs) using F-factor replicons and origins of replication as a cloning vehicle.

Several precautions need to be taken when using size selection. Kolchinsky *et al.* (1994) showed that high EDTA concentrations caused small molecular weight DNA to be trapped in the compression zone, resulting decrease in the average insert size.

YACs carrying soybean DNA were confirmed by ethidium bromide staining of pulse field gel separated yeast karyotypes (Fig. 9.5). Some YAC colonies contained more than one YAC. These gels were plotted and probed with vector

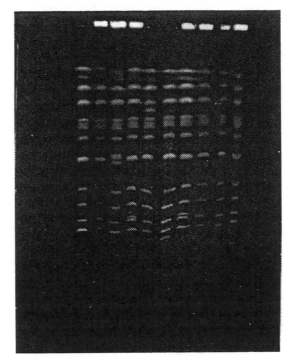

Fig. 9.5. Pulse-field generated karyotype of yeast carrying soybean insert YACs. The chromosomes range in molecular size from 225 kb to 2000 kb. The gel was stained with ethidium bromide. Several YACs are seen; for example lane 5 from the left shows a large YAC as the fourth largest chromosome. Lane 3 from the left has two YACs of nearly identical size. These can be separated after mitotic segregation. Further analysis is required to determine whether such YACs are derived from each other by deletion or rearrangement. (Photograph courtesy of Roel Funke, Plant Molecular Genetics, University of Tennessee, Knoxville.)

DNA, soybean genomic DNA and chloroplast DNA. The hybridization signal from the vector probe showed chromosomes of similar intensity and mobility as the ethidium bromide visualization. The soybean genomic probe gave a pattern of different intensities, as some YACs clearly harbour more repeated DNA than others. However, the size class for all YACs detected by soybean DNA hybridization was identical to that described for the other two detection methods. YACs containing highly repeated DNA regions may be the first candidates for fluorescent *in situ* hybridization (FISH) as the target sizes are increased considerably over single copy DNA probes.

Terminal clones of soybean YAC inserts were generated by vectorette ligation (Ashikawa *et al.*, 1992). Clones were tested for the presence of repeated DNA sequences by hybridization to digested soybean DNA. Two out of four tested clones detected repeated DNA, while the other two detected single bands. None tested so far were polymorphic between *G. soja* and *G. max.* Interestingly, terminal clones from YACs known to contain large amounts of repeated DNA were single copy, suggesting that there is no direct correlation.

At present we are extending this library to become useful for gene isolation. We propose to screen the library in pooled groups using PCR, and to dot blot the clones onto Nylon filters for direct hybridization.

DNA Amplification Fingerprinting Applied to Soybean Genomes

Molecular mapping of genes requires a large number of markers. While the repertoire of RFLPs is sufficient to produce a map, one can notice many gaps. Indeed, the region around the *nts* gene is rather bare of RFLP markers, other than pUTG-132a. It is therefore important to use other methods to detect molecular variability.

Short single primers of arbitrary sequence amplify soybean DNA and give multibanded patterns, when separated by polyacrylamide gel electrophoresis (Caetano-Anollés *et al.*, 1991b, 1992). Figure 9.6 shows the single primer amplification process, and demonstrates its superficial similarity to the PCR pro-

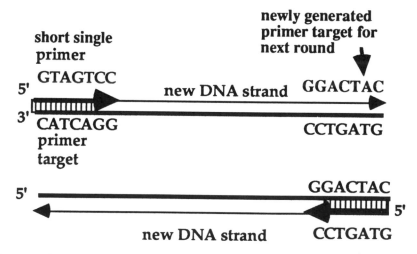

Fig. 9.6. The DAF amplicon. Single short oligonucleotide primers anneal to single stranded DNA generated by temperature dissociation of double stranded target DNA and are extended from the 3′ end by thermostable DNA polymerase. At the newly generated 3′ terminus of the replicated DNA strand a new complimentary primer target site is generated to facilitate another round of amplification after temperature dissociation.

cedure. However, in contrast to PCR, no prior sequence knowledge is needed as the primer has an arbitrary sequence. Furthermore, the primer is short and single. [N.B. DAF is similar, but distinct in several features such as primer–template ratios and primer length, to two other single primer amplification methods developed concurrently by Williams *et al.* (1990) and Welsh and McClelland (1990).] Detection of DAF products is efficiently and routinely achieved by a high sensitivity silver staining method (Bassam *et al.*, 1991; Caetano-Anollés and Gresshoff, 1994), although other separation technologies such as capillary electrophoresis (Pat Williams, Gaithersburg, Maryland, unpublished results), agarose gel electrophoresis (Kolchinsky *et al.*, 1993), Phast-System^{TM} (Pharmacia) (Baum *et al.*, 1994; Caetano-Anollés and Bassam, unpublished data), and automated DNA sequencer have been demonstrated.

A survey of 25 primers, eight nucleotides long, used to amplify DNA from *Glycine soja* and *G. max* showed a high frequency of amplification polymorphisms. On average, about 1.5 polymorphisms per primer were detected (Prabhu and Gresshoff, 1994). Amplification polymorphisms (AFLPs) were also detected between commercial cultivars of soybean. The majority of AFLPs had a molecular size of 100 to 300 basepairs, but can range up to 800 bp. An eight-nucleotide long primer (called an 8-mer) used in a reaction with *G. max* DNA and Stoffel fragment DNA polymerase Perkin-Elmer Corp.) gave about 25 scorable and repeatable bands. This frequency of scorable marker is four to five times higher than observed for RAPD reactions with soybean (Williams *et al.*, 1990).

Gresshoff and Mackenzie (1994) demonstrated that separate DNA isolations, amplifications, and gel manipulations produced only a minute error in band presence and migration. Intensity of staining varied, but did not affect the scoring of the major bands. The major source of error was assigned to the ability to measure band migration from the well, rather than the experimental design and handling. Additional error may come from altered DNA concentration. It is difficult to measure DNA accurately at such low concentrations. Dilutions are usually used, making errors more likely. We now try to confirm promising amplification polymorphisms in repeated runs, even varying the sample DNA concentration (A. Kolchinsky, Knoxville, unpublished data). In general, we find that low, but consistent DNA concentrations (5 ng per 25 μl reaction volume) together with a high (3 μM primer concentration) are needed for the initial DNA amplification to achieve well-staining and informative DNA patterns. Cycle number should not exceed 35. Quality of DNA, purity of primers and DNA polymerase also need to be confirmed. It is wise to conduct 'dummy' runs, which contain all amplification components except template DNA.

Most of the AFLPs generated in soybean by short primers were inherited in a dominant Mendelian fashion (Prabhu and Gresshoff, 1994). Lack of codominance is seen as a weakness of DAF and RAPD markers (Williams *et al.*, 1990) as heterozygotes cannot be distinguished from dominant homozygotes.

However, conversion of a DAF polymorphism either to an RFLP probe or a sequence characterized amplified region (SCAR) would overcome this problem and detect heterozygotes. Alternatively, we have shifted to the analysis of recombinant inbred lines (RILs; Lark *et al.*, 1993). In these, heterozygotes are nearly eliminated giving segregation ratios for most molecular markers of 1:1 (Prabhu and Gresshoff, 1994). Thus, all samples are numerically valid. Not all DAF polymorphisms may generate polymorphic SCARs or RFLP probes; however, the utility and ease of using PCR primers to detect a polymorphisms make this approach worthwhile.

Three DAF markers were mapped to the RIL map of Lark *et al.* (1993). Close linkage to flanking markers were detected (Prabhu and Gresshoff, 1994). It is possible that DAF markers, being generated by a different paradigm, can fill more gaps and combine linkage groups in soybean molecular maps.

As a first step of converting an amplified band to a cloned molecular marker, Weaver *et al.* (1994) isolated silver stained DAF bands from polyacrylamide gels, reamplified them prior to cloning, and subsequent Southern blot analysis of both DAF gels and genomic DNA. Likewise Kolchinsky *et al.* (1993) separated DAF products by agarose electrophoresis, excised and eluted bands and cloned these to obtain a probe for Southern blotting.

About one quarter of the DAF generated polymorphisms showed non-Mendelian inheritance. Some of these cases followed maternal inheritance patterns and presumably are polymorphisms generated from chloroplast or mitochondrial genomes. However, some DAF markers gave paternal inheritance patterns, which require further study, and may suggest yet undiscovered mechanisms in single primer DNA amplification (Prabhu and Gresshoff, 1994).

While there is a significant number of polymorphisms distinguishing the nts382 mutant of *G. max* from *G. soja* to generate a map around the *nts* locus, it was impossible to detect a difference between the parent line Bragg and its EMS-mutagenesis derived mutants nts382 and nts1007 (Caetano-Anollés *et al.*, 1993). Out of a total of 470 scored DAF bands, all were identical for the three near-isogenic lines. This demonstrated two points. First, the DAF method, if executed carefully, gives reliable results, even with separated DNA preparations and genotypes (cf. Gresshoff and MacKenzie, 1994). Second, the EMS mutagenesis that produced the supernodulation mutants did not cause major genomic changes. The mutant and parent line therefore are equivalent to near-isogenic lines (NILs) commonly used in genome analysis after periods of repeated backcrossing to a recurrent parent line.

When the template DNA of the three isolines was predigested with restriction nucleases, that cut four base sequences, then amplified with the same short DAF 8-mer primers, several AFLPs were detected. Segregation analysis of these bands in a subpopulation of F_2 plants derived from a *G. soja* × *G. max* (nts382) cross showed a strong cotransmission (100% for some) with the *nts* locus. Further studies are needed to confirm the nuclear assignment of these amplified regions. It is unclear why and how predigestion causes the increased detection of molecular polymorphisms. The same effect was observed with banana and fungal DNA. It is possible that predigestion removes a major population of soybean DNA from the amplification pool, if it is cut into small pieces, insufficient for DAF. One presumes that repeated DNA would be preferentially lost, leaving more unique DNA sequence available for amplification.

These polymorphic bands distinguishing the supernodulators and wild-type soybean are being cloned. The cloned fragments need to be tested for their ability to detect an RFLP. Recombinant plants in the *nts* region will be tested for their genotype relative to the DAF derived clones; in other words, they will be mapped to generate a more densely filled map region around the *nts* locus.

Telomeres and Satellites

Telomeres and satellite DNA are integral parts of eukaryotic genomes (Kolchinsky and Gresshoff, 1993, 1994). Their origin, distribution and function are still open to speculation. As part of our research into the soybean genome, we developed methods to investigate these apparently structural elements. We discovered a major satellite DNA (SB92) as well as the fact that telomere associated sequences (TAS) vary between cultivars of soybean, permitting the mapping of them and fingerprinting of soybean germplasm.

In other plants and animals, telomeres have an unequal distribution of nucleotides. This allowed the synthesis of putatively strand-specific primers. Single telomere-related primers $(TTTAGGG)_4$ (GT-primer) and $(CCCTAAA)_4$ (AC-primer) were used to 'drive' a PCR-reaction with soybean DNA. The reaction, run in stringent conditions, generated a reproducible pattern of bands. These fingerprints differed for several leguminous plants; they also distinguished several lines of soybean (*Glycine max*) and accessions of the corresponding wild subspecies *G. soja*. The polymorphic bands were co-dominant in heterozygous progeny and therefore were scored in segregating populations. One of the polymorphic PCR products was mapped to the end of linkage group E of the USDA/ARS soybean RFLP map (Kolchinsky and Gresshoff, unpublished data). Five PCR-products generated by single telomeric primers were cloned and sequenced. They contained a clear G- and T-rich strand and numerous degenerate homologies to the consensus plant telomeric repeat CCCTAAA in both orientations. It appears that telomere-related primers amplify predominantly telomere-associated sequences, an important source of molecular markers for the ends of linkage groups.

Digestion of soybean genomic DNA with several restriction nucleases revealed a common DNA fragment only 92 bp in length (Kolchinsky and Gresshoff, 1995). Sequencing of several cloned copies of the fragment revealed a new DNA satellite species (SB92), which exists in soybean in internally variant forms. About 15% sequence variation was detected for the ten copies sequenced, giving rise to the restriction sites as detected by digestion. The satellite represents 0.7% of the soybean genome and is not homologous to PFGE separated telomeric fragments. It is likely that the satellite is centromeric. An identical satellite was used in FISH to paint soybean chromosomes (P. Keim, Flagstaff, Arizona; personal communication). SB92 exists in four foci, apparently located near centromeric regions.

Genomic Analysis of Cell Cycle Genes Related to Nodulation

Nodulation requires new cell divisions in tissues and cells that are committed to other cellular fates. Many nodule-specific proteins may present the consequence of early differentiation events and may not be directly involved in the sequence of required biochemical steps leading to the functional nodule. Progressively, it is being discovered that DNA sequences equivalent to nodulins are expressed in non-symbiotic tissue, supporting the view that the symbiotic processes may be duplications and alteration of pre-existing plant capabilities.

A normal root maintains mitosis in the root apex, while pericycle cells are resting, until lateral root development occurs. Likewise, cortical cells only induce cell division after inoculation with *Bradyrhizobium*. Autoregulation of nodulation slows and terminates renewed cell divisions in the cortex, yet pericycle and apex activity continues. It was therefore important to initiate an analysis of the cell cycle related proteins. It is understood that there are many other proteins involved in the early nodulation steps, so that the analysis of cyclin B, p34 protein kinase (*cdc2* gene product), and the early nodulin enod40 merely functions as indicators of early cell cycle regulation associated with nodulation.

The induction of cell division and the maintenance of a meristem are plant processes which have been recruited into the nodulation ontogeny. Building on the data base generated from yeast cells, it was possible to isolate genomic clones homologous to the p34 protein kinase encoded by the *cdc2* gene of yeast as well as a homologue to a cyclin B like gene. The original probes for these studies were cDNA clones provided by Dr D.P.S. Verma (Ohio State University) and Dr H. Kouchi (Tsukuba, Japan). Two classes of cdc2 clones were discovered (N. Taranenko, unpublished data). One possessed an internal *Xho*I restriction site (class I), similar to the one seen in Verma's cDNA clone. Sequence analysis also demonstrated homology to the untranslated 5′ region of the cDNA (Miao *et al.*, 1993). The other clone lacked the *Xho*I site. The class II clone may represent a truncated gene. The 5′ region of both genes revealed extensive sequence similarity to the cDNA clone. Several introns were discovered in the class I clone.

Both cDNA (Hata *et al.*, 1991) and genomic clones of cyclin B from soybean (Deckert *et al.*, 1994) have been isolated. Among two cDNA clones, one (*S13-7*) was truncated at 5' end, but the second clone represented a full cDNA clone. These clones share 95% identity at both the nucleotide and amino acid level; however, the nucleotide homology was only 75% in the 3' non-coding region.

A genomic clone partially homologous to the early nodulin enod40 cDNA was isolated through PCR primers (Kruchinina and Gresshoff, unpublished data). The putative promoter site and the first part of the gene, homologous to the published cDNA sequence were found. All three genomic clones were used as an indicator of early cell division activity in soybean root associated with nodulation.

Expression analysis of cdc2, cyclinB and enod40 in different plant tissues (leaf, root, stem, nodule, expanding root zone) and different soybean symbiotic mutants (supernodulation mutant nts382 and non-nodulating mutant nod139; see Gresshoff, 1993a) revealed very similar patterns. Northern blotting of RNA from different supernodulation and non-nodulation mutants either in inoculated or uninoculated roots showed that cyclinB, cdc2 and enod40 were expressed in uninoculated roots. No expression was found in leaves or mature root regions. Inoculation with *Bradyrhizobium* increased RNA levels within 24 hours. This effect was increased and maintained at a higher level for longer in the supernodulation mutant nts382. Mutant nod139, which is characterized by normal isoflavone excretion (Mathews *et al.*, 1989c; Sutherland *et al.*, 1990) and *Bradyrhizobium* colonization and attachment, but fails to have cell divisions (Mathews *et al.*, 1989a, 1992) showed no increase in the RNA level detected with any of the cell division related probes. This demonstrated that the mutational blockage in nod139 [controlled by gene *rj6* (Mathews *et al.*, 1989b; Pracht *et al.*, 1993)] is very early in the signal recognition cascade. Genetic studies suggest that two loci are responsible for the recessive phenotypes, perhaps reflecting the ancestral genome duplication in soybean. Interestingly, some soybean varieties such as Bragg, are already homozygously recessive at the *rj5* locus, resulting in apparent single gene inheritance of the mutant *rj6* locus. Gresshoff and Landau-Ellis (1994) reported that crosses of nod139 with *G. soja* give a 15:1 segregation ratio in the F_2.

Conclusions and Perspectives

Soybean genome analysis has progressed rapidly following a hiatus. Already, 15 to 20 years ago, it was possible to characterize soybean DNA for genome complexity and repeated DNA (Gurley *et al.*, 1979). Soybean was one of the first plants for which a transposon-like element was demonstrated by molecular means (see Vodkin and Rhodes, 1985). Vodkin *et al.* (1983) found a transposable element like sequence in an inactivated lectin gene of soybean.

While the soybean genome is larger (about 1000 Mb) than that of *Arabidopsis* (120 Mb) and rice (460 Mb), it is one of the smaller genomes of legumes with agricultural potential and a scientific database. Alternative model legumes have been proposed, because of their smaller diploid genome and ease of transformation by *Agrobacterium* (see Jiang and Gresshoff, 1993 for *Lotus japonicus* review). Research with the model legumes *Lotus japonicus* and *Medicago truncatula* is progressing in several laboratories and has produced some valuable information. However, the agricultural application of seed legumes such as soybean and pea make it essential that, despite biological difficulties, molecular tools are further developed to investigate and understand mechanisms relating to plant performance, adaptation, and growth.

The construction of the first RFLP map of soybean in late 1988 by Shoemaker and associates, coupled with the pachytene karyotype elucidation (Singh and Hymowitz, 1988), and *Agrobacterium tumefaciens* and biolistic (microprojectile) transformation of soybean (Hinchee *et al.*, 1988; McCabe *et al.*, 1988) provided essential advances. In 1991, the first symbiotic gene was mapped onto the Iowa State University map, and in late 1993 the first RFLP linkage group was placed onto a pachytene chromosome by trisomic segregation analysis (Kollipara *et al.*, 1994).

While transformation of soybean is still a slow process, the recent demonstration of *A. rhizogenes* transformed roots in soybean (Bond, 1993; Cheon *et al.*, 1993; Miao and Verma, 1993) may facilitate an experimental tool to test gene regulation in root tissues. Antisense disruption of resident gene expression may allow the study of orphan genes (Cheon *et al.*, 1993).

In this context it is important to be careful as a symbiotic phenotype may be caused by the transformation process rather than the transgene in antisense configuration. Bond (1993; see also Gresshoff, 1993a) transformed roots of soybean with *Agrobacterium rhizogenes* strain K599 and found that 15% of the transgenic roots produced nodules with altered morphology and nitrogen fixing ability. Since strain K599 contains a resident Ri plasmid harbouring *rol* genes as well as the plasmid carrying the reporter and selection genes, it is likely that some roots are transformed only with the reporter plasmid (gus^+, rol^-), some only with the Ri (gus^-, rol^+), and some with both plasmids or the region within the T-DNA border sequences (gus^+, rol^+). Furthermore, we have evidence that inactivation of *gus* can occur in growing roots (Chian and Gresshoff, unpublished data). The problem is increased, as soybean does not demonstrate a clear-cut hairy root phenotype, as was observed for alfalfa (Beach and Gresshoff, 1988). It is plausible that *rol* genes slow nodule initiation and development. This may have resulted in nodules unable to fix nitrogen efficiently, possibly because the transition in nodule morphology from determinate to indeterminate, meristematic type altered temporal and spatial relationships (Gresshoff, 1993a,b; Bond *et al.*, in preparation).

It is hoped that further advances in cell culture techniques, for example, using protoplasts as well as microprojectile bombardment of immature embryos (Finer and McMullen, 1991; Bond *et al.*, 1992), will be applied to make gene transfer technology more routine for soybean. At present, gene tagging by insertion mutagenesis is still a dream for soybean researchers.

The construction of YAC clones, telomere analysis, RIL availability, satellite DNA arrangements, and the analysis of a growing number of genomic regions containing cDNA homologous sequences as described in this review will surely accelerate the analysis of nodulation and differentiation related genes.

Knowledge of the genomic nature of symbiosis controlling genes will reveal whether functional homologous genes exist in other legumes and other non-nodulated plants. The potential analysis of such genes to determine their function in a non-symbiotic root may demonstrate which genetic steps are absent in non-nodulating plants. It is possible that genes used in the nitrogen-fixing symbiosis are the evolutionary derivatives of genes used for normal cellular and organismic processes. If this is the case, it is likely that similar genes are present in other plants. The molecular genetic understanding of the differences between symbiotic and non-symbiotic plants, and the subsequent genetic modification of such functions in both legume and non-legume may be one avenue for the eventual extension of the nodulation abilities to other plants. While such goals are 'holy grails', it is significant that the research in that direction is repeatedly yielding results which are of significance to plant biology and genetics in general.

Acknowledgements

Research was supported by the Human Frontiers Science Programme, the Tennessee Soybean Promotion Board, the American Soybean Association (USB), Agrigenetics Advanced Science Company, NIH Superfund, USDA CSRS project TEN0046, NSF US-Nepal program, and the Ivan Racheff Endowment. The staff of the Racheff Chair is thanked for providing unpublished information as well as continued research enthusiasm.

References

Akao, S. and Kouchi, H. (1992) A supernodulating mutant isolated from soybean cultivar Enrei. *Soil Science and Plant Nutrition* 38, 183–187.

Ashikawa, I., Katayose, Y., Momma, T., Miyazaki, A., Saji, S. Shimizu, T., Tanoue, H., Umehara, Y., Yasukochi, Y., Yoshino, K. and Kurata, N. (1992) Physical mapping of rice genome. *Rice Genome* 1, 8–9.

Bassam, B.J., Caetano-Anollés, G. and Gresshoff, P.M. (1991) A fast and sensitive silver-staining for DNA in polyacrylamide gels. *Analytical Biochemistry* 196, 80–83.

Baum, T.J., Gresshoff, P.M., Lewis, S.A. and Dean, R.A. (1994) Characterization and phylogenetic analysis of four root-knot nematode species using DNA amplification fingerprinting and automated polyacrylamide gel electrophoresis. *Molecular Plant Microbe Interaction* 7, 39–47.

Beach, K. and Gresshoff, P.M. (1988) Characterization and culture of *Agrobacterium rhizogenes* transformed roots of forage legumes. *Plant Science* 57, 73–81.

Blackhall, N.W., Hammatt, N. and Davey, M.R. (1991) Analysis of variation in the DNA content of *Glycine* species: A flow cytometric study. *Soybean Genetics Newsletter* 18, 194–200.

Bond, J.E. (1993) Transformation as a tool to study the genetics of nodulation in *Glycine max*. PhD dissertation. The University of Tennessee, Knoxville, Tennessee.

Bond, J.E., McDonnell, R., Finer, J. and Gresshoff, P.M. (1992) Construction and use of a low cost micro-projectile gene gun for gene transfer in plants. *Tennessee Farm and Home Science* 162, 4–14.

Buzzell, R.I., Buttery, B.R. and Ablett, G. (1990) In: Gresshoff, P.M., Roth, L.E., Stacey, G. and Newton, W.E. (eds), *Nitrogen Fixation: Achievements and Objectives*. Routledge, Chapman and Hall, New York, 726 pp.

Caetano-Anollés, G. and Gresshoff, P.M. (1991) Plant genetic control of nodulation. *Annual Review of Microbiology* 45, 345–382.

Caetano-Anollés, G. and Gresshoff, P.M. (1992) Plant genetic suppression of the nonnodulation phenotype of *Rhizobium meliloti* host range *nodH* mutants: gene-for-gene interaction in the alfalfa-*Rhizobium* symbiosis? *Theoretical and Applied Genetics* 84, 624–632.

Caetano-Anollés, G. and Gresshoff, P.M. (1994) Staining nucleic acids with silver: an alternative to radioisotopic and fluorescent labeling. *Promega Notes* 45, 13–18.

Caetano-Anollés, G., Paparozzi, E.T. and Gresshoff, P.M. (1991a) Mature nodules and root tips control nodulation in soybean. *Journal of Plant Physiology* 137, 389–396.

Caetano-Anollés, G., Bassam, B.J. and Gresshoff, P.M. (1991b) DNA amplification fingerprinting using very short arbitrary oligonucleotide primers. *Bio/Technology* 9, 553–557.

Caetano-Anollés, G., Joshi, P.A. and Gresshoff, P.M. (1991c) Spontaneous nodules induce feedback suppression of nodulation in alfalfa. *Planta* 183, 77–82.

Caetano-Anollés, G., Bassam, B.J. and Gresshoff, P.M. (1992) Primer-template interactions during *in vitro* amplification with short oligonucleotides. *Molecular and General Genetics* 235, 157–165.

Caetano-Anollés, G., Bassam, B.J. and Gresshoff, P.M. (1993) Enhanced detection of polymorphic DNA by multiple arbitrary amplicon profiling of endonuclease digested DNA: identification of markers linked to the supernodulation locus of soybean. *Molecular and General Genetics* 241, 57–64.

Calvert, H.E., Pence, M.K., Pierce, M., Malik, N.S.A. and Bauer, W.D. (1984) Anatomical analysis of the development and distribution of *Rhizobium* infections in soybean roots. *Canadian Journal of Botany* 62, 2375–2383.

Carroll, B.J., McNeil, D.L. and Gresshoff, P.M. (1984) Breeding soybeans for increased nodulation in the presence of external nitrate. In: Ghai, B.S. (ed.) *Symbiotic Nitrogen Fixation* (I). USG Publishers, Ludhiana, pp. 43–50.

Carroll, B.J., McNeil, D.L. and Gresshoff, P.M. (1985a) Isolation and properties of soybean mutants which nodulate in the presence of high nitrate concentrations. *Proceedings of the National Academy of Sciences, USA* 82, 4162–4166.

Carroll, B.J., McNeil, D.L. and Gresshoff, P.M. (1985b) A supernodulation and nitrate tolerant symbiotic (nts) soybean mutant. *Plant Physiology* 78, 34–40.

Carroll, B.J., McNeil, D.L. and Gresshoff, P.M. (1986) Mutagenesis of soybean (*Glycine max* (L.) Merr.) and the isolation of non-nodulating mutants. *Plant Science* 47, 109–119.

Cheon, C.-I., Lee, N.-G., Siddique, A.-B.-M., Bal, A.K. and Verma, D.P.S. (1993) Roles of plant homologs of Rap1p and Rap7p in the biogenesis of the peribacteroid membrane, a subcellular compartment formed (*de novo*) during root nodule symbiosis. *EMBO Journal* 12, 4125–4135.

Deckert, J., Taranenko, N. and Gresshoff, P.M. (1994) Cell cycle genes and their plant homologues. In: Plant Genome Analysis. *Current Topics in Plant Molecular Biology* vol. 3, CRC Press, Boca Raton, Florida, pp. 169–195.

Delves, A.C., Carroll, B.J. and Gresshoff, P.M. (1988) Genetic analysis and complementation studies on a number of mutant supernodulating soybean lines. *Journal of Genetics* 67, 1–8.

Delves, A.C., Mathews, A., Day, D.A., Carter, A.S., Carroll, B.J. and Gresshoff, P.M. (1986) Regulation of the soybean-*Rhizobium* symbiosis by shoot and root factors. *Plant Physiology* 82, 588–590.

Duc, G. and Messager, A. (1989) Mutagenesis of pea (*Pisum sativum* L.) and the isolation of mutants for nodulation and nitrogen fixation. *Plant Science* 60, 207–213.

Finer, J.J. and McMullen, M.D. (1991) Transformation of soybean via particle bombardment of embryogenic suspension culture tissue. *InVitro Cellular and Developmental Biology* 27, 175–182.

Funke, R.P. and Kolchinsky, A. (1994) Plant yeast artificial chromosome libraries and their use: status and some strategic considerations. In: Plant Genome Analysis. (ed. P.M. Gresshoff). *Current Topics in Plant Molecular Biology* vol. 3, CRC Press, Boca Raton, 111–119.

Funke, R.P., Kolchinsky, A. and Gresshoff, P.M. (1993) Physical mapping of a region in the soybean (*Glycine max*) genome containing duplicated sequences. *Plant Molecular Biology* 22, 437–446.

Funke, R.P., Kolchinsky, A. and Gresshoff, P.M. (1994) High EDTA concentrations cause entrapment of small DNA molecules in the compression zone of pulsed field gels, resulting in smaller than expected insert sizes in YACS prepared from size selected DNA. *Nucleic Acids Research* 22, 2708–2709.

Ganal, M.W., Young, N.D. and Tanksley, S.D. (1989) Pulse field gel electrophoresis and physical mapping of large DNA fragments in the Tm-2a region of chromosome 9 in tomato. *Molecular and General Genetics* 215, 395–400.

Gibson, S.I. and Somerville, C. (1993) In: Koncz, C., Chua, N.H. and Schell, J. (eds) *Methods in* Arabidopsis *Research.* World Scientific, Singapore, pp. 119–143.

Gremaud, M.F. and Harper, J.E. (1989) Selection and initial characterization of partially nitrate tolerant nodulation mutants of soybean. *Plant Physiology* 89, 169–173.

Gresshoff, P.M. (1993a) Molecular genetic analysis of nodulation genes in soybean. *Plant Breeding Reviews* 11, 275–318.

Gresshoff, P.M. (1993b) Plant function in nodulation and nitrogen fixation in legumes. In: Palacios, R., Mora, J. and Newton, W.E. (eds) *New Horizons in Nitrogen Fixation.* Kluwer Academic Publishers, Dordrecht, The Netherlands, pp. 31–42.

Gresshoff, P.M. and Caetano-Anollés, G. (1992) Systemic regulation of nodulation in legumes. In: Gresshoff, P.M. (ed.) *Current Topics in Plant Molecular Biology.* CRC Press, Boca Raton, Florida, pp. 87–100.

Gresshoff, P.M. and Landau-Ellis, D. (1994) In: Gresshoff, P.M. (ed.) Plant Genome Analysis. *Current Topics in Plant Molecular Biology* vol. 3, CRC Press, Boca Raton, Florida, pp. 97–112.

Gresshoff, P.M. and MacKenzie, A. (1994) Low experimental variability of DNA profiles generated by arbitrary primer based amplification (DAF) of soybean. *Chinese Journal of Botany* 6, 1–6.

Gresshoff, P.M. and Rolfe, B.G. (1978) Viability of *Rhizobium* bacteroids isolated from soybean nodule protoplasts. *Planta* 142, 329–333.

Gurley, W.B., Hepburn, A.G. and Key. J.L. (1979) Sequence organization of the soybean genome. *Biochimica et Biophysica Acta* 561, 167–183.

Hata, S., Kouchi, H., Suzuka, I. and Ishi, T. (1991) Isolation and characterization of cDNA clones for plant cyclins. *EMBO Journal* 10, 2681–2688.

Hinchee, M.A.W., Conner-Ward, D.V., Newell, C.A., McDonnell, R.E., Sato, S.J., Gasser, C.S., Fischoff, D.A., Re, D.B., Fraley, R.T. and Horsch, R.B. (1988) Production of transgenic soybean plants using *Agrobacterium*-mediated DNA transfer. *Bio/Technology* 6, 915–922.

Hirsch, A., Bhuvaneswari, T.V., Torrey, J.E. and Bisseling, T. (1989) Early nodulins are induced in alfalfa root outgrowths elicited by auxin transport inhibitors. *Proceedings of the National Academy of Sciences, USA* 86, 1244–1248.

Jacobsen, E. and Feenstra, W.J. (1984) A new pea mutant with efficient nodulation in the presence of nitrate. *Plant Science Letters* 33, 337–344.

Jiang, Q. and Gresshoff, P.M. (1993) *Lotus japonicus*: a model plant for structure-function analysis in nodulation and nitrogen fixation. In: Gresshoff, P.M. (ed.) Plant Responses to the Environment. *Current Topics in Plant Molecular Biology* vol. 2, CRC Press, Boca Raton, Florida, pp. 97–110.

Joshi, P.A., Caetano-Anollés, G., Graham, E.T. and Gresshoff, P.M. (1991) Ontogeny and ultrastructure of spontaneous nodules in alfalfa (*Medicago sativum*). *Protoplasma* 162, 1–11.

Keim, P., Diers, B.W., Olson, T.C. and Shoemaker, R.C. (1990) RFLP mapping in soybean: association between marker loci and variation in quantitative traits. *Genetics* 126, 735–742.

Kolchinsky, A. and Gresshoff, P.M. (1993) Direct labeling of telomeres. *Genome* 36, 224–229.

Kolchinsky, A. and Gresshoff, P.M. (1994) Plant telomeres as molecular markers. In: Gresshoff, P.M. (ed.) *Current Topics in Plant Molecular Biology* vol. 3. CRC Press, Boca Raton, Florida, pp. 113–124.

Kolchinsky, A. and Gresshoff, P.M. (1995) A major satellite DNA of soybean is a 92-base pairs tandem repeat. *Theoretical and Applied Genetics* 90, 621–626.

Kolchinsky, A. Funke, R.P. and Gresshoff, P.M. (1993) DAF-amplified fragments can be used as markers for DNA from pulse field gels. *Biotechniques* 14, 400–403.

Kolchinsky, A., Funke, R.P. and Gresshoff, P.M. (1994) Dissecting molecular mechanisms of nodulation: taking a leaf from *Arabidopsis*. *Plant Molecular Biology* 26, 549–552.

Kolchinsky, A., Landau-Ellis, D. and Gresshoff, P.M. (1995) Genome stability of a diploidized DNA region around the supernodulation (*nts-1*) locus of soybean (in review).

Kollipara, K.P., Ahmad, F., Burridge, J.A., Xu, S.J., Singh, R.J. and Hymowitz, T. (1994) Gene mapping in soybean with primary trisomics. Abstracts. Second Biannual Conference on Cellular and Molecular Biology of Soybean. Athens, Georgia, p. 6.

Landau-Ellis, D. and Gresshoff, P.M. (1992) Supernodulating soybean mutant alleles *nts382* and *nts1007* show no recombination with the same RFLP marker supporting complementation data. *Molecular Plant Microbe Interactions* 5, 428–429.

Landau-Ellis, D. and Gresshoff, P.M. (1994) The RFLP molecular marker closely linked to the supernodulation locus of soybean contains three inserts. *Molecular Plant Microbe Interactions* 7, 423–433.

Landau-Ellis, D., Angermüller, S.A., Shoemaker, R. and Gresshoff, P.M. (1991) The genetic locus controlling supernodulation co-segregates tightly with a cloned molecular marker. *Molecular and General Genetics* 228, 221–226.

Lark, K.G., Weisemann, J.M., Mathews, B.F., Palmer, R., Chase, K. and Macalma, T. (1993) A genetic map of soybean (*Glycine max* L.) using an intraspecific cross of two cultivars: 'Minsoy' and 'Noir I'. *Theoretical and Applied Genetics* 86, 901–906.

Legocki, R. and Verma, D.P.S. (1979) A nodule-specific plant protein (Nodulin-36) from soybean. *Science* 205, 190–193.

Leyser, H.M.O., Lincoln, C.A., Timpte, C., Lammer, T. and Estelle, M. (1993) *Arabidopsis* auxin-resistance gene AXR1 encodes a protein related to ubiquitin-activating enzyme E1. *Nature* 364, 161–163.

Marcker, A., Lund. M, Jensen, E.O. and Marcker, K.A. (1984) Transcription of the soybean leghemoglobin genes during nodule development. *EMBO Journal* 3, 1691–1695.

Martin, G.B., Brommonschenkel, S.H., Chunwongse, J., Frary, A., Ganal, M.W., Spirey, R., Wu, T., Earle, E.D. and Tanksley, S.D. (1993) Map-based cloning of a protein kinase gene conferring disease resistance in tomato. *Science* 262, 1432–1435.

Mathews, A., Carroll, B.J. and Gresshoff, P.M. (1989a) Development of *Bradyrhizobium* infections in supernodulating and non-nodulating mutants of soybean (*Glycine max* (L.) Merrill). *Protoplasma* 150, 40–47.

Mathews, A., Carroll, B.J. and Gresshoff, P.M. (1989b) A new recessive gene conditioning non-nodulation in soybean. *Journal of Heredity* 80, 357–360.

Mathews, A., Kosslak, R.M., Sengupta-Gopalan, C., Appelbaum, E.R., Carroll, B.J. and Gresshoff, P.M. (1989c) Biological characterization of root exudates and extracts from nonnodulating and supernodulating soybean mutants. *Molecular Plant Microbe Interactions* 2, 283–290.

Mathews, A., Carroll, B.J., Gresshoff, P.M. (1990) The genetic interaction between non-nodulation and supernodulation in soybean: an example of developmental epistasis. *Theoretical and Applied Genetics* 79, 125–130.

Mathews, A., Carroll, B.J. and Gresshoff, P.M. (1992) Studies on the root control of non-nodulation and plant growth of non-nodulating mutants and a supernodulating mutant of soybean (*Glycine max* (L.) Merr.) *Plant Science* 83, 35–43.

McCabe, D.E., Swain, W.F., Martinell, B.J. and Christou, P. (1988) Stable transformation of soybean (*Glycine max*) by particle acceleration *Bio/Technology* 6, 923–926.

Mellor, R.B. and Rosendahl, L. (1994) A soybean peribacteroid space component, riboflavin, represses daidzein-induced common *nod*-gene expression in *Bradyrhizobium japonicum*. *Journal of Plant Physiology* 144, 34–37.

Meyer, A.D. and Mellor, R.B. (1993) NOD-active compounds in soya nodules. *Journal of Plant Physiology* 142, 57–60.

Miao, G.-H. and Verma, D.P.S. (1993) Soybean nodulin 26 gene – encoding a channel protein – is expressed only in the infected cells of nodules and is regulated differently in roots of homologous and heterologous plants. *Plant Cell* 5, 781–794.

Miao, G.-H., Hong, Z. and Verma. D.P.S. (1993) Two functional soybean genes encoding p34^{cdc2} protein kinases are regulated by different plant developmental pathways. *Proceedings of the National Academy of Sciences, USA* 90, 943–947.

Moore, G., Gale, M.D., Kurata, N and Flavell, R.B. (1993) Molecular analysis of small grain legume genomes: status and prospects. *Bio/Technology* 11, 584–585.

Nutman, P.S. (1968) Symbiotic effectiveness in nodulated red clover: V. The *n* and *d* factors for ineffectiveness. *Heredity* 23, 537–551.

Paran, I. and Michelmore, R.W. (1993) Development of reliable PCR markers linked to downy mildew resistance genes in lettuce. *Theoretical and Applied Genetics* 85, 985–993.

Park, S.J. and Buttery, B.R. (1988) Nodulation mutants of white bean (*Phaseolus vulgaris* L,) induced by ethyl-methane sulphonate. *Canadian Journal of Plant Science* 68, 199–202.

Pierce, M. and Bauer, W.D. (1993) A rapid regulatory response governing nodulation in soybean. *Plant Physiology* 73, 286–290.

Prabhu, R.R and Gresshoff, P.M. (1994) Inheritance of polymorphic markers generated by DNA amplification fingerprinting and their use as genetic markers in soybean. *Plant Molecular Biology* 26, 105–116.

Pracht, J.E., Nickell, C.D. and Harper, J.E. (1993) Rj_5 and Rj_6 genes controlling nodulation in soybean. *Crop Science* 33, 711–713.

Sanchez, F., Padilla, J.E., Pérez, H. and Lara, M. (1991) Control of nodulin genes in root-nodule development and metabolism. *Annual Review of Plant Physiology and Plant Molecular Biology* 42, 507–528.

Schwartz, D.C. and Cantor, C.R. (1984) Separation of yeast chromosome-sized DNAs by pulsed field gradient gel electrophoresis. *Cell* 37, 67–75.

Shoemaker, R.C., Guffy, R.D., Lorenzen, L. and Specht, J.E. (1992) Molecular genetic mapping of soybean: Map utilization. *Crop Science* 32, 1091–1098.

Shoemaker, R.C., Lorenzen, L.L., Diers, B.W. and Olson, T.C. (1994) Genome mapping and agriculture In: Gresshoff, P.M. (ed.) Plant Genome Analysis. *Current Topics in Plant Molecular Biology* vol. 3, chapter 1. CRC Press, Boca Raton, Florida, pp. 1–10.

Singh, R.J. and Hymowitz, T. (1988) The genomic relationship between *Glycine max* (L.) Merr. and *Glycine soja* (Sieb. and Zucc.) as revealed by pachytene chromosome analysis. *Theoretical and Applied Genetics* 76, 705–711.

Stacey, G., Burris, R.H. and Evans, H.J. (1992) *Biological Nitrogen Fixation.* Chapman and Hall, New York.

Sutherland, T.D., Bassam, B.J., Schuller, L.J. and Gresshoff, P.M. (1990) Early nodulation signals in wild type and symbiotic mutants of soybean (*Glycine max* (L.) Merr. cv. Bragg). *Molecular Plant Microbe Interactions* 3, 122–128.

Verma, D.P.S. (1992) Signals in root nodule organogenesis and endocytosis of *Rhizobium. Plant Cell* 4, 373–382.

Vodkin, L.O. and Rhodes, P.R. (1985) Movable genetic elements found in soybean. *Agricultural Research* 33, 6–7.

Vodkin, L.O., Rhodes, P.R. and Goldberg, R.B. (1983) A lectin gene insertion has the structural features of a transposable element. *Cell* 34, 1023–1031.

Weaver, K., Caetano-Anollés, G., Gresshoff, P.M. and Callahan, L.M. (1994) Isolation and cloning of DNA amplification products from silver stained polyacrylamide gels. *BioTechniques* 16, 226–227.

Welsh, J. and McClelland, M. (1990) Fingerprinting genomes using PCR with arbitrary primers. *Nucleic Acids Research* 18, 7213–7218.

Wicking, C. and Williamson, B. (1991) From linked marker to gene. *Trends in Genetics* 7, 288–290.

Williams, L.F. and Lynch, D.L. (1954) Inheritance of a non-nodulating character in the soybean. *Agronomy Journal* 46, 28–29.

Williams, J.G.K., Kubelik, A.R., Livak, K.J., Rafalski, J.A. and Tingey, S.V. (1990) DNA polymorphisms amplified by arbitrary primers are useful genetic markers. *Nucleic Acids Research* 18, 6531–6535.

Improvement of Soybean for Nitrogen Fixation: Molecular Genetics of Nodulation

10

A.J. Delauney[1] and D.P.S. Verma[2]

[1] *Department of Biology, University of the West Indies, Cave Hill Campus, Barbados:* [2] *Plant Biotechnology Center, Ohio State University, Columbus, Ohio 43210, USA.*

Introduction

Due to their ability to fix nitrogen, nodulated legume plants can thrive in soils that are completely devoid of nitrogen, and thus many cultivated grain legumes, including soybean, benefit from the *Rhizobium* symbiosis, producing nitrogen-rich grains. However, this important trait has not been well selected in breeding programmes, due largely to the fact that breeding has been carried out in nitrogen-rich soils. The growing demand for soybean in protein-rich diets is likely to increase world soybean production. This should occur by harnessing the full potential of symbiotic nitrogen fixation and not by simply adding more nitrogen fertilizer as has been the case with corn and many other crops.

Our knowledge of symbiotic nitrogen fixation has increased significantly during the last 20 years. As more information is generated on the molecular genetics of the symbiosis, it is becoming increasingly reasonable to expect that improvements in symbiotic nitrogen fixation may be achievable by genetic manipulation of the host plant and/or the endosymbiont. Attention has focused primarily on the latter due to the relative ease of genetic analysis and manipulation in bacteria. This has led to the production of rhizobia that fix nitrogen with increased efficiency due, for example, to recycling of hydrogen (which otherwise is wasted during nitrogen fixation). However, Hup$^+$ rhizobial isolates which double the relative nitrogen fixing efficiency of Hup$^-$ isolates show no significant increase in plant growth and yield (Fuhrmann, 1992). This suggests that the efficiency of nitrogen fixation by the bacteria may not be an important factor limiting nitrogen input to the host plant (Streeter, 1993).

219

The relative difficulty of genetically transforming legume plants has to some extent hindered attempts at improving symbiotic efficiency by direct genetic alteration of the host plant. However, transformation and regeneration of several legume species, including soybean, has been reported recently, and methods for the production of transgenic plants by microprojectible bombardment have been developed (Christou *et al.*, 1992; see Chapter 11). Thus, transformation is no longer a major obstacle. A more serious hindrance is the incomplete characterization of the plant genes and metabolic and developmental pathways required for an effective symbiosis. A prerequisite to the genetic manipulation of the host to improve symbiotic nitrogen fixation is the identification of essential nodule-active genes, and elucidation of their function and mode of regulation.

Symbiotic nitrogen fixation at the molecular level in soybean has been well characterized. This review focuses on key events in the establishment and maintenance of a functioning root nodule, and identifies genes and processes which should be targeted for manipulation with a view to improving symbiotic nitrogen fixation, and thus ultimately crop productivity, in soybean.

Nodule Organogenesis

Molecular signals in the initiation of nodule morphogenesis

An exchange of molecular signals (Fisher and Long, 1992; Hirsch, 1992; Verma, 1992) between the host plant and the rhizobia in the soil leads to the initiation of nodulation. This communication process begins when flavonoids exuded by the seedling root (Peters and Verma, 1990), chemotactically attract the rhizobia towards the root and also activate the *Rhizobium* nodulation (*nod*) genes. The common *nod* genes (*nod ABCIJ*) occur in all rhizobial strains and are essential for nodulation. Host-specific *nod* genes which occur in particular strains influence the efficiency of nodulation on different hosts (Fisher and Long, 1992).

Expression of the common *nod* genes leads to the synthesis of an oligosaccharide (3–5 residues) of N-acetyl-glucosamine residues which is chemically modified by the products of the host-specific *nod* genes to produce a functional Nod factor. These chemical modifications include the addition of C_{16}–C_{18} unsaturated fatty acid to the non-reducing end of the oligosaccharide, and may also involve the addition of sulphate, on the reducing end of the sugar residues (Denarie and Roche, 1992). The functional lipo-oligosaccharide Nod factors serve as the primary signal in initiating nodule morphogenesis. Following chemotaxis, rhizobia bind to the freshly emerged root hairs and within 6–18 h, the tip of the root hair undergoes marked deformations, often curling to form characteristic 'shepherd's crook' structures. When applied to compatible hosts, purified Nod factors cause root hair deformation and also elicit division of cells in the root cortex leading to formation of nodule primordia.

The large size and complexity of the Nod factor precludes a role as a diffusible signal to trigger cortical cell division; it is therefore likely that the Nod factor activates a second signal which is transmitted to the internal cortical cells. The Nar phenotype (**N**odulation in the **A**bsence of **R**hizobium) exhibited by certain alfalfa varieties (Truchet *et al.*, 1989; Caetano-Anolles *et al.*, 1992) confirms that a signal, independent of Nod factor does stimulate cortical cell division during nodule organogenesis. The nature of this signal is unknown, but it may involve changes in hormonal concentrations and transport (Libbenga and Bogers, 1974; Long and Cooper, 1988; Hirsch *et al.*, 1989), concentration of internal root flavonoids (Hirsch, 1992) and membrane potential (Ehrhardt *et al.*, 1992). The secondary signal may then act to promote mitosis in certain cortical cells presumed to be arrested in the G2 phase of cell division. However, recent observations (Yang *et al.*, 1994) suggest that the histone H4 gene is activated as early as 6 h after application of Nod factor in pea and therefore these cells may not be arrested in the G2 phase of the cell cycle. Cyclin genes, the activity of which increases in concert with cell proliferation in soybean nodules (Hata *et al.*, 1991) may also be involved in initiating cortical cell division.

Avoidance of host defence responses

The rhizobial symbiosis resembles a refined pathogenic interaction (Djordjevic *et al.*, 1987) in which the legume host selectively lowers its protective barriers to accommodate a controlled invasion by *Rhizobium*. Thus, for example, soybean roots infected with compatible *Bradyrhizobium japonicum* express a distinct subset of chalcone synthase and phenylalanine ammonia lyase enzymes that are different from the enzymes expressed as part of a defence response in pathogenic or ineffective symbiotic interactions (Estabrook and Sengupta-Gopalan, 1991). This correlates with the relatively low level of the phytoalexin, glyceollin I, produced by soybean roots in symbiotic interactions compared to pathogenic interactions (Schmidt *et al.*, 1992). Furthermore, in ineffective symbioses, the host plant can mount a vigorous chemical attack against the *Rhizobium* 'pathogen' (Werner *et al.*, 1984; 1985; Parniske *et al.*, 1991), presumably mediated by the expression of genes that are normally silenced in effective symbioses. The lysis of bacteria in the vicinity of the host cell nucleus in bacteria-induced ineffective nodules (Werner *et al.*, 1985), resembles a conditional mutation in the host imposed by reducing the expression of the *rab7* gene in soybean (Cheon *et al.*, 1993) indicating a gene-for-gene type interaction observed in many plant-pathogen interactions.

It is not known how rhizobia succeed in suppressing the host defence response. As noted by Fisher and Long (1992) the Nod factor may be evolutionarily related to chitin fragments and chemical modifications, such as acylation, acetylation, sulphation etc. of the basic oligosaccharide structure help identify rhizobia as symbionts rather than pathogens. Other features of the bacterial cell surface are involved in identifying the symbiotic nature of rhizobia. Rhizobial mutants with altered surface exo- and lipo-polysaccharides (Noel, 1992) abort nodule development at different stages. Even in nodulation by effective wild-type rhizobia, the frequency of infection events that successfully result in nodule formation is extremely low as many infection threads are aborted in the root hairs. The abortion of infection threads is apparently due to a hypersensitive response by the plant and the production of phytoalexins (Hirsch, 1992). Truche and his colleague has demonstrated histochemical accumulation of enzymes involved in phytoalexin production (Vasse et al., 1993).

Following endocytosis, rhizobia are compartmentalized within a membrane called the peribacteriod membrane (PBM) when they enter the infected cells from the infection thread. Thus the bacteria are kept out of direct contact with the host cell cytoplasm. Failure to form this membrane compartment or its premature disintegration as seen with certain bacterial mutants, renders the association pathogenic and provokes a defence reaction from the host (Werner et al., 1985). The mechanism by which rhizobia evade the host plant defences is therefore a fruitful area for further study.

Early nodulin gene expression

The nodule is a highly specialized organ in which numerous plant genes are exclusively expressed. Soon after the initial attachment of Rhizobium to the root hairs, the synthesis of several proteins, not present in uninoculated root hairs, is induced. These proteins have been detected in infected root hairs of pea (Gloudemans et al., 1989) and cowpea (Krause and Broughton, 1992), but with the exception of two early nodulins (see below), they have not yet been characterized at the genetic level. Most of the so called early nodulin genes encode (hydroxy) proline-rich proteins which are thought to be located in the cell wall on the basis of their homology to cell wall proteins (Franssen et al., 1992). In pea, two early nodulins have been ascribed functions related to the infection process. The PsENOD12 gene is expressed in root hairs, cortical cells and nodule cells containing the growing infection threads, as well as in several cell layers ahead of the infection thread where cells are preparing for infection thread penetration (Scheres et al., 1990a.). Another infection related early nodulin in pea, PsENOD5, is only present in cells containing the growing infection thread (Scheres et al., 1990b), and is thought to be a component both of the infection thread plasma membrane and the peribacteriod membrane. Both PsENOD12 and PsENOD5 are detectable in infected root hairs (Scheres et al., 1990 a,b).

Other early nodulin genes are involved in nodule morphogenesis rather than the infection process. Examples include the soybean GmENOD2 and

GmENOD13 genes which are expressed in the nodule parenchyma (van de Wiel *et al.*, 1990a; Franssen *et al.*, 1992). Due to their likely location in the cell wall, the ENOD2 and ENOD13 proteins are thought to contribute to the special morphology of the parenchyma cells that enable them to form the O_2 diffusion barrier.

The expression of some of the root hair-specific and early nodulin genes is inducible by the application of Nod factors from compatible bacteria. In pea root hairs, transcription of the RH-42 gene is stimulated by *R. leguminosarum* bv. *viciae* exudates containing Nod factors (Gloudemans *et al.*, 1989). Purified Nod factors from rhizobia also induce the expression of PsENOD12 in pea root hairs. It has previously been noted that this nodulin is expressed in cells preparing for growth of the infection thread. Although Nod factors, in the absence of live rhizobia, do not induce formation of infection threads, they do induce the formation of pre-infection thread structures (van Brussel *et al.*, 1992) which are characterized by radially aligned cytoplasmic bridges traversing the central vacuole of cells in the outer root cortex.

Biogenesis of the PBM compartment

The rhizobia enter the infected cells by endocytosis involving the budding-off of the plasma membrane at the tip of the infection thread, followed by fusion with new membrane vesicles to form the fully differentiated PBM. The term 'symbiosome' (Roth *et al.*, 1988) has been aptly applied to the unit comprising the bacteroid, the PBM and the peribacteriod space on the inside of the PBM. The term emphasizes the organellar nature of the functional unit with its resemblance to the mitochondria and choloroplasts that originated from endosymbiotic events. Enclosed in the PBM, the rhizobia are 'internalized' in the plant cell even though they remain physiologically outside the host cytoplasm (Verma 1990).

The compartmentalization of several thousands of bacteroids, each within its own PBM, in the infected cells of nodules necessitates copious membrane synthesis in these cells; in fact, infected cells of soybean nodules contain almost 30 times more membrane in the form of PBM than plasma membrane (Verma *et al.*, 1978). The presence of an intact PBM plays a critical role in mitigating host defence responses against the invading bacteria. The PBM also regulates all exchanges of metabolites between the host cell cytoplasm and the endosymbiont. These functions are mediated in part by the many nodulins located specifically in the PBM (Verma and Fortin, 1988). Considering this crucial position of the PBM as the interface between the symbionts, genetic alteration of its structure and properties may allow for improvement in symbiotic efficiency.

In order to form large quantities of PBM and to target specific nodulins to this subcellular compartment, the vesicular transport machinery of the host has to be activated and directed for synthesis of PBM. Although the vesicular transport of proteins is not completely understood, small GTP-binding proteins in yeast and mammals (Pryer *et al.*, 1992) are thought to play a key role. Homologues of these proteins have been indentified in plants (Bednarek and Raikhel, 1992; Verma *et al.*, 1994a) and the role of some of these proteins in PBM biogenesis has recently been investigated (Cheon *et al.*, 1993). The expression of soybean and moth bean *Rab1* and *Rab7* genes increased 3- and 12-fold, respectively, during nodulation, coincidentally with membrane proliferation in nodules. Reducing the expression of Rab1p in soybean nodules using antisense RNA indicated that endocytosis of bacteroids occurred normally but the proliferation of the membrane system was inhibited causing disintegration of the initially formed PBM. Many bacteroids were consequently released into the central vacuole which had persisted in the infected cells, contrary to its disappearance in infected cells of wild type nodules.

Antisense RNA-inhibition of Rab7p affected maturation of the PBM, resulting in fewer PBM compartments and accumulation of late endosome autophagic structures in infected cells, a situation found in certain ineffective nodules (Werner *et al.*, 1985). These data confirm the involvement of the Rab subfamily of small GTP-binding proteins in the vesicular transport system in PBM biogenesis.

The Functional Root Nodule

Although the bacteroids are directly responsible for reducing N_2 to NH_4^+, the legume plant contributes significantly towards the process by creating an appropriate physiological environment for the activity of the nitrogenase enzyme, delivering a high flux of photosynthate-derived carbon substractes to support bacteroid respiration, and providing the enzymatic machinery for the efficient assimilation and transport of NH_4^+. Thus, the nodule is very active in both nitrogen and carbon metabolism. The presence of abundant nitrogenous and carbon metabolites in the cytoplasm of infected cells inevitably raises its osmotic pressure, some four-fold compared to that in root cells (Verma *et al.*, 1978) and this necessitates an active osmoregulatory mechanism to maintain the metabolic activity of nodule cells (see below).

Many of the host cell metabolic functions are mediated by late nodulins and other nodule-active proteins. Numerous late nodulin genes have been cloned from soybean (Delauney and Verma, 1988). Although the functions of only a few nodulins such as leghaemoglobin, glutamine synthetase, uricase and sucrose synthase are known (see Verma and Delauney, 1988; Nap and Bisseling, 1990). Gene cloning has recently contributed greatly to identifying and characterizing several enzymes that are active in key metabolic pathways in nodules described below.

Nitrogen assimilation in nodules

NH_4 produced by the bacteroids is transferred to the host cell cytoplasm where it is assimilated via the coupled activities of glutamine synthetase (GS) and glutamate synthase (GOGAT; Atkins, 1991; Schubert, 1986). NH_4^+ is first incorporated into the amide position of glutamate forming glutamine in a reaction catalysed by GS; glutamine then reacts with 2-oxoglutarate in a reductive amination reaction catalysed by GOGAT. DNA clones for GS and GOGAT have been characterized in a variety of legumes. In alfalfa (Dunn *et al.*, 1988), kidney bean (Gebhardt *et al.*, 1986) and soybean (Sengupta-Gopalan and Pitas, 1986), nodule-specific GS genes are developmentally induced, whereas in pea nodules (Tingey *et al.*, 1987) the activity of the root-type GS gene is enhanced. In soybean, a significant part of the increase in GS activity is also due to induction of the root-type gene by NH_4^+ (Hirel *et al.*, 1987; Miao *et al.*, 1991). A nodule-specific GS cDNA has recently been cloned from *Vigna aconitifolia* (Lin *et al.*, 1995). An alfalfa nodule GOGAT cDNA has also been characterized (Gregarson *et al.*, 1993). The enzyme is not nodule-specific but its activity is enhanced in nodules.

The subsequent assimilatory route of reduced nitrogen diverges in different legumes. Legumes of temperate origin (including pea, lupin, broadbean, alfalfa and clovers) export fixed nitrogen from the nodule in the form of the amides, asparagine and, to a lesser extent, glutamine. Soybean and other legumes of tropical origin (such as kidney bean, pigeon pea, cowpea, winged bean, and moth bean) export fixed nitrogen mostly as the ureides, allantoin an allantoic acid.

Amide synthesis

Asparagine is synthesized principally by the amidation of aspartate in the glutamine-dependent asparagine synthetase reaction (Atkins, 1991; Schubert, 1986), although a limited amount of asparagine is synthesized by the direct incorporation of ammonia into the amide group. The precursor, aspartate, is synthesized from oxaloacetate (OAA) by the action of aspartate aminotransferase (AAT). The high levels of OAA required are generated by the oxidation of sugars via the TCA cycle but a significant amount is also synthesized by dark CO_2 fixation catalysed by phosphenolpyruvate carboxylase (PEPC). The activities of these three key enzymes increase with the onset of N fixation in nodules of lupin and alfalfa (Vance and Stade, 1984). These enzymes have been purified from nodules of various legumes and cDNA clones have recently been isolated (see Vance and Heichel, 1991). Nodule-specific isozymes of AAT have been detected at the protein level (Vance *et al.*, 1988) but this has not been confirmed at the gene level.

Ureide synthesis

The ureides, allantoin and allantoic acid are formed in nodules of tropical legumes by oxidative catabolism of *de novo* synthesized purine nucleotides. Accordingly, the *de novo* purine synthesis pathway is highly induced in these nodules following the commencement of nitrogen fixation. This pathway for nitrogen assimilation is considered to be more energy efficient as compared to that using amides (Schubert, 1986).

The de novo purine synthesis pathway   *De novo* purine synthesis occurs via a complex pathway: ten enzymatic steps are required to convert the activated ribose precursor, 5-phosphoribosyl-1-pyrophosphate (PRPP) to the purine nucleotide, inosine 5-monophosphate (IMP). Our understanding of this pathway is derived largely from studies on bacterial (Ebbole and Zalkin, 1987) and animal (Henikoff, 1987) systems. However, recent characterizations of nodule cDNA clones in moth bean and soybean, as well as a number of *Arabidopsis* clones, have shed light on the organization and regulation of *de novo* purine synthesis in plants. Purine synthesis cDNAs characterized to date from plants include *Arabidopsis* clones encoding 5'-phosphoribosyl-5-aminoimidazole ribonucleotide (AIR) synthetase (Senecoff and Meagher, 1993; Schnorr *et al.*, 1994), glycinamide ribonucleotide (GAR) synthetase and GAR transformylase (Schnorr *et al.*, 1994), moth bean clones encoding AIR carboxylase, 5'-amino-imidazole-4-N-succinocarboxamide ribonucleotide (SAICAR) synthetase (Chapman *et al.*, 1994) and *Vigna* glutamine PRPP amidotransferase (PRAT) as well as a soybean PRAT (Kim *et al.* 1995). All these clones have been isolated by genetic complementation of auxotrophic *Escherichia coli* mutants (see Delauney and Verma, 1990a), and it is expected that clones encoding the remaining enzymes in plants will be rapidly isolated using this or similar techniques.

   Structural analysis of these cDNAs confirm that the enzymatic steps in plants are identical to those in animals and bacteria, but there are differences in the ways the enzymes are organized. In bacteria, all the enzyme activities occur on separate polypeptides (Ebbole and Zalkin, 1987) whereas several enzyme activities are associated into different multifunctional polypeptides in animals (Henikoff, 1987; Aimi *et al.*, 1990a,b; Schild *et al.*, 1990). In plants, the enzymes so far characterized are monofunctional polypeptides except the AIR carboxylase which combines two separate activities encoded by bacterial *purE* and *purK* genes into a single polypeptide; the PurK domain is absent in the avian enzyme.

   The plant enzymes are translated with an N-terminal presequence signalling transit to chloroplasts in leaves (Senecoff and Meagher, 1993; Schnorr *et al.*, 1994) or plastids in nodules (Chapman *et al*, 1994; Kim *et al.*, 1995). This is consistent with the earlier localization of the purine enzymes in plastids (Schubert, 1986; Atkins, 1991).

Regulation of purine synthesis in nodules   Regulation of the *de novo* purine synthetic pathway is thought to be exerted at the first enzymatic step catalysed by PRAT. This enzyme has been shown to be subject to end-product inhibition in bacteria (Messenger and Zalkin, 1979), animals (Holmes *et al.*, 1973) and plants (Reynolds *et al.*, 1984). Recently, transcriptional regulation of the PRAT gene in moth bean nodules was investigated by determining RNA levels at different stages of nodule development and under different physiological conditions (Kim *et al.*, 1995). PRAT mRNA concentrations were found to increase steadily in parallel with N fixation activity as nodules matured from 13 days to 23 days. Levels of PRAT mRNA in Fix $^-$ nodules were also found to be high (approximately 60% of the levels in normal Fix $^+$ nodules) suggesting that this pathway is induced prior to and independent of nitrogen fixation.

Taken together, the data suggest that control of *de novo* purine synthesis in nodules involves regulation of PRAT activity at the transcriptional level which is responsive to both nodule developmental and metabolic cues (Kim *et al.*, 1995). Such regulation probably acts in concert with feedback regulation of the PRAT enzyme (Reynolds *et al.*, 1984).

Oxidation of purines to ureides   IMP, the end-product of the *de novo* purine synthesis pathway, is exported from the plastids and metabolized in the infected cells by cytosolic IMP dehydrogenase, 5′-nucleotidase (phosphatase) and xanthosine nucleosidase first to xanthosine monophosphate, then to xanthosine and finally xanthine (Atkins, 1991). The cellular localization of the next enzyme in the pathway, xanthine dehydrogenase (XDH), is somewhat controversial. There have been contradictory reports that XDH is located in the cytosol of infected cells (Triplett, 1985) and in the peroxisomes of uninfected cells (Nguyen, 1986). A more recent study by Triplett's group (Datta *et al.*, 1991) localized the enzyme predominantly in the cytosol of uninfected cells although substantial concentrations of XDH were also detectable in the infected cells. The data from this study suggest that xanthine, produced in the infected cells, may be transported to the uninfected cells for further catabolism. However, the presence of high concentrations of XDH in the infected cells (Triplett, 1985; Datta *et al.*, 1991) implies that xanthine may undergo conversion to uric acid by a NADH-dependent dehydrogenase, rather than an oxidase (Triplett *et al.*, 1982), avoiding a step requiring molecular O_2 which is present in low concentration in the nodule.

The next enzyme in the pathway, uricase (urate oxidase), is located in the peroxisomes of uninfected cells (Nguyen *et al.*, 1985), implying that uric acid or xanthine must be transported from the infected cells to be further metabolized in the uninfected cells. Uricase, a tetramer of 35 kDa subunits (nodulin-35), is encoded by a nodule specific gene whose expression is under evelopmental control (Nguyen *et al.*, 1985). Uric acid is oxidized to allantoin by uricase in a reaction utilizing free O_2 and this may explain the localization of uricase in the uninfected cells in an organelle containing catalase which generates O_2 from H_2O_2.

Allantoin is hydrolysed to allantoic acid by allantoinase, located in the smooth ER of the uninfected cells (Hanks *et al.*, 1981). Fixed nitrogen is transported away from the nodules via this system in the form of allantoin and allantoic acid.

Amide versus ureide export

Ureides are apparently more efficient forms of N transport than amides both in terms of the carbon requirement for skeletons in assimilating NH_4^+ and the energetics of synthesis (Schubert, 1986). It has been experimentally demonstrated that cowpea, a ureide exporter, requires as little as 1.4 g C per g N fixed, which is almost three times less than the minimum requirement in lupin, an amide exporter (Atkins, 1991). The interpretation of these calculations must be tempered by the obvious fact that the comparison involves different legumes, differing in several characteristics other than the nature of the exported N solute, such as rates of H_2 recycling by the bacteroids, nodule morphology and PEPC activity. Also, the efficiency with which N and C are subsequently utilized by the plant factors into estimates of the overall C/N budgets of ureide and amide-exporters.

Modern techniques of genetic engineering may enable the manipulation of either ureide or amide production in the same legume. This would allow more definitive comparisons of metabolic symbiotic efficiency if plants could be engineered to adopt the more favourable metabolic pathway. In a recent genetic analysis of the regulation of metabolic flux along the ureide and amide pathways in moth bean, we inhibited uricase activity using an antisense uricase gene construct (Lee *et al.*, 1993). This resulted in a reduction in ureide synthesis and the appearance of N deficiency symptoms in the plant, similar to results obtained in cowpea when XDH was inhibited by allopurinol (Triplett, 1985; Atkins *et al.*, 1988). This incapacity of ureide-exporting nodules to export fixed N as amides suggested that these nodules are developmentally controlled to export only ureides. However, it may be argued that the blocking of uricase, a terminal enzyme in ureide synthesis, would not prevent large quantities of glutamine from being funnelled into the purine and ureide synthesis pathway leading to the accumulation of uric acid. A more critical step might be to inhibit the very first committed reaction of purine synthesis catalysed by PRAT. Eliminating all metabolic flux through the *de novo* purine pathway might provide a more reliable indication of the ability of ureide exporters to utilize alternative assimilatory and transport routes. Moth bean and soybean nodule PRAT cDNA clones have recently been characterized (Kim *et al.*, 1995) and antisense-mediated inhibition of PRAT activity may shed light on the form of N solutes exported from such transgenic nodules.

Carbon metabolism in nodules

The nodule constitutes a very strong sink for carbon substrates. Accordingly, with the onset of nitrogen fixation, there is a rapid increase in sink activity (Streeter, 1980) as carbon compounds are delivered to the nodule to serve as substrates for bacteroidal respiration as well as provide skeletons for assimilation of the fixed nitrogen.

Photosynthate, transported into nodule primarily as sucrose, is used to support N fixation (Reibach and Streeter, 1983). Sucrose can be hydrolysed by two enzymes, invertase and sucrose synthase, both of which are active in legume nodules. Sucrose synthase is probably more important in sucrose hydrolysis since its activity, but not that of invertase, correlates with nitrogenase activity (Anthon and Emerich, 1990). A nodule-enhanced protein previously identified as nodulin-100 was extensively characterized and shown to be a subunit of sucrose synthase (Thummler and Verma, 1987). The soybean nodule enzyme was recently shown to be concentrated mainly in the cytoplasm of uninfected cells in the central infected region of the nodule. The enzyme is not present in high concentration in the bulk of the cortex but is relatively abundant in the vascular endodermis and in the cortical cells close to the vascular bundles (Gordon *et al.*, 1992). The cellular localization of sucrose synthase and the apparent arrangement of uninfected cells in continuous files extending from the vascular bundles towards the centre of the infected region (Gordon *et al.*, 1992) suggest that the uninfected cells may provide channels through which sucrose from the phloem is hydrolysed and translocated into the central region of the nodule. The cellular localization of sucrose synthase is consistent with the suggestion (Day and Copeland, 1991) that the production of C_4-dicarboxylic acids from sucrose (see below) occurs in the uninfected cells with the subsequent transfer of these organic acids to the infected cells. Further corroboratory evidence derives from the demonstration that infected cells isolated from soybean nodules readily take up malate but are impermeable to sucrose and poorly permeable to hexoses (Li *et al.*, 1990).

UDP-glucose and fructose produced from sucrose hydrolysis are further metabolized by glycolytic enzymes which are highly active in the nodule cytosol (Kouchi *et al.*, 1988; Copeland *et al.*, 1989). There is evidence (Vance and Rustin, 1984) that, in plants, a major branch point in glycolysis proceeds from PEP to OAA via the PEPC reaction. OAA can then enter the TCA cycle or, more likely, be reduced by malate dehydrogenase (MDH) yielding malate as a primary product of glycolysis in plants. Malate may be oxidized in plant

mitochondria via the MDH reaction, but under the microaerobic conditions of
the nodule may be further reduced to fumarate and succinate (Vance and Heichel, 1991). It has been demonstrated using labelled sucrose that the major fate
of sucrose transported to nodules is conversion to organic acids (Reibach and
Streeter, 1983).

Reductive synthesis for these dicarboxylic acids may be a major adaptation of nodule carbohydrate metabolism to the prevailing microaerobic conditions. Reductive carbon flow to malate and succinate under microaerobic and
anaerobic conditions has been noted in a variety of plants and nodules (Vance
and Heichel, 1991). Interestingly, in *Selenastrum minutum*, the anaerobic shift
to reductive pathways to C_4-dicarboxylic acids increased in response to
NH_4^+. This may be instructive to the mode of regulation of the reductive pathway in nodules.

There is now overwhelming evidence (Day and Copeland, 1991; Streeter,
1991) that C_4-dicarboxylic acids constitute the major, if not the only carbon
substrates used to fuel nitrogenase activity in the bacteroids. The bacteroids
generally have low levels of glycolytic enzymes for the catabolism of sugars
(Streeter, 1991) whereas malate is readily catabolized via bacteroidal malic enzyme and MDH (Day and Copeland, 1991); consequently, mutants unable to
utilize glucose and fructose still form effective nodules whereas mutants in-
capable of taking up malate, fumarate and succinate or deficient in some steps
of the TCA cycle, form ineffective nodules (see Streeter, 1991; Vance and Heichel, 1991). Glycolysis may be an adaptation of host metabolism to the microaerobic environment of the nodule. The preferential utilization of organic
acids, rather than sugars, as respiratory substrates by the bacteroids represent a matching bacterial adaptation to the symbiotic state.

The requirement for substantial amounts of photosynthate in the nodule
raises the question of whether N fixation is limited by supply of photosynthate. After a comprehensive review of the available, sometimes contradictory experimental data, Vance and Heichel (1991) have persuasively
argued that N fixation is limited not by the delivery of photosynthate but
rather by carbon utilization within the nodule. That excess carbohydrate is
delivered to the nodule is substantiated by the fact that starch accumulates
to 'luxury' levels in nodule cells and is metabolized only when external re-
serves are depleted. Similarly, the accumulation of large quantities of poly-β-
hydroxybutyrate by bacteroids suggest that carbon substrates delivered to
them exceed their respiratory capacity in a microaerobic environment (Day
and Copeland, 1991).

It has become clear that a major factor limiting carbon utilization in nodules is the low concentration of O_2 available which restricts oxidative catabolism in both host mitochondria (Rawsthorne and LaRue, 1986; Suganuma
et al., 1987) and bacteroids (McDermott *et al.*, 1989). However, since a low O_2
environment is essential for the activity of the O_2-labile nitrogenase enzyme,
the modifications to carbon utilization pathways operative in both the host

cytoplasm and the bacteroids may be a necessity for the symbiotic state even if these pathways are energetically inefficient.

Osmoregulation in nodules

Symbiotic nitrogen fixation has long been known to be very sensitive to the imposition of water stress. The regulation of osmotic potential gradients within the nodule has been proposed as a mechanism controlling O_2 permeability (Purcell and Sinclair, 1994), and O_2 concentration is in turn a major modulator of nodule metabolism and N fixation. In a recent thought-provoking review (Streeter, 1993), the availability of water for export of nitrogenous solutes from the nodules was proposed as a critical factor limiting the efficiency of symbiotic N fixation. Though this hypothesis has not been directly substantiated by experimental data, there is suggestive evidence that water relations in the nodule play a pivotal role in regulating nodule function and efficiency.

Nodule cells contain high concentrations of sugars, amino acids and organic acids as a result of the active C and N metabolic pathways operating in these cells. The abundance of these osmotically active metabolites in both the infected and uninfected cells of the central region of the nodule raise the osmoticum in the infected cells four- to five-fold higher than in root cells (Verma *et al.*, 1978) and is expected to impose osmotic stress on the nodule cells. Thus, quite distinct from the imposition of drought and salinity stresses which nodules may occasionally experience as a result of climatic conditions, it is likely that osmotic stress is a normal condition with which infected cells in the nodule have to contend. Accordingly, the active employment of osmoregulatory mechanisms may be necessary for optimum functioning of this tissue.

The suggestion that the bacteroids are subject to osmotic stress is given credence by their active synthesis and accumulation of α-α-trehalose (Streeter, 1985), a known osmoprotectant in a variety of organisms (Crowe *et al.*, 1984). The induction of trehalose synthesis may also be related to low O_2 concentration (see Hoelzle and Streeter, 1990), further indicative of a link between osmoregulation and the concentration of O_2 in the nodule.

Many plans respond to osmotic stress by accumulating organic osmolytes, the most common being proline and glycine betaine (see Delauney and Verma, 1993). Soybean nodules are very active in proline biosynthesis and in particular, the activity of pyrroline-5-carboxylate reductase (P5CR), the enzyme responsible for converting P5C to proline was found to be much higher in nodules than in most other animal and plant tissues (Kohl *et al.*, 1988). This led to the suggestion that the elevated P5CR activity in nodules is used to generate the high levels of $NADP^+$ required for the synthesis of the purine precursor ribose-5-phosphate, via the pentose phosphate pathway. The validity of this proposal is debatable since a special role for P5CR in regulating ureide synthesis was not substantiated by experimental data (Kohl *et al.*, 1990). On the other hand, exogenous proline supplied to nodulated soybean roots was found to stimulate acetylene reduction activity to the same extent as did succinate, consistent with a role for proline as an energy source for nitrogen fixation (Zhu *et al.*, 1992).

In view of the widespread use of proline as an osmoregulatory solute in plants (Delauney and Verma, 1993), and given the likelihood that certain nodule cells may be continuously subjected to osmotic stress, it seems plausible that a crucial role of proline synthesis is in fact an involvement in osmoregulation. Indeed, imposition of salt or drought stress on nodulated alfalfa and soybean plants caused a marked accumulation of proline in nodules (Fougére *et al.*, 1991; Kohl *et al.*, 1991) reflecting an osmoregulatory role for proline in nodule tissue.

cDNA clones encoding the two enzymes involved in converting glutamate to proline, P5C synthetase (P5CS; Hu *et al.*, 1992) and P5CR (Delauney and Verma, 1990b), have been isolated from legume nodules and characterized. RNAs for both enzymes are inducible by exposure of uninoculated roots to salt stress, consistent with an osmoregulatory function for proline in nodules.

It would be desirable to gain a better understanding of osmoregulation in nodules in view of the indications that the availability of water affects O_2 diffusion in nodules (Purcell and Sinclair, 1994) and may constrain the efficiency of symbiotic nitrogen fixation (Streeter, 1993). This may be achieved by overexpression of P5CS under the control of the leghaemoglobin gene promoter.

Regulation of Nodule-active Genes

Regulated plant genes may generally be categorized as either tissue-specific or inducible. This distinction, however, becomes somewhat blurred when applied to genes that are expressed in nodules since the nodule is itself an inducible organ; hence, all nodule-active genes are ultimately inducible by rhizobia. Nevertheless, it has become common practice to distinguish between nodule-specific (nodulin) genes (Legocki and Verma, 1980) whose expression, responsive to rhizobial or nodule developmental signals, is exclusive to the nodule, and inducible genes whose expression, though enhanced in nodules, is also detectable in roots and other tissues.

Multiple signals are involved in regulating gene expression in nodules. The inducible genes are obviously activated by different metabolites or conditions while among the nodulin genes, the marked variation in temporal activation of the early and late genes immediately suggests the involvement of at least two different sets of signals. In fact, since the classification of nodulin genes into the broad 'early' and 'late' subgroups, suggestive of two timepoints for the activation of nodulin genes, is an oversimplification, there are clearly additional signals operating. Within each of these subgroups, there are differences in the temporal induction of particular genes as well as in the nature of the signals eliciting their expression (Verma and Delauney, 1988; Scheres *et al.*, 1990b).

Regulation of gene expression in nodules also involves the specific silencing of certain genes that are active in roots. Candidate genes in this category include those whose products constitute part of the plant's defence against pathogens, and genes encoding enzymes which perform specialized root functions that are not required in the nodule or which might even be disruptive to the unique balance of metabolic flux in the nodule. This phenomenon has not been subject to detailed investigation and very little is known about the underlying mechanisms.

Signals involved in regulation of plant gene expression in nodules

Bacterial factors

As discussed on pp. 220–221, the Nod factors, when applied to compatible legume roots, are able to form nodule-like structures. Accordingly, some of the early nodulin genes are induced in root hairs (Gloudemans *et al.*, 1989; Scheres *et al.*, 1990a,b). For this close-range gene activation in the root hairs, the Nod factor itself may be the main signal eliciting gene expression. During infection, expression of the same PsENOD12 gene in nodule primordia and root cortex cells involves long-range activation by the rhizobia (Scheres *et al.*, 1990a); in this case, secondary plant signals, almost certainly involving plant hormones (see Hirsch, 1992), may be the inducers. The likely involvement of phytohormones is supported by findings that early nodulin genes are expressed in pseudonodules elicited by auxin transport inhibitors (Hirsch *et al.*, 1989; van de Wiel *et al.*, 1990a) and cytokinins (Long and Cooper, 1988; Dehio and de Bruijn, 1992). Similarly, a cloned cytokinin (zeatin) biosynthesis gene partially complemented a *nod* ABC $^-$ *R. meliloti* mutant, resulting in empty alfalfa nodules in which MsENOD2 was expressed (Cooper and Long, 1994).

The Nod factors are not the only bacterial components involved in activating gene expression in the host plant. This is borne out by the demonstration that *R. leguminosarum* bv. *viciae* exudate containing Nod factors stimulates expression of the root hair-specific RH-44 gene but not that of the RH-42 gene (Gloudemans *et al.*, 1989) implying that rhizobial components, other than Nod factors, are required for RH-42 gene expression.

The *Rhizobium* extracellular matrix contains acidic exopolysaccharides (EPS), lipopolysaccharides (LPS) and cyclic β-1,2-glucans encoded by the *exo*, *lps* and *ndv* genes, respectively (see Hirsch, 1992). Mutations in these genes frequently cause the formation of empty nodules which either lack infection threads, contain infection threads that abort prematurely, or are defective in bacterial release from the infection threads and thus contain no intracellular bacteria (Long, 1989; Noel, 1992). Expression of early nodulin genes and different subsets of late nodulin genes has been detected in such empty nodules (Dickstein et al., 1988; van de Wiel et al., 1990a,b; Verma and Delauney, 1988). Nap and Bisseling (1990) have suggested that the products of the exo, lps and ndv genes may act as avoidance determinants which enable Rhizobium to avoid provoking defensive responses in the host plant. Mutations that unmask such determinants result in the plant's defence mechanisms being activated and nodule development aborted. This view is supported by evidence that infection threat abortion and premature disintegration of the symbiosome membrane are associated with hypersensitive-like responses by the plant (Puhler et al., 1991; Werner et al., 1985). By contrast, the results of Battisti et al. (1992) suggest that EPS may function as a specific elicitor of certain steps in nodule development. These workers showed that low molecular weight succinoglycan (an acidic exopolysaccharide), applied to alfalfa roots 24 h prior to inoculation with R. meliloti exo mutants, restored nodule invasion by these mutants.

Nodule developmental signals

The use of numerous Rhizobium mutants that elicit nodule-like structures, blocked at various stages of development, has suggested that nodule organogenesis occurs in a series of phases, each characterized by the expression of different subsets of nodulin genes (see Verma and Delauney, 1988; Franssen et al., 1992). Although, as indicated in the previous section, the invading rhizobia may produce many signals that regulate host gene expression, it is possible that plant gene expression is also responsive to signals generated according to a nodule developmental programme within the plant itself. However, until such plant organogenetic signals are identified and characterized, it is difficult to distinguish them from primary signals originating directly from the bacteria.

Nodule cell metabolites

The highly active carbon and nitrogen metabolic pathways in nodules produce high concentrations of metabolites that could conceivably serve as inducers of genes whose products are required for the utilization of these substances. The best characterized example involves the stimulation of GS gene expression by NH_4^+ in soybean nodules. Fusion of soybean GS gene promoter with the β-glucuronidase (GUS) reporter gene led to root-specific expression of GUS in transgenic Lotus corniculatus and tobacco plants (Miao et al., 1991). Treatment with NH_4^+ increased GUS expression in L. corniculatus roots but not in tobacco suggesting that legumes may contain specific transcriptional factors that mediate induction of this gene by NH_4^+. Because a reduction in n-GS activity in Phaseolus nodules can be correlated with both high and low nitrogenase activities, Sanchez *et al.* (1991) suggest that GS expression may be modulated not by NH_4^+ availability *per se* but by the carbon/nitrogen balance within the nodule. However, our data (Hirel *et al.*, 1987; Miao *et al.*, 1991) suggest that one of the soybean GS genes expressed in nodules is directly regulated by NH_4^+ supplied exogenously or from symbiotic nitrogen fixation. Induction is affected specifically by NH_4^+ as other sources of fixed nitrogen, i.e. glutamine, asparagine and nitrate, did not stimulate GS-GUS expression in transgenic *Lotus* nodules (Miao *et al.*, 1991).

The induction of GS activity by NH_4^+ is not difficult to rationalize as NH_4^+ is quite cytotoxic, and since copious amounts of NH_4^+ are secreted by the bacteroids into the cytosol of the infected cell, it is essential that this NH_{4+} is efficiently assimilated. GS is the main enzyme involved in the assimilation of NH_4^+ and thus NH_4^+-enhanced GS expression ensures that NH_4^+ does not accumulate to toxic levels. NH_4^+ might also be involved in stimulating the transcription of another gene involved in its assimilation. We have shown that the PRAT gene which encodes the rate-limiting enzyme in *de novo* purine synthesis (see p. 227) is inducible by NH_4^+ (Kim *et al.*, 1995), though in this case, glutamine, an immediate precursor of purine synthesis, is a more effective inducer.

Another nodule metabolite which may be involved in regulation of gene activity is haem. Haem is the prosthetic group of leghaemoglobin (Lb), supplied by the bacterium. Thus a *R. meliloti hemA* mutant, defective in the first step of haem biosynthesis, induces Fix⁻ nodules on alfalfa (Leong *et al.*, 1982). Haem may also regulate (reduce) the activity of sucrose synthase in soybean nodules at the enzyme level (Thummler and Verma, 1987) but an involvement in the regulation of sucrose synthase gene expression has not been investigated.

The majority of late nodulin genes are induced just prior to the onset of nitrogen fixation and are also expressed in nodules elicited by Fix⁻ *Rhizobium* mutants, suggesting that their expression is responsive to rhizobial or developmental signals independent of nitrogenase activity (Verma and Delauney, 1988). Frequently, however, the levels of nodulin gene expression in Fix⁻ nodules are reduced to varying degrees. This may be an indication that although the products of nitrogen fixation do not serve as the primary signals for activating nodulin gene expression (except in a few cases such as the soybean GS gene), they may be required for full induction or for maintaining the expression of these genes. Alternatively, the generally reduced levels of gene expression in Fix⁻ nodules may simply be a manifestation of the pleiotropic effects of an ineffective symbiosis in which the bacterium is essentially a parasite.

Physiological factors

Cells in the infection zone of the nodule are unusual, if not unique, in containing an extremely low concentration of free oxygen and a very high osmoticum (see pp. 230–231). By analogy with osmoregulated and microaerobically regulated genes in a variety of organisms, it might be expected that these physiological conditions are involved in regulating gene expression in the nodule. Expression of the *nif* and *fix* genes in the bacteroids is known to be regulated by O_2 via a complex cascade system (dePhilip *et al.*, 1990; Gilles-Gonzalez *et al.*, 1991). However, in a direct investigation of nodulin gene induction by low O_2 concentration Govers *et al.* (1985) were unable to detect the activation of pea nodulin genes by microaerobiosis. Nevertheless, there is evidence that the activities of purine nucleosidase and uricase in soybean callus increase under microaerobic conditions (Larsen and Jochimsen, 1987). Furthermore, sucrose synthase genes in maize are inducible by anaerobiosis, raising the possibility that the sucrose synthase (nodulin-100) gene in nodules may be similarly regulated.

In view of the importance of osmoregulation in the nodule, it is conceivable that the high osmoticum in some cells may be involved in the regulation of gene activity. Although this possibility has not been fully investigated, there are indications that the proline biosynthesis genes in the nodule may be regulated by osmotic potential. Consistent with this view, soybean genes encoding the enzymes, P5CS and P5CR, that catalyse the synthesis of proline from glutamate are inducible by exposure to high salt conditions (Delauney and Verma, 1990; Hu *et al.*, 1992).

Improvement of Soybean Symbiosis by Genetic Manipulation of the Host

Although the functions of most nodulin genes remain to be ascertained and many processes in the nodule are incompletely understood, our knowledge of the hosts' contribution to the symbiosis at the genetic and biochemical level has increased significantly over the last decade. In particular, a comprehensive picture of the key pathways of carbon and nitrogen metabolism in the nodule is now available. With the genes encoding several of the enzymes in these pathways having been cloned, it is now possible to contemplate genetic manipulation of the major metabolic pathways in the nodule with a view to increasing the efficiency of symbiotic nitrogen fixation. Moreover, as the genetics of *Rhizobium* has advanced, it is now possible to judicially select host/ *Rhizobium* combinations for optimum symbiosis under specific soil conditions. The latter also plays an important role in optimization of this process. Thus, a three-way solution is needed for developing specific germplasm and a suitable inoculum that yields a high nitrogen-fixing soybean crop in a particular agroclimatic condition.

Increasing carbon delivery and catabolism in the nodule

The high demand for carbon substrates in the nodule raises the possibility that the supply of photosynthate to this organ may be inadequate to maximally support bacteroidal respiration and provide skeletons for the assimilation of fixed nitrogen, in which case increasing the delivery of sucrose to the nodule through genetic engineering may result in increased nitrogen fixation activity.

The factors which govern the partitioning of photoassimilate to different sink organs are not fully understood. Nevertheless, it may be feasible to alter the sink activity of the nodule by genetic manipulation of its sucrose hydrolytic activities (cf. Sonnewald *et al.*, 1994). However, before embarking on such attempts at genetic engineering, the question of whether symbiotic nitrogen fixation is indeed constrained by the supply of photosynthate needs to be critically examined. Although it has been suggested that the availability of photosynthetically derived carbon substrates may limit N fixation, a preponderance of the experimental evidence indicates that legume nodules do not generally have excess N fixation capacity that can be activated by increasing carbon supply to the nodule (Vance and Heichel, 1991).

If carbon delivery to the nodule exceeds demand, might other aspects of carbon metabolism constrain nodule function? It has been suggested that limitations on the conversion of sucrose to C_4-dicarboxylic acids and the exchange of metabolites across the symbiosome membrane may reduce the efficiency of nitrogen fixation (Vance and Heichel, 1991). Reductive synthesis of dicarboxylic acids occurs in the absence of 'normal' mitochondrial function and is therefore inefficient in terms of energy generation. Indeed, Vance and Heichel (1991) suggest that the large carbon requirement of symbiosis may be due to limited production of ATP by the fermentative pathways utilized for C_4-dicarboxylate synthesis. However, attempts to redirect dicarboxylate synthesis through energetically efficient oxidative mitochondrial pathways by genetic engineering are likely to be nonproductive since the reductive pathways appear to be necessitated by the microaerobic environment of the nodule.

There is some evidence pointing to the role of the symbiosome membrane in regulating symbiotic efficiency: peanut nodules formed by *Rhizobium* strain 32H1 were shown to have much higher nitrogen fixing activity than cowpea nodules elicited by the same strain (Sen and Weaver, 1980). A striking structural difference between the two types of nodules is that the peanut symbiosomes contain single bacteroids whereas cowpea symbiosomes contain multiple bacteroids, suggesting that the symbiosome membrane surface area per bacteroid may limit the availability of respiratory substrates to the bacteroids (Streeter, 1991). However, the evidence is not very compelling in view of other significant differences between peanut and cowpea, in particular peanut nodules being predominantly asparagine exporters whereas cowpea nodules are ureide exporters (Atkins, 1991).

Stimulation of nitrogen assimilation

It may be possible to increase symbiotic efficiency by genetically enhancing the assimilation of NH_4^+ by the host plant or increasing the export of the assimilation products from the nodule. The latter would require increasing the translocation of solutes into the xylem, a complex process which may be dependent on the import of water in the phloem and the general availability of water in the nodule (Streeter, 1993). Furthermore, the water status of the nodule appears to be directly involved in regulating O_2 permeability in the nodule (Streeter, 1993; Purcell and Sinclair, 1994) which in turn impacts on multiple aspects of nodule metabolism. This complex interdependence of physiological and metabolic processes is far from being fully understood, and thus enhancement of solute export is unlikely to be amenable to genetic manipulation in the immediate future.

Pathways leading to the assimilation of fixed nitrogen in tropical legumes have been elucidated and several genes encoding key enzymes in these pathways have recently been cloned. This may allow metabolic engineering strategies aimed at altering the assimilatory pathways of NH_4^+ leading to increased symbiotic efficiency. Confidence in the efficacy of this approach derives from studies in which exposure of alfalfa and soybean nodules to the GS inhibitor, tabtoxinine-β-lactam, induced an approximate doubling in N fixation in nodules and overall plant growth (Knight and Langston-Unkefer, 1988). The precise mechanism underlying these phenomena is not clear but it does suggest that alteration of the N-assimilatory pathways in nodules provides a possible means of enhancing the efficiency of nodule function.

Key enzymes in N assimilation in tropical legumes which might be targeted for genetic engineering include GS and PRAT. Since GS is already a highly expressed enzyme in nodules, comprising as much as 2% of total cytosolic protein, there may be limited scope for engineering substantial increases in GS concentration, though based on the data of Knight and Langston-Unkefer (1988), a decrease in root GS activity and an increase in the nodule form of GS may be desirable. This could be achieved by inhibition of GS activity using specific antisense GS constructs. Increases in PRAT activity with a view to stimulating purine and ureide synthesis in nodules may be readily achievable by overexpression of the PRAT gene in nodule plastids or targeting the PRAT gene product to plastids. Given the elaborate cellular and organelle compartmentalization of the *de novo* purine and ureide synthetic pathway, it is likely that additional enzymes in different organelles also play ancillary regulatory roles, and such enzymes are additional potential targets for genetic engineering. The identity of these additional enzymes is likely to be revealed as more information is generated on the regulation of purine and ureide synthesis.

Enhancement of osmotolerance

As proposed above (pp. 231–232), some nodule cells are continuously subjected to osmotic stress as is evidenced by their adoption of active osmoregulatory measures, and genetic engineering of increased osmotolerance in these nodule cells can be expected to result in more efficient cell function. Presently, the most promising strategies to genetically engineer greater osmoprotection in plant cells are focused on achieving the accumulation of solutes such as glycine betaine (McCue and Hanson, 1990), polyols (Tarczynski *et al.*, 1993) and proline (Delauney and Verma, 1993). Genes encoding the rate-limiting enzymes in the appropriate biosynthetic pathways have been cloned and preliminary experiments have already demonstrated limited increases in osmotolerance using these strategies (Tarczynski *et al.*, 1993; Kavi Kishore *et al.*, 1995). To be applied effectively in the nodule, it will probably be necessary to target osmolyte accumulation to a particular cell type, most likely to the infected cells.

A close interplay of osmotic potential gradients and O_2 permeability in the nodule occurs (Purcell and Sinclair, 1994). Thus, any alteration of osmotic gradients is likely to affect O_2 availability which in turn has widespread effects on nodule metabolism, not least the activity of the nitrogenase enzyme itself. The multiple pleiotropic effects of altering osmotic relationships within the nodule make it extremely difficult to predict the net impact of such adjustments. This is, however, an area requiring active investigation as a possible route for the enhancement of symbiotic efficiency.

Concluding Remarks

Currently available tools of plant biotechnology afford considerable scope of manipulation of plant gene expression and cellular metabolism. Recent progress in our comprehension of nodule organogenesis and function, and in the characterization of important nodule-active genes has enabled various metabolic pathways to be targeted for genetic manipulation. Thus, in the legume–*Rhizobium* symbiosis, there is potential for genetic engineering of the host plant as a means of improving the efficiency of symbiotic nitrogen fixation.

The desire to realize this objective is, in many cases, constrained by our incomplete understanding of the relevant physiological and metabolic processes and their interdependence. We have attempted within the confines of available information, to identify processes whose manipulation might lead to improvement of the host symbiont. Our analysis suggests that the delivery of photosynthate to the nodule, carbon metabolism in that organ and transport to the bacteroids are not suitable targets for genetic manipulation. On the other hand, stimulation of the assimilation of fixed N appears to be a more promising approach towards enhancing symbiotic efficiency, although there are uncertainties about the location of the rate-limiting steps in the sequence of processes between NH_4^+ production and the export of nitrogenous solutes from the nodule. Genetically enhancing osmoprotective mechanisms in the nodules may lead to enhanced nitrogen fixation and crop yield, particularly under drought and salinity stress.

Genetic manipulation of O_2 concentration in the nodule has not been discussed in detail because our knowledge of the multiple mechanisms underlying the regulation of O_2 levels is very incomplete. However, it is abundantly clear that many processes in the nodule are O_2-limited and several metabolic pathways apparently reflect adaptations to the microaerobic environment of the nodule. A full understanding of the effects of low O_2 concentration on nodule metabolism will undoubtedly help identify factors limiting nodule efficiency. Antagonistic demands for O_2 lie at the heart of nodule biochemistry: the bacteroids have a high O_2 requirement for aerobic respiration whereas activity of the O_2-labile nitrogenase enzyme requires a microaerobic milieu. The necessity for precise control over O_2 concentration is therefore obvious. When the complexity of the control mechanisms is unravelled, it may be possible to perform sophisticated manipulations whereby increased levels of O_2 are supplied to the O_2-limited machinery of the nodule while still ensuring the protection of the nitrogenase enzyme. This may ultimately afford the greatest scope for enhancing the efficiency of symbiotic N fixation.

Finally, simple genetic selection of host–microsymbiont combinations to produce germplasm able to provide high yield in nitrogen-deficient soils is a worthwhile pursuit which has not received deserved attention. In this regard it is interesting to note that super-nodulating plant mutants, isolated using such selection, do not necessarily perform better but are instructive in dissecting other rate-limiting step(s) in this process.

Acknowledgements

Recent work in Verma's laboratory was supported by grants from the NSF and the USDA.

References

Aimi, J., Badylak, J., Williams, J., Chen, Z., Zalkin, H. and Dixon, J. (1990a) Cloning of a cDNA encoding adenylosuccinate lyase by functional complementation in *Escherichia coli. Journal of Biological Chemistry* 265, 9011–9014.

Aimi, J., Qiu, H., Williams, J., Zalkin, H. and Dixon, J. (1990b) *De novo* purine nucleotide biosynthesis: Cloning of human and avian cDNAs encoding the trifunctional glycinamide ribonucleotide synthetase–aminoimidazole ribonucleotide synthetase–glycinamide ribonucleotide transformylase by functional complementation in *E. coli. Nucleic Acids Research* 18, 6665–6672.

Anthon, G. and Emerich, D. (1990) Developmental regulation of enzymes of sucrose and hexose metabolism in effective and ineffective soybean nodules. *Plant Physiology* 92, 346–351.

Atkins, C. (1991) Ammonia assimilation and export of nitrogen from the legume nodule. In: Dilworth M.D. and Glenn, A. (eds) *Biology and Biochemistry of Nitrogen Fixation.* Elsevier, Amsterdam, pp. 293–319.

Atkins, C., Shelp, B., Stover, P. and Pate, J. (1984) Nitrogen nutrition and the development of biochemical functions associated with nitrogen fixation and ammonia assimilation of nodules on cowpea seedlings. *Planta* 162, 327–333.

Atkins, C., Storer, P. and Pate, J. (1988) Pathways of nitrogen assimilation in cowpea nodules studied using $^{15}N_2$ and allopurinol. *Plant Physiology* 86, 204–207.

Battisti, L., Lara, J.C. and Leigh, J.A. (1992) Specific oligosaccharide form of the *Rhizobium meliloti* exopolysaccharide promotes nodule invasion in alfalfa. *Proceedings of the National Academy of Sciences, USA* 89, 5625–5629.

Bednarek, S.Y. and Raikhel, N.V. (1992) Intracellular trafficking of secretory proteins. *Plant Molecular Biology* 20, 133–150.

Caetano-Anollés, G., Joshi, P. and Gresshoff, P.M. (1992) Nodulation in the absence of *Rhizobium.* In: Gresshoff, P.M. (ed.) *Current Topics in Plant Molecular Biology.* CRC Press, Boca Raton, Florida, pp. 61–70.

Chapman, K., Delauney, A., Kim, J. and Verma, D.P.S. (1994) Structural organization of *de novo* purine biosynthesis enzyme in plants: 5-aminoimidozole ribonucleotide carboxylase and 5-aminoimidazole-4-N-succinocarboxamide ribonucleotide synthetase cDNAs from *Vigna aconitifolia. Plant Molecular Biology* 24, 389–395.

Cheon, C.-I., Lee, N.-G., Siddique, A.-B.M., Bal, A.K. and Verma. D.P.S. (1993) Roles of plant homologs of Rab1p and Rab7p in the biogenesis of the peribacteroid membrane, a subcellular compartment formed *de novo* during root nodule symbiosis. *EMBO Journal* 12, 4125–4135.

Christou, P., McCabe, D., Swain, W. and Russell, D. (1992) Legume transformation. In: Verma, D.P.S. (ed.) *Control of Plant Gene Expression.* CRC Press, Boca Raton, Florida, pp. 547–564.

Cooper, J.B., Long, S.R. (1994) Morphogenetic rescue of *Rhizobium meliloti* nodulation mutants by trans-zeatin secretion. *Plant Cell* 6, 215–225.

Copeland, L., Vella, J. and Hong, Z.Q. (1989) Enzymes of carbohydrate metabolism in soybean nodules. *Phytochemistry* 28, 57–61.

Crowe, J., Crowe, L. and Chapman, D. (1984) Preservation of membranes in anhydrobiotic organisms: The role of trehalose. *Science* 223, 701–703.

Datta, D.B., Triplett, E.W. and Newcomb, E.H. (1991) Localization of xanthine dehydrogenase in cowpea root nodules – implications for the interaction between cellular compartments during ureide biogenesis. *Proceedings of the National Academy of Sciences, USA* 88, 4700–4702.

Day, D. and Copeland, L. (1991) Carbon metabolism and compartmentation in nitrogen-fixing legume nodules. *Plant Physiology and Biochemistry* 29, 185–201.

Dehio, C. and de Bruijn, F.J. (1992) The early nodulin gene *SrEnod2* from *Sesbania rostrata* is inducible by cytokinin, *Plant Journal* 2, 117–128.

Delauney, A.J. and Verma, D.P.S. (1988) Cloned nodulin genes for symbiotic nitrogen fixation. *Plant Molecular Biology Reporter* 6, 279–285.

Delauney, A.J. and Verma, D.P.S. (1990a) Isolation of plant genes by heterologous complementation in *Escherichia coli*. In: Schilperoort, R.A., Gelvin, S.B. and Verma, D.P.S. (eds) *Plant Molecular Biology Manual*. Kluwer, Dordrecht, pp. A14: 1–23.

Delauney, A.J. and Verma, D.P.S. (1990b) A soybean gene encoding pyrroline-5-carboxylate reductase was isolated by functional complementation in *Escherichia coli* and is found to be osmoregulated. *Molecular and General Genetics* 211, 299–305.

Delauney, A.J. and Verma, D.P.S. (1993) Proline biosynthesis and osmoregulation in plants. *Plant Journal* 4, 215–223.

Denarie, J. and Roche, P. (1992) *Rhizobium* nodulation signals. In: Verma, D.P.S. (ed.) *Molecular Signals in Plant-Microbe Interactions*. CRC Press, Boca Raton, Florida, pp. 295–324.

dePhilip, P., Batut, J. and Boistard, P. (1990) *Rhizobium meliloti* FixL is an oxygen sensor and regulates *R. meliloti nifA* and *fixK* genes differently in *Escherichia coli*. *Journal of Bacteriology* 172, 4255–4262.

Dickstein, K., Bisseling, T., Reinhold, V.N. and Ausubel, F.M. (1988) Expression of nodule-specific genes in alfalfa root nodules blocked at an early stage of development. *Genes and Development* 2, 677–687.

Djordjevic, M., Gabriel, D. and Rolfe, B. (1987) *Rhizobium* – the refined parasite of legumes. *Annual Review of Phytopathology* 25, 145–168.

Dunn, K., Dickstein, R., Feinbaum, R., Burnett, B.K., Peterman, T.K., Thoids, G., Goodman, H.M. and Ausubel, F.M. (1988) Developmental regulation of nodule-specific genes in alfalfa root nodules. *Molecular Plant-Microbe Interactions* 1, 66–75.

Ebbole, D. and Zalkin, H. (1987) Cloning and characterization of a 12 gene cluster from *Bacillus subtilis* encoding nine enzymes for *de novo* purine nucleotide synthesis. *Journal of Biological Chemistry* 262, 8274–8587.

Ehrhardt, D.W., Atkinson, E.M. and Long, S.R. (1992) Depolarization of alfalfa root hair membrane potential by *Rhizobium meliloti* Nod factors. *Science* 256, 998–1000.

Estabrook, E.M. and Sengupta-Gopalan, C. (1991) Differential expression of phenylalanine ammonia-lyase and chalcone synthase during soybean nodule development. *Plant Cell* 3, 299–308.

Fisher, R. and Long, S. (1992) *Rhizobium* plant signal exchange. *Nature* 357, 655–660.

Fougere, F., Rudulier, D. and Streeter, J. (1991) Effects of salt stress on amino acid, organic acid and carbohydrate composition of roots bacteroids and cytosol of alfalfa. *Medicago sativa* L. *Plant Physiology* 96, 1228–1236.

Franssen, H., Vijn, I., Yang, W.C. and Bisseling, T. (1992) Developmental aspects of the *Rhizobium*-legume symbiosis. *Plant Molecular Biology* 19, 89–107.

Fuhrmann, J. (1992) Symbiotic effectiveness of indigenous soybean bradyrhizobia as related to serological, morphological, rhizobitoxine, and hydrogenase phenotypes. *Applied and Environmental Microbiology* 56, 224–229.

Gebhardt, C., Oliver, J.E., Forde, B.G., Saarelainen, R. and Miflin, B.J. (1986) Primary structure and differential expression of glutamine synthetase genes in nodules, roots and leaves of *Phaseolus vulgaris*. *EMBO Journal* 5, 1429–1435.

Gilles-Gonzales, M.A., Ditta, G.S. and Helinski, D.R. (1991) A haemoprotein with kinase activity encoded by the oxygen sensor of *Rhizobium meliloti*. *Nature* 150, 170–172.

Gloudemans, T., Bhuvaneswari, T., Moerman, M., Brussel, T., Kammen, A. and Bisseling, T. (1989) Involvement of *Rhizobium leguminosarum* nodulation genes in gene expression in pea root hairs. *Plant Molecular Biology* 12, 157–167.

Gordon, A.,Thomas, B. and Reynolds, P. (1992) Localization of sucrose synthase in soybean root nodules. *New Phytologist* 122, 35–44.

Govers, F., Goudemans, T., Moerman, M., Kammen, A. and Bisseling, T. (1985) Expression of plant genes during the development of pea root nodules. *EMBO Journal* 4, 861–867.

Gregarson, R.G., Miller, S.S., Twary, S.N., Gantt, J.S. and Vance, C.P. (1993) Molecular characterization of NADH-dependent glutamate synthase from alfalfa nodules. *Plant Cell* 5, 215–226.

Hanks, J., Tolbert, N. and Schubert, K. (1981) Localization of enzymes of ureide biosynthesis in peroxisomes and microsomes of nodules. *Plant Physiology* 68, 65–69.

Hata, S., Kouchi, H., Suzuka. I. and Ishii, T. (1991) Isolation and characterization of cDNA clones for plant cyclins. *EMBO Journal* 10, 2681–2688.

Henikoff, S. (1987) Multifunctional polypeptides for *de novo* purine synthesis, *Bio Essays* 6, 8–13.

Hirel, B., Bouet, C., King, B., Layzell, D., Jacobs, F. and Verma, D.P.S. (1987) Glutamine synthetase genes are regulated by ammonia provided externally or by symbiotic nitrogen fixation. *EMBO Journal* 6, 1167–1171.

Hirsch, A. (1992) Developmental biology of legume nodulation. *New Phytologist* 122, 211–237.

Hirsch. A., Bhuvaaneswari,T.,Torrey, J. and Bisseling,T. (1989) Early nodulin genes are induced in alfalfa root outgrowths elicited by auxin transport inhibitors. *Proceedings of the National Academy of Sciences, USA* 86, 1244–1248.

Hoelzle, I. and Streeter, J. (1990) Increased accumulation of trehalose in rhizobia cultured under 1% oxygen. *Applied and Environmental Microbiology* 90, 3213–3215.

Holmes, E., Mcdonald, J., Mccord, J.,Wyngaardm, J. and Kelley,W. (1973) Human glutamine phosphoribosylpyrophosphate amidotransferase. *Journal of Biological Chemistry* 248, 144–150.

Hu, C.-A.A., Delauney, A.J. and Verma, D.P.S. (1992) A bifunctional enzyme (Δ^1-pyrroline-5-carboxylate synthetase) catalyzes the first two steps in proline biosynthesis in plants. *Proceedings of the National Academy of Sciences, USA* 89, 9354–9358.

Krause, A. and Broughton,W.J. (1992) Proteins associated with root-hair deformation and nodule initiation in *Vigna unguiculata*. *Molecular Plant–Microbe Interactions* 5, 96–103.

Kavi Kishore, P., Hong, Z., Miao, G.-H., Hu, C.-A. and Verma, D.P.S. (1995) Overexpression of Δ^1-pyrroline-5-carboxylate synthetase increases proline production and confers osmotolerance in transgenic plants 1089, 1387–1394.

Kim, J.H., Delauney, A. and Verma, D.P.S. (1995) Control of *de novo* purine biosynthesis genes in ureide-producing legumes: Induction of glutamine phosphoribosylpyrophosphate amidotransferase gene and characterization of its cDNA from soybean and *Vigna*. *Plant Journal* 7, 77–86.

Knight, T. and Langston-Unkefer, P. (1988) Enhancement of symbiotic dinitrogen fixation by a toxin-releasing plant pathogen *Science* 241, 951–954.

Kohl, D., Schubert, K., Carter, M., Hagendorn, C. and Shearer, G. (1988) Proline metabolism in N_2-fixing root nodules: Energy transfer and regulation of purine synthesis. *Proceedings of the National Academy of Sciences, USA* 85, 2036–2040.

Kohl, D., Lin, J., Shearer, G. and Schubert, K. (1990) Activities of the pentose phosphate pathway and enzymes of proline metabolism in legume root nodules. *Plant Physiology* 94, 1258–1264.

Kohl, D., Kennelly, J., Zhu, Y., Schubert, K. and Shearer, G. (1991) Proline accumulation, nitrogenase C_2H_2 reducing activity and activities of enzymes related to proline metabolism in drought stressed soybean nodules. *Journal of Experimental Botany* 42, 831–837.

Kouchi, H., Fukai, K., Katagiri, H., Minamisawa, K. and Tajima, S. (1988) Isolation and enzymological characterization of infected and uninfected cell protoplasts from root nodules of *Glycine max*. *Physiologica Plantarum* 73, 372–334.

Lance, C. and Rustin, P. (1984) The central role of malate in plant metabolism. *Physiologie Vegetale* 22, 625–641.

Larsen, K. and Jochimsen, B.U. (1987) Appearance of purine-catabolizing enzymes in *fix*[+] and *fix*[–] root nodules on soybean and effect of oxygen on the expression of the enzymes in callus tissue. *Plant Physiology* 85, 452–456.

Lee, N., Stein, B., Suzuki, H. and Verma, D.P.S. (1993) Expression of antisense nodulin-35 RNA in *Vigna aconitifolia* transgenic root nodules retards peroxisome development and affects nitrogen availability to the plant. *Plant Journal* 3, 599–606.

Legocki, R. and Verma, D. (1980) Identification of nodule-specific host proteins (nodulins) involved in the development of *Rhizobium*-legume symbiosis. *Cell* 20, 153–163.

Leong, S., Ditta, G. and Helinski, D. (1982) Heme biosynthesis in *Rhizobium*. *Journal of Biological Chemistry* 257, 8724–8730.

Li, Y., Ou Yang L-J., Quinnel, R.G., Udvardi, M.K. and Day, D.A. (1990) Malate transport and metabolism in infected cells of soybean nodules. In: Gresshoff, P.M., Roth, L.E., Stacey, G. and Newton, W.E. (eds) *Nitrogen Fixation: Achievements and Objectives*. Chapman and Hall, New York, p. 753.

Libbenga, K. and Bogers, R. (1974) Root-nodule morphogenesis. In: Quispel, A. (ed.) *The Biology of Nitrogen Fixation*. North-Holland Publishing Co., Amsterdam, pp. 430–472.

Lin, Z., Miao, G.-H and Verma, D.P.S. (1994) A cDNA sequence encoding glutamine synthetase is preferentially expressed in nodules of *Vigna aconitifolia*. *Plant Physiology* 107, 279–280.

Long, S.R. and Cooper, J. (1988) Overview of symbiosis. In: Palacios, R. and Verma, D.P.S. (eds) *Molecular Genetics of Plant-Microbe Interaction*. APS Press, Minnesota.

Long, S.R. (1989) *Rhizobium*-legume nodulation: life together in the underground. *Cell* 56, 203–214.

McCue, K. and Hanson, A. (1990) Drought and salt tolerance towards understanding and application. *Trends in Biotechnology* 8, 358–362.

McDermott, T., Griffith, S., Vance, C. and Graham, P. (1989) Carbon metabolism in

Bradyrhizobium japonicum bacteroids. *FEBS Microbiological Reviews* 63, 327–340.

Messenger, L. and Zalkin, H. (1979) Glutamine phosphosphoribosylpyrophosphate amidotransferase from *Escherichia coli*. *Journal of Biological Chemistry* 254, 382–392.

Miao, G.-H., Hirel, B., Marsolier, M.C., Ridge, R.W. and Verma, D.P.S. (1991) Ammonia-regulated expression of a soybean gene encoding cytosolic glutamine synthetase in transgenic *Lotus corniculatus*. *Plant Cell* 3, 11–22.

Nap, J.-P. and Bisseling, T. (1990) The roots of nodulins. *Physiologia Plantarum* 79, 407–414.

Nguyen, J. (1986) Plant xanthine dehydrogenase: Its distribution, properties and function. *Physiologie Vegetale* 24, 263–281.

Nguyen, T., Zelechowska, M., Foster, V., Bergmann, H. and Verma, D.P.S. (1985) Primary structure of the soybean nodulin-35 gene encoding uricase II localized in the peroxisomes of uninfected cells of nodules. *Proceedings of the National Academy of Sciences, USA* 82, 5040–5044.

Noel, K.D. (1992) Rhizobial polysaccharides required in symbioses with legumes. In: Verma, D.P.S. (ed.) *Molecular Signals in Plant-Microbe Communications*. CRC Press, Boca Raton, Florida, pp. 341–357.

Parniske, M., Ahlborn, B. and Werner, D. (1991) Isoflavonoid-inducible resistance to the phytoalexin glyceollin in soybean rhizobia. *Journal of Bacteriology* 173, 3432–3439.

Peters, N. and Verma, D. (1990) Phenolic compounds as regulators of gene expression in plant-microbe interaction. *Molecular Plant–Microbe Interactions* 3, 4–8.

Pryer, N.K., Wuestenhube, L. and Schekman, R. (1992) Vesicle-mediated protein sorting. *Annual Review of Biochemistry* 61, 471–516.

Puhler, A., Arnold, W., Buendia-Claveria, A., Kapp, D., Keller, M., Niehaus, K., Quant, J., Roxlau, A. and Weng, W. (1991) The role of *Rhizobium meliloti* exopolysaccharides EPSI and EPSII in the infection process of alfalfa nodules. In: Hennecke, H. and Verma, D.P.S. (eds) *Advances in Molecular Genetics of Plant-Microbe Interactions*. Kluwer Academic Publishers, Dordrecht, pp. 189–194.

Purcell, L. and Sinclair, T.R. (1994) An osmotic hypothesis for the regulation of oxygen permeability in soybean nodules. *Plant Cell Environment* 17, 837–843.

Rawsthorne, S. and LaRue, T. (1986) Preparation and properties of mitochondria from cowpea nodules. *Plant Physiology* 81, 1092–1096.

Reibach, P. and Streeter, J. (1983) Metabolism of ^{14}C-labeled photosynthate and distribution of enzymes of glucose metabolism in soybean nodules. *Plant Physiology* 72, 634–640.

Reynolds, P., Blevins, D., Randall, D. (1984) 5-Phosphoriboslypyrrophosphate amidotransferase from soybean root nodules: Kinetic and regulatory properties. *Biochimica et Biophysica Acta* 229, 623–631.

Roth, L., Jeon, K. and Stacey, G. (1988) Homology in endosymbiotic systems: The term 'symbiosome'. In: Palacios, R. and Verma, D.P.S. (eds) *Molecular Genetics of Plant-Microbe Interactions*. The American Phytopathological Society Press, St. Paul, Minnesota, pp. 220–225.

Sanchez, F., Padilla, J.E., Perez, H. and Lara, M. (1991) Control of nodulin genes in rootnoule development and metabolism. *Annual Review of Plant Physiology* 42, 507–528.

Scheres, B., van de Wiel, C., Zalensky, A., Horvath, B., Spaink, H., van Eck, H., Zwartkruis, F., Wolters, A., Gloudemans, T., van Kammen, A. and Bisseling, T. (1990a) The ENOD12 gene product is involved in the infection process during the pea rhizobium interaction. *Cell* 60, 281–294.

Scheres, B., van Englen, F., van der Knaap, E., van de Wiel, C., van Kammen, A. and Bisseling, T. (1990b) Sequential induction of nodulin gene expression in the developing pea nodule. *Plant Cell* 2, 687–700.

Schild, D., Brake, A., Kiefer, M., Young, D. and Barr, P. (1990) Cloning of three human multifunctional *de novo* purine biosynthetic genes by functional complementation of yeast mutations. *Proceedings of the National Academy of Sciences, USA* 7, 2916–2920.

Schmidt, P.E., Parniske, M. and Werner, D. (1992) Production of the phytoalexin glyceollin-I by soybean roots in response to symbiotic and pathogenic infection. *Botanica Acta* 105, 18–25.

Schnorr, K., Nygaard, P. and Laloue, M. (1994) Molecular characterization of *Arabidopsis thaliana* cDNAs encoding three purine biosynthetic enzymes. *Plant Journal* 6, 113–121.

Schubert, K. (1986) Products of biological nitrogen fixation in higher plants: synthesis, transport, and metabolism. *Annual Review of Plant Physiology* 37, 539–574.

Scott, D. and Farnden, K. (1976) Amonia assimilation in lupin nodules. *Nature* 263, 703–705.

Sen, D. and Weaver, R. (1980) Nitrogen fixing activity of rhizobial strain 32H1 in peanut and cowpea nodules. *Plant Science Letters* 18, 315–318.

Senecoff, J. and Meagher, R. (1993) Isolating the *Arabidopsis thaliana* genes for *de novo* purine synthesis by suppression of *Escherichia coli* mutants. *Plant Physiology* 102, 387–399.

Sengupta-Gopalan, S. and Pitas, J. (1986) Expression of nodule-specific glutamine synthetase genes during nodule development in soybeans. *Plant Molecular Biology* 7, 189–199.

Sonnewald, U., Lerchl, J., Zrenner, R. and Frommer, W. (1994) Manipulation of sink-source relations in transgenic plants. *Plant Cell Environment* 17, 649–658.

Sprent, J.I. (1981) Functional evolution in some papilionoid root nodules. In: Polhill, R.M. and Raven, P.H. (eds) *Advances in Legume Systematics*. Royal Botanical Gardens, Kew, pp. 671–676.

Streeter, J. (1993) Translocation – a key factor limiting the efficiency of nitrogen fixation in legume nodules. *Physiologia Plantarum* 87, 616–623.

Streeter, J.G. (1980) Carbohydrates in soybean nodules. II Distribution of compounds in seedlings during the onset of nitrogen fixation. *Plant Physiology* 66, 471–476.

Streeter, J.G. (1985) Accumulation of a α-trehalose by *Rhizobium* bacteria and bacteroids. *Journal of Bacteriology* 164, 78–84.

Streeter, J.G. (1991) Transport and metabolism of carbon and nitrogen in legume nodules. *Advances in Botanical Research* 18, 129–187.

Suganuma, N., Kitou, M. and Yamaoto, Y. (1987) Carbon metabolism in relation to cellular organization of soybean root nodules and respiration of mitochondria aided by leghemoglobin. *Plant Cell Physiology* 28, 113–122.

Tarczynski, M., Jensen, R. and Bohnert, H. (1993) Stress protection of transgenic tobacco by production of the osmolyte mannitol. *Science* 259, 508–510.

Thummler, F. and Verma, D. (1987) Nodulin-100 of soybean is the subunit of sucrose synthase regulated by the availability of free heme in nodules. *Journal of Biological Chemistry* 262, 14730–14736.

Tingey, S.V., Walker, E.L. and Coruzzi, G.M. (1987) Glutamine synthetase genes of pea encode distinct polypeptides which are differentially expressed in leaves, roots and nodules. *EMBO Journal* 6, 1–9.

Triplett, E.W. (1985) Intracellular nodule localization and nodule specificity of xanthine dehydrogenase in soybean. *Plant Physiology* 77, 1004–1009.

Truchet, G., Barker, D.G., Camut, S., Debilly, F., Vasse, J. and Huguet, T. (1989) Alfalfa nodulation in the absence of *Rhizobium*. *Molecular and General Genetics* 219, 65–68.

Van Brussel, A.A.N., Bakhuizen, R., van Spronsen, P.C., Spaink, H.P., Tak, T., Lugtenberg, B.J.J. and Kijne, J.W. (1992) Induction of pre-infection thread structures in the leguminous host plant by mitogenic lipo-oligosaccharides of *Rhizobium*. *Science* 257, 70–72.

Van de Wiel, C., Scheres, B., Franssen, H., Vanlierop, M.J., Vanlammeren, A., Vankammen, A. and Bisseling, T. (1990a) The early nodulin transcript ENOD2 is located in the nodule parenchyma (inner cortex) of pea and soybean root nodules. *EMBO Journal* 9, 1–7.

Van de Wiel, C., Norris, J., Bochenek, B., Dickstein, R., Bisseling, T. and Hirsch, A. (1990b) Nodulin gene expression and ENOD1 localization in effective nitrogen fixing and ineffective bacteria free nodules of alfalfa. *Plant Cell* 2, 1009–1017.

Vance, C.P. and Stade, S. (1984) Alfalfa root nodule carbon dioxide fixation II. Partial purification and characterization of root nodule phosphoenolpyruvate carboxylase, *Plant Physiology* 75, 261–264.

Vance, C., Griffith, S., Miller, S. and Egli, M. (1988) Plant regulated aspects of nodulation and nitrogen fixation. *Plant Cell Environment* 11, 413–427.

Vance, C. and Heichel, G. (1991) Carbon in N_2 fixation limitations or exquisite adaptation. *Annual Reviews of Plant Physiology and Plant Molecular Biology* 42, 373–392.

Vasse, J., De Billy, F. and Truchet, G. (1993) Abortion of infection during the *Rhizobium meliloti*-alfalfa symbiotic interaction is accompanied by a hypersensitive reaction. *Plant Journal* 4, 555–566.

Verma, D.P.S. (1990) Endosymbiosis of *Rhizobium*: Internalization of the extracellular compartment and metabolite exchange. In: Gresshoff, P. and Steacy, G. (eds) *Nitrogen Fixation: Achievements and Objectives*. Chapman and Hall, New York, pp. 235–237.

Verma, D. (1992) Signals in root nodule organogenesis and endocytosis of *Rhizobium*. *Plant Cell* 4, 373–382.

Verma, D.P.S. and Delauney, A. (1988) Root nodule symbiosis: Nodulins and nodulin genes. In: Verma, D.P.S. and Goldberg, R.B. (eds) *Temporal and Spatial Regulation of Plant Genes*. Springer-Verlag, Vienna, pp. 169–199.

Verma, D.P.S. and Fortin, M.G. (1988) Nodule development and formation of the endosymbiotic compartment. In: Vasil, I.K. (ed.) *The Molecular Biology of Nuclear Genes*. Academic Press, New York, pp. 329–353.

Verma, D., Kazain, V., Zogbie, V. and Bal, A. (1978) Isolation and characterization of the membrane envelope enclosing the bacteroids in soybean root nodules. *Journal of Cell Biology* 78, 919–939.

Verma, D.P.S., Cheon, C.-I. and Hong, Z. (1994) Small GTP-binding proteins and membrane biogenesis in plants. *Plant Physiology* 106, 1–6.

Werner, D., Morschel, E., Korat, R., Meilor, R. and Bassarab, S. (1984) Lysis of bacteroids in the vicinity of the host cell nucleus in an ineffective (fix ⁻) root nodule of soybean (*Glycine max*). *Planta* 16, 8–16.

Werner, D., Mellor, R., Hahn, M. and Grisbach, H. (1985) Soybean root response to symbiotic infection glyceollin I accumulation in an ineffective type of soybean nodules with an early loss of the peribacteroid membrane. *Zeitschrift für Naturforschung* 40, 179–181.

Yang, W.C., de Blank, C., Meskiene, I., Hirt, H., Bakker, J., van Kammen, A., Franssen, H. and Bisseling, T. (1994) Rhizobium nod factors reactivate the cell cycle during infection and nodule primordia formation, but the cell cycle is only completed in primordium formation. *Plant Cell* 6, 1415–1426.

Zhu, Y., Shearer, G. and Kohl, D. (1992) Proline fed to intact soybean plants influences acetylene reducing activity and content and metabolism of proline in bacteroids. *Plant Physiology* 98, 1020–1028.

Soybean Transformation: Technologies and Progress

J.J. Finer[1], T.-S. Cheng[1][*] and D.P.S. Verma[2]

[1] *Horticulture and Crop Sciences Department, Ohio Agricultural Research and Development Center, Ohio State University, Wooster, Ohio 44691, USA:* [2] *Plant Biotechnology Center, Ohio State University, Columbus, Ohio 43210, USA.*

Introduction

Since the first report of transformation and plant regeneration of tobacco (Horsch *et al.*, 1985), it was assumed that transformation of all plants would become routine in a few years. Unfortunately, soybean and many other crop plants have provided some of the greatest challenges to transformation efforts. Although soybean transformation has been reported in 1988 by two independent laboratories using two completely different protocols (Hinchee *et al.*, 1988; McCabe *et al.*, 1988), soybean transformation is far from routine in most laboratories. In fact, very few laboratories have even been able to repeat the work that was described in these first reports. This is not to say that these reports were incomplete; rather, other laboratories working in soybean transformation have simply not put the required efforts into properly setting up and executing soybean transformation experiments. The inefficiencies, variability and various other problems associated with soybean transformation have necessitated a refinement of the existing systems and the development of new methodologies for gene introduction in soybean.

Success with soybean transformation requires an understanding of both tissue culture and transformation processes. Although soybean tissue culture is the subject of Chapter 6, it will be briefly reviewed here as it relates to transformation. The following portions of this chapter will cover the various transformation systems and prospects for development of new systems.

---

* Present address: National Tainan Teachers' College, Tainan, Taiwan, Republic of China.

Soybean Tissue Culture

Without doubt, the single factor that has limited transformation of soybean is the response of soybean to tissue culture manipulations. For transformation to be efficient and successful, DNA must be introduced into cells that give rise to plants or germ-line tissue. In many cases, it is difficult to target these tissues. In other cases, DNA can be introduced into regenerable and responsive soybean cells, but lack of fertility often occurs in the regenerated plants. In order to control tissue culture-related limitations, plant regeneration in soybean should first be understood.

Soybean can regenerate via the following two distinct processes; shoot morphogenesis and somatic embryogenesis (Barwale *et al.*, 1986). Shoot morphogenesis is the process of shoot meristem organization and development. Shoots grow out from a source tissue and are excised and rooted to obtain an intact plant. During somatic embryogenesis, an embryo (similar to the zygotic embryo), containing both shoot and root axes, is formed from somatic plant tissue (tissue other than germ-line). An intact plant rather than a rooted shoot results from the germination of the somatic embryo. Shoot morphogenesis and somatic embryogenesis are very different processes and the specific route of regeneration is primarily dependent on the explant source and media used for tissue culture manipulations. Although the systems are different, some commonalities do exist. Both systems show cultivar-specific responses where some lines are more responsive to tissue culture manipulations than others. A line that is highly responsive in shoot morphogenesis may not generate many somatic embryos. Lines that produce large numbers of embryos during an 'induction' step may not give rise to rapidly-growing proliferative cultures. Essentially, the optimum tissue culture conditions should be defined for each soybean line, regardless of the method of regeneration. In addition to line-specific responses, proliferative cultures can be observed with both systems. Proliferation is beneficial for both systems as it allows a single, transformed cell to multiply to the point that it will contribute to germ-line tissue. To gain an understanding of the benefits and limitations of each system, shoot morphogenesis and somatic embryogenesis in soybean first need to be defined.

Shoot morphogenesis was first reported by Wright *et al.* (1986). They described a system whereby shoots were obtained *de novo* from cotyledonary nodes of soybean seedlings. The shoot meristems were formed subepidermally and morphogenic tissue could proliferate on a medium containing benzyl adenine (BA). This 'cot node' system can be used for transformation if the subepidermal, multicellular origin of the shoots is recognized and proliferative cultures are utilized. The idea here is to target the tissue that will give rise to the new shoots and proliferate those cells within the meristematic tissue to lessen problems associated with chimerism. Tissue targeting can be difficult as the shoot meristem-forming cells are not very accessible. Formation of chimeras, resulting from transformation of only a single cell in a meristem, are problematic if the transformed cell is not adequately proliferated and does not give rise to germ-line tissue. Once the cot node system is well understood and reproduced satisfactorily, it can be used as one target tissue for soybean transformation.

Somatic embryogenesis in soybean was first reported by Christianson *et al.* (1983). They described a system in which embryogenic tissue was initially obtained from the zygotic embryo axis. These embryogenic cultures were proliferative but the repeatability of the system was low and the origin of the embryos was not reported. Later histological studies of a different proliferative embryogenic soybean culture showed that proliferative embryos were of apical or surface origin with a small number of cells contributing to embryo formation (Finer, 1988). In this study, cells on the apical surface of older embryos gave rise to new embryo initials. However, soybean somatic embryos may not always be surface-derived. The origin of primary embryos (the first embryos derived from the initial explant) is dependent on the explant tissue and the auxin levels in the induction medium (Hartweck *et al.*, 1988). With proliferative embryonic cultures, single cells or small groups of surface cells of the 'older' somatic embryos form the 'newer' embryos. Embryogenic cultures can also be used successfully for transformation if the origin of the embryos is recognized and the biological limitations of proliferative embryogenic cultures are understood. Biological limitations include the difficulty in developing proliferative embryogenic cultures and reduced fertility problems (culture-induced variation) associated with plants regenerated from long-term proliferative embryogenic cultures. Some of these problems are accentuated in prolonged cultures.

Transformation Methods: Overview

Genes can be introduced into soybean as well as any other plant, using either biological or physical methods. While *Agrobacterium* provides the only biological vector for transformation of soybean, numerous physical methods for DNA introduction exist.

In *Agrobacterium*-mediated transformation, the bacterium is inoculated directly on the responsive tissues and a portion of its DNA is transferred to the genome of the target cells. In order for this biological vector to be useful, the bacterium must deliver the DNA to the proper regenerable target tissues and must then be eliminated. The main problems with use of *Agrobacterium* are tissue and host incompatibilities. Although soybean is a suitable host for *Agrobacterium*, it is clearly not as responsive to infection and subsequent DNA transfer as many other dicot plants. In addition, embryos and embryogenic tissues are typically not very susceptible to *Agrobacterium* infection. Host and tissue-type incompatibilities between *Agrobacterium* and soybean have been partially overcome by using modified highly-virulent *Agrobacterium* strains (Hansen *et al.*, 1994) and acetosyringone: an inducer of the DNA transfer process (Stachel *et al.*, 1985).

Physical methods for DNA introduction include particle bombardment (Sanford, 1988), protoplast electroporation (Shillito *et al.*, 1985), electroporation of intact tissue (D'Halluin *et al.*, 1992), and silicon carbide whiskers (Kaeppler *et al.*, 1990). Several other physical methods exist but these have not shown good consistency in tissue response and efficiency. Of these physical methods for DNA delivery, only particle bombardment has led to the consistent recovery of transgenic plants and progeny (McCabe *et al.*, 1988; Finer and McMullen, 1991). The main problems with the physical methods is that DNA must be specifically targeted to cells that are both transformation- and regeneration-competent, and the DNA integration patterns in transformed tissue can be quite complex. *Agrobacterium*-mediated transformation, on the other hand, is relatively inexpensive and results in more controlled integration events. At present, the choice of the transformation method for soybean must ultimately be based on the efficiency of transformation as efficiencies are still quite low. In the end, the soybean transformation system that will be most commonly used may be a modification of an existing method, a completely new method, or a hybrid between two or more systems (e.g. particle bombardment with *Agrobacterium*; Bidney *et al.*, 1992).

Soybean Transformation via *Agrobacterium*

Large efforts have been made to develop an efficient *Agrobacterium tumefaciens*-mediated transformation system for genetic improvement of soybean (for recent reviews see Christou, 1991, 1992). The first successful transformation experiments in soybean utilized wild-type *Agrobacterium*, which resulted in tumour formation directly on soybean plants (Pedersen *et al.*, 1983). In 1987, Baldes *et al.* (1987) reported transformation of soybean following cocultivation of soybean protoplast with *Agrobacterium*. Although soybean transformation was demonstrated in these first two reports, transgenic plants were not recovered. The first transgenic soybean plants were recovered using a disarmed *Agrobacterium* strain (Hinchee *et al.*, 1988). Cotyledonary node explants from the cultivar Peking were inoculated with *Agrobacterium* which conferred kanamycin and glyphosate resistance and expressed β-glucuronidase (GUS). After inoculation, the explants were placed on a medium containing BA, which induced shoot morphogenesis, and kanamycin for selection for kanamycin resistance. A few months later, 6% of the selected regenerated plantlets tested positive for either GUS expression and/or glyphosate tolerance. Genetic analysis of progeny produced from two of these plants demonstrated cosegregation of kanamycin resistance and either GUS expression or glyphosate tolerance in a 3:1 ratio. Therefore, this transformation resulted in insertion of introduced DNA in a single site, displaying Mendelian inheritance.

This landmark discovery in soybean transformation confirmed that soybean plans can be transformed using *Agrobacterium* if a cultivar is utilized (Peking in this case) that is both susceptible to infection and responsive to tissue culture manipulations. The three most critical parameters in developing this soybean transformation protocol were: (i) use of a cultivar susceptible to *Agrobacterium* infection; (ii) development of the regeneration system from soybean cotyledons; and (iii) enriching for transformed tissue by kanamycin selection. A modification to this *Agrobacterium*-mediated transformation was reported (Townsend, personal communication) where a commercial soybean line was transformed. The main improvements in transformation efficiency came from acetosyringone application, proper wounding of the target tissue, and refinement of the kanamycin selection system. Hopefully, full details of this procedure will be made available to others working in the area. The target tissue for all of this transformation work was *de novo* shoots that lie below the surface of this tissue. *Agrobacterium* worked in these systems because of its ability to move somewhat and target certain cells within the plant.

Transformation of germinating seeds of soybean with *Agrobacterium* has also been reported (Chee *et al.*, 1989). Inoculation of the plumule, cotyledonary node, and adjacent cotyledonary tissues of germinating soybean seeds with *Agrobacterium* conferring kanamycin resistance generated sixteen soybean plants that showed some expression of the introduced DNA. This gave a primary transformation frequency of about 0.7%. Unfortunately, only one tenth of these primary transgenic plants yielded progeny that were also transgenic. Southern hybridization analysis and PCR were used to demonstrate transformation events from the primary transformants and progeny. *Agrobacterium*-mediated germinating seedling transformation does avoid the requirement for tissue culture manipulations but the inefficiency of the system makes it impractical for most applications.

Transgenic primary soybean plants have also been recovered through *Agrobacterium*-mediated transformation of immature cotyledon tissues (Parrott *et al.*, 1989). *Agrobacterium*, which contained a binary plasmid encoding a 15 kD zein gene under regulatory control of the phaseolin promoter, was cocultivated with immature cotyledons of soybean. With this cotyledon culture system, embryos and not shoots were obtained. Three transgenic plants were recovered but the progeny from all of these plants did not contain the introduced DNAs, indicating that the primary transformed plants were apparently all chimeric. As the origin of primary embryos in soybean can be subepidermal and multicellular (Hartweck *et al.*, 1988), one would expect to obtain chimeric plantlets using this system. Possibly, the use of proliferative embryogenic cultures, where the embryo originates from a single epidermal cell, would be beneficial for *Agrobacterium*-mediated transformation work.

Agrobacterium rhizogenes-*mediated transformation*

One of the approaches to obtain transgenic tissue from soybean that avoids regeneration problems is to use *Agrobacterium rhizogenes*-induced hairy roots. These roots develop normally and can be used for nodulation and nitrogen fixation studies. The chimeric plants containing transgenic roots and wild type shoots can be easily obtained. Cheon *et al.* (1993) recently used this system successfully and demonstrated that it is a good experimental tool to study expression of nodule-expressed plant genes. In this case, a binary vector containing a foreign gene in tandem with a reporter gene was introduced in wild type *A. rhizogenes* strain and hairy roots were developed on hypocotyls of young seedlings. These roots were monitored using GUS assay, and all wild type hairy roots as well as primary roots were cut. The plant was supported on transgenic root(s) which were then nodulated by an appropriate strain of *Rhizobium*. Nodules formed on such plants were functional and were able to support the nitrogen requirement of the plant. This system saves significant time for studies on nodulation and nitrogen fixation (for more details see Chapter 10).

Particle Bombardment-mediated Transformation

The development of particle bombardment techniques (Klein *et al.*, 1987; Sanford, 1988) has led to the recovery of transgenic plants from a number of agronomically important crops, such as maize (Fromm *et al.*, 1990; Gordon-Kamm *et al.*, 1990), cotton (Finer and McMullen, 1990; McCabe and Martinell, 1993), rice (Christou *et al.*, 1991; Cao *et al.*, 1992), wheat (Vasil *et al.*, 1992; Weeks *et al.*, 1993) and soybean (McCabe *et al.*, 1988; Finer and McMullen, 1991). The main benefit of this method over other methods is that DNA is physically delivered into plant cells and the biological incompatibilities associated with use of *Agrobacterium* can be avoided. In addition, intact plant tissues can serve as the target, and production and manipulation of protoplasts is not necessary. Use of particle bombardment has become widespread because of its simplicity (Vain *et al.*, 1993) and both transient and stable transformants have been obtained (Sanford *et al.*, 1993).

The basis of particle bombardment is the acceleration of small DNA-coated particles (~ 1 μm gold, tungsten, or platinum) towards plant cells. After penetration through the cell wall by the particles, the DNA dissociates from the particles and integrates into the chromosome. The original particle gun utilized ignition of gunpowder to generate the energy to accelerate the particles (Klein *et al.*, 1987). Since then, many different versions of the particle gun have been built (Christou *et al.*, 1988; Sautter *et al.*, 1991; Finer *et al.*, 1992; Takeuchi *et al.*, 1992) and used for plant transformation. The major modification to the gunpowder-driven device is the use of air, gas (nitrogen or helium), or electrical discharge to propel the particles. Full utilization of particle bombardment is still limited due to the high cost and complexity of the commercial device. The development of an inexpensive and efficient particle bombardment device which is simple to build and operate (Finer *et al.*, 1992; Vain *et al.*, 1993) should greatly aid in the distribution and utilization of this technology for soybean as well as other crops.

Soybean transformation via particle bombardment

The first report of soybean transformation via particle bombardment utilized callus tissue, which could not regenerate plants (Christou *et al.*, 1988). Plasmid DNA containing the neomycin phosphotransferase II (NPTII) coding region under regulatory control of the CaMV 35S promoter and the NOS polyadenylation region was introduced into soybean callus via an electric arc discharge particle gun (Christou *et al.*, 1988). Kanamycin-resistant calli were recovered and NPTII enzyme assays and Southern hybridization analysis demonstrated both the expression of the foreign DNA and its stable integration into the soybean genome. Although no transgenic soybean plants were obtained in this first report, this work demonstrated the utility of particle bombardment for soybean transformation.

Kanamycin-resistant transgenic soybean plants were subsequently obtained using the method to introduce DNA-coated particles into shoot meristematic tissue of soybean (McCabe *et al.*, 1988). Although the authors claimed that shoot transformation avoided tissue culture manipulations, the shoot apex was treated like cotyledonary nodes and multiple shoots were proliferated prior to whole plant regeneration. Most of the transformed shoots from bombarded meristems were chimeric for the introduced gene due to the nature of the target tissue and the method for proliferation and regeneration. Although chimerism is viewed as undesirable, chimeric plants can yield transgenic progeny if the introduced DNA is in germ-line tissues in the primary plant. In subsequent reports (Christou *et al.*, 1989; Christou, 1990; Yang and Christou, 1990), Mendelian inheritance of transgenes, the recovery of both homozygous and heterozygous T1 plants, and cell-type specific expression of the GUS gene in soybean were described.

Shoot apex transformation is very labour-intensive because the meristematic tissue is difficult to target and, without selection, a large number of potentially transformed shoots must be regenerated and analysed. In order to lessen the number of plants that need to be regenerated, efforts were made to predict germ-line transformation events based on a correlation of the expression patterns of the GUS marker gene with transmission of the transgene to progeny (Christou and McCabe, 1992). Through these early-screening efforts, the number of regenerated plants that were grown to maturity for progeny analysis was tremendously reduced. In spite of this elaborate work and the large numbers of reports that resulted from it, recovery of transgenic soybean through shoot apex bombardment remains labour-intensive and has not been repeated by any other group (Sato *et al.*, 1993).

Other approaches to particle bombardment-mediated transformation of soybean have yielded one system that has good potential for more widespread use. Recovery of transgenic soybean plants via particle bombardment of embryogenic suspension cultures has been reported by several different laboratories (Finer and McMullen, 1991; Sato et al., 1993; Parrott et al., 1994). This system is dependent on the development of embryogenic suspension cultures of soybean (Finer and Nagasawa, 1988) where the embryogenic tissue proliferates from surface tissues of older somatic embryos (Finer, 1988; Finer and McMullen, 1991). Surface proliferation makes this system more efficient for transformation because the embryonic cells on the surface of proliferating clumps of embryogenic tissue can be easily targeted. In addition, selection for transformed tissue using antibiotic or herbicide resistance is possible with these suspension cultures and large numbers of potentially transformed plants do not have to be screened for the presence of the introduced DNAs. Although bombardment of embryogenic suspension culture tissue does have clear advantages over shoot apex bombardment, there are some problems with this system that are related to the tissue culture manipulations. The two main problems are the large effort required to establish the culture and, by the time the cultures are established, the plants regenerated from these cultures can display culture-induced variation (see Chapter 6) such as in partial or complete sterility. Rapid establishment of embryogenic suspension cultures of soybean is possible and transgenic plants recovered from these cultures after particle bombardment do not appear to have severely reduced fertility as was observed with older cultures (W. Parrott, personal communication).

In a comparison of the response of the shoot apex and embryogenic suspension cultures to particle bombardment resulting in the recovery of transgenic plants, Sato et al. (1993) demonstrated that embryogenic suspension cultures were clearly more responsive. In this report, although about 30% of the bombarded shoot tips displayed GUS-positive sectors, none of the regenerants continued to express GUS in regenerated plants. On the other hand, bombarded embryogenic suspension culture tissues produced an average of four independent transgenic lines per bombardment and the GUS-positive embryos regenerated into GUS-positive plants. In both regeneration systems, the DNA-coated particles were only capable of penetrating the first two cell layers. The different transformation efficiencies of these two systems are related to accessibility of the target cells that are capable of forming whole plants. A particle bombardment device that will permit efficient shoot apex transformation by controlling the depth of particle penetration is not yet commercially available.

Cotransformation in soybean

Cotransformation is the simultaneous introduction of two or more separate plasmids. This allows a selectable marker gene to be introduced along with a target gene on a separate plasmid. The advantage of cotransformation is that selectable and multiple nonselectable genes can be mixed without the need to place all genes on a single plasmid. Cotransformation of two different plasmids into callus tissue of soybean was first reported by Christou and Swain (1990). In order to study the transformation and recombination process in soybean, twelve different plasmids were introduced into embryogenic cultures via particle bombardment (Hadi and Finer, unpublished). The DNAs used for cotransformation included ten plasmids containing RFLP markers for maize and two separate plasmids encoding hygromycin resistance and β-glucuronidase. Southern hybridization analyses of stably transformed soybean tissue revealed that all 12 of the plasmids could be taken up and incorporated, with no preferential integration of any one of the plasmids. Plasmid amplification may have occurred in some clones, concatemer formation (indicative of homologous recombination) was observed to a limited extent, and ligation of plasmid fragments was also occurring. This study showed that interplasmid illegitimate recombination is very efficient in soybean and probably occurs prior to integration of the foreign DNAs.

Soybean Transformation Using Electroporation of Protoplasts

Electroporation has been used to stably transform a number of plants including soybean (Christou *et al.*, 1987; Lin *et al.*, 1987). In these first reports of soybean transformation via electroporation, protoplasts were isolated from either zygotic embryos or immature cotyledons. In both cases, stable transformations and integration of foreign DNA into the nuclear genome of soybean was shown. However, transgenic soybean plants were not obtained from these studies due to the difficulty in regenerating plants from soybean protoplasts.

The first report of soybean plant regeneration from protoplasts utilized immature cotyledons of soybean, cv. Heilong 26 as a protoplast source (Wei and Xu, 1988). Based on this work, regeneration of soybean from protoplasts of other cultivars was reported (Dhir *et al.*, 1991c). Production of transgenic soybean via protoplast electroporation (Dhir *et al.*, 1991b, 1992) is not well established. Although production of transgenic soybean was reported, this work could not be repeated.

For efficient recovery of transgenic soybean plants via protoplast electroporation, the following three components need to be available; an efficient gene transfer method, an effective selection scheme, and a reliable tissue culture regeneration system. The main benefit of electroporation-mediated gene transfer is that large numbers of transformants can be obtained and the protoplast selection system allows production of nonchimeric cell lines. However, until a truly reliable and reproducible soybean protoplast regeneration system is developed, production of transgenic soybean via protoplast transformation may be extremely difficult.

Limitations and Future Prospects

Since the first reports of soybean transformation using both the particle gun (McCabe et al., 1988) and Agrobacterium (Hinchee et al., 1988), production of transgenic soybean has been rated from 'near-to-impossible' to 'routine'. Perhaps, the most fair assessment at present is 'inefficient but consistent'. The procedures described for soybean transformation in this chapter have been proven and the molecular evidence for production of transgenic soybean is clear. The three most commonly used procedures (Agrobacterium with cotnodes, particle gun with shoot apex, particle gun with embryogenic suspension cultures) are extremely complex and technically demanding, and many of the laboratories that are attempting soybean transformation may not have adequate facilities to perform this type of work. Laboratories that are successfully performing soybean transformation have put effort into studying and understanding soybean tissue culture and transformation responses. We have provided embryogenic suspension cultures to a number of different laboratories that are interested in soybean transformation via bombardment of embryogenic suspension culture tissue. The most difficult aspect of embryogenic suspension culture transformation is not the maintenance but the establishment of suspension cultures. Even after providing established suspension cultures, soybean transformation has not become routine or standard, and many laboratories have difficulties simply maintaining these cultures. The day that soybean transformation becomes as routine as tobacco (or even maize) may be approaching but there is still much work to be done.

Acknowledgements

The authors wish to gratefully acknowledge Dr Glenn Collins and Dr Wayne Parrott for helpful discussions in the area of soybean transformation. This work was supported by a grant from the American Soybean Association/United Soybean Board to JF and DPSV, and State and Federal funds appropriated to OSU/OARDC. Work in Verma's laboratory was supported by NSF and USDA grants. Mention of trademark or proprietary products does not constitute a guarantee or warranty of the product by OSU/OARDC, and nor does it imply approval to the exclusion of other products that may also be suitable. (OARDC Journal Article No. 20-95.)

References

Baldes, R., Moos, M. and Geider, K. (1987) Transformation of soybean protoplasts from permanent suspension cultures by cocultivation with cells of *Agrobacterium tumefaciens*. *Plant Molecular Biology* 9, 135–145.

Barwale, U.B., Kerns, H.R. and Widholm, J.M. (1986) Plant regeneration from callus cultures of several soybean genotypes via embryogenesis and organogenesis. *Planta* 167, 473–481.

Bidney, D., Scelonge, C., Martich, J., Burrus, M., Sims, L. and Huffman, G. (1992) Microprojectile bombardment of plant tissues increases transformation frequency by *Agrobacterium tumefaciens*. *Plant Molecular Biology* 18, 301–313.

Cao, J., Duan, X., McElroy, D. and Wu, R. (1992) Regeneration of herbicide resistant transgenic rice plants following microprojectile mediated transformation of suspension culture cells. *Plant Cell Reports* 11, 586–591.

Chee, P.P., Fober, K.A. and Slightom, J.L. (1989) Transformation of soybean (*Glycine max*) by infecting germinating seeds with *Agrobacterium tumefaciens*. *Plant Physiology* 91, 1212–1218.

Cheon, C.-I., Lee, N.-G., Siddique, A.B.M., Bal, A.K. and Verma, D.P.S. (1993) Roles of plant homologs of Rab1p and Rab7p in biogenesis of peribacteroid membrane, a subcellular compartment formed *de novo* during root nodule symbiosis, *EMBO Journal* 12, 4125–4135.

Christianson, M.L., Warnick, D.A. and Carlson, P.S. (1983) A morphogenetically competent soybean suspension culture. *Science* 222, 632–634.

Christou, P. (1990) Morphological description of transgenic soybean chimeras created by the delivery, integration and expression of foreign DNA using electric discharge particle acceleration. *Annals of Botany* 66, 379–386.

Christou, P. (1991) From crown gall to field testing: Historical review of soybean transformation. *Soybean Genetics Newsletter* 18, 201–209.

Christou, P. (1992) *Genetic Engineering and* in vitro *Culture of Crop Legumes*. Technomic Publishing, Ancaster, Pennsylvania.

Christou, P. and McCabe, D.E. (1992) Prediction of germ-line transformation events in chimeric Ro transgenic soybean plantlets using tissue-specific expression patterns. *Plant Journal* 2, 283–290.

Christou, P. and Swain, W.F. (1990) Cotransformation frequences of foreign genes in soybean cell cultures. *Theoretical and Applied Genetics* 79, 337–341.

Christou, P., Murphy, E. and Swain, W.F. (1987) Stable transformation of soybean by electroporation and root formation from transformed callus. *Proceedings of the National Academy of Sciences, USA* 84, 3962-3966.

Christou, P., McCabe, D.E. and Swain, W.F. (1988) Stable transformation of soybean callus by DNA-coated gold particles. *Plant Physiology* 87, 671–674.

Christou, P., Swain, W.F., Yang, N. and McCabe, D.E. (1989) Inheritance and expression of foreign genes in transgenic soybean plants. *Proceedings of the National Academy of Sciences, USA* 86, 7500–7504.

Christou, P., Ford, T.L. and Kofron, M. (1991) Production of transgenic rice (*Oryza sativa* L.) plants from agronomically important Indica and Japonica varieties via electric discharge particle acceleration of exogenous DNA into immature zygotic embryos. *Bio/Technology* 9, 957–962.

D'Halluin, K., Bonne, E., Bossut, M., Beuckeleer, M.D. and Leemans, J. (1992) Transgenic maize plants by tissue electroporation. *Plant Cell* 4, 1495–1505.

Dhir, S.K., Dhir, S. and Widholm, J.M. (1991a) Plantlet regeneration from immature cotyledon protoplasts of soybean (*Glycine max* L.) *Plant Cell Reports* 10, 39–43.

Dhir, S.K., Dhir, S., Sturtevant, A.P. and Widholm, J.M. (1991b) Regeneration of transformed shoots from electroporated soybean (*Glycine max* (L.) Merr.) protoplasts. *Plant Cell Reports* 10, 97–101.

Dhir, S.K., Dhir, S., Hepburn, A. and Widholm, J.M. (1991c) Factors affecting transient gene expression in electroporated *Glycine max* protoplasts. *Plant Cell Reports* 10, 106–110.

Dhir, S.K., Dhir, S., Savka, M.A., Belanger, F., Kriz, A.L., Farrand, S.K. and Widholm, J.M. (1992) Regeneration of transgenic soybean (*Glycine max*) plants from electroporated protoplasts. *Plant Physiology* 99, 81–88.

Finer, J.J. (1988) Apical proliferation of embryogenic tissue of soybean [*Glycine max* (L.) Merrill]. *Plant Cell Reports* 7, 238–241.

Finer, J.J. and McMullen, M.D. (1990) Transformation of cotton (*Gossypium hirsutum* L.) via particle bombardment. *Plant Cell Reports* 8, 586–589.

Finer, J.J. and McMullen, M.D. (1991) Transformation of soybean via particle bombardment of embryogenic suspension culture tissue. *InVitro Cellular and Developmental Biology* 27P, 175–182.

Finer, J.J. and Nagasawa, A. (1988) Development of an embryogenic suspension culture of soybean (*Glycine max* Merrill.). *Plant Cell, Tissue and Organ Culture* 15, 125–136.

Finer, J.J., Vain, P., Jones, M.W. and McMullen, M.D. (1992) Development of the particle inflow gun for DNA delivery to plant cells. *Plant Cell Reports* 11, 323–328.

Finer, J.J., Finer, K.R. and Santarem, E.R. (In Press) Physical methods for plant cell transformation. In: *The Encyclopedia of Molecular Biology.* VCH Publishers, Weinheim, Germany.

Fromm, M.E., Morrish, F., Armstrong, C., Williams, R., Thomas, J. and Klein, T.M. (1990) Inheritance and expression of chimeric genes in the progeny of transgenic maize plants. *Bio/Technology* 8, 833–839.

Gordon-Kamm, W.J., Spencer, T.M., Mangano, M.L., Adams, T.R., Daines, R.J., Start, W.G., O'Brien, J.V., Chambers, S.A., Adams, W.R.J., Willetts, N.G., Rice, T.B., Mackey, C.J., Krueger, R.W., Kausch, A.P. and Lemaux, P.G. (1990) Transformation of maize cells and regeneration of fertile transgenic plants. *Plant Cell* 2, 603–618.

Hansen, G., Das, A. and Chilton, M.D. (1994) Constitutive expression of the virulence genes improves the efficiency of plant transformation by *Agrobacterium*. *Proceedings of the National Academy of Sciences, USA* 91, 7603–7607.

Hartweck, L.M., Lazzeri, P.A., Cui, D., Collins, G.B. and Williams, E.G. (1988) Auxin orientation effects on somatic embryogenesis from immature soybean cotyledons. *InVitro Cellular and Developmental Biology* 24, 821–828.

Hinchee, M.A.W., Connor-Ward, D.V., Newell, C.A., McDonnell, R.E., Sato, S.J., Gasser, C.S., Fischhoff, D.A., Re, D.B., Fraley, R.T. and Horsch, R.B. (1988) Production of transgenic soybean plants using *Agrobacterium* mediated DNA transfer. *Bio/Technology* 6, 915–922.

Horsch, R.B., Fry, J.E., Hoffman, N.L., Eicholtz, D., Rogers, S.G. and Fraley, R.T. (1985) A simple and general method for transferring genes into plants. *Science* 227, 1229–1231.

Kaeppler, H.F., Gu, W., Somers, D.A., Rines, H.W. and Cockburn, A.F. (1990) Silicon carbide-mediated DNA delivery into plant cells. *Plant Cell Reports* 9, 415–418.

Klein, T.M., Wolf, E.D., Wu, R. and Sanford, J.C. (1987) High velocity microprojectiles for delivering nucleic acids into living cells. *Nature* 327, 70–73.

Lin, W., Odell, J.T. and Schreiner, R.M. (1987) Soybean protoplast culture and direct gene uptake and expression by cultured soybean protoplasts. *Plant Physiology* 84, 856–861.

McCabe, D.E. and Martinell, B.J. (1993) Transformation of elite cotton cultivars via particle bombardment of meristems. *Biotechnology* 11, 596–598.

McCabe, D.E., Swain, W.F., Martinell, B.J. and Christou, P. (1988) Stable transformation of soybean (*Glycine max*) by particle acceleration. *Bio/Technology* 6, 932–926.

Parrott, W.A., Hoffman, L.M., Hildebrand, D.F., Williams, E.G. and Collins, G.B. (1989) Recovery of primary transformants of soybean. *Plant Cell Reports* 7, 615–617.

Parrott, W.A., All, J.N., Adang, M.J., Bailey, M.A. and Boerma, H.R. (1994) Recovery and evaluation of soybean (*Glycine max* [L.] Merr.) plants transgenic for a *Bacillus thuringiensis* var. *kurstaki* insecticidal gene. *InVitro Cellular and Developmental Biology* 30, 144–149.

Pedersen, H.C., Christiansen, J. and Wyndaele, R. (1983) Induction and *in vitro* culture of soybean crown gall tumors. *Plant Cell Reports* 2, 201–204.

Sanford, J.C. (1988) The biolistic process. *Trends in Biotechnology* 6, 299–302.

Sanford, J.C., Smith, F.D. and Russell, J.A. (1993) Optimizing the biolistic process during different biological applications. *Methods in Enzymology* 217, 483–509.

Sato, S., Newell, C., Kolacz, K., Tredo, L., Finer, J. and Hinchee, M. (1993) Stable transformation via particle bombardment in two different soybean regeneration systems. *Plant Cell Reports* 12, 408–413.

Sautter, C., Waldner, H., Neuhaus-Url, G., Galli, A., Neuhaus, G. and Potrykus, I. (1991) Micro-targeting: High efficiency gene transfer using a novel approach for the acceleration of micro-projectiles. *Bio/Technology* 9, 1080–1085.

Shillito, R.D., Saul, M.W., Paszkowski, J., Mueller, M. and Potrykus, I. (1985) High efficiency direct gene transfer to plants. *Bio/Technology* 3, 1099–1103.

Stachel, S.E., Messens, E., Van Montagu, M. and Zambryski, P. (1985) Identification of the signal molecules produced by wounded plant cells which activate the T-DNA transfer process in *Agrobacterium tumefaciens*. *Nature* 318, 624–629.

Takeuchi, Y., Dotson, M. and Keen, N.T. (1992) Plant transformation: a simple particle bombardment device based on flowing helium. *Plant Molecular Biology* 18, 835–839.

Vain, P., McMullen, M.D. and Finer, J.J. (1993) Osmotic treatment enhances particle bombardment mediated transient and stable transformation of maize. *Plant Cell Reports* 12, 84–88.

Vasil, V., Brown, S.M., Re, D. and Vasil, I. (1992) Herbicide resistant fertile transgenic wheat plants obtained by micro projectile bombardment of regenerable embryogenic callus. *Bio/Technology* 10, 667-674.

Weeks, J.T., Anderson, O.D. and Blechl, A.E. (1993) Rapid production of multiple independent lines of fertile transgenic wheat (*Triticum aestivum*). *Plant Physiology* 102, 1077–1084.

Wei, Z. and Xu, Z. (1988) Plant regeneration from protoplasts of soybean (*Glycine max* L.). *Plant Cell Reports* 7, 348–351.

Wright, M.S., Koehler, S.M., Hinchee, M.A. and Carnes, M.G. (1986) Plant regeneration by organogenesis in *Glycine max*. *Plant Cell Reports* 5, 150–154.

Yang, N. and Christou, P. (1990) Cell type specific expression of a CaMV 35S-GUS gene in transgenic soybean plants. *Developmental Genetics* 11, 289–293.

Index

Abnormal embryos 116
Acceptability in food 137, 144, 149
Accessions 1, 5–9
Acetohydroxamic acid 110
Acetosyringone 252, 253
Acetyl CoA carboxylase 130–131
cv. Acme 107
Aconitase 117
Activator elements 70, 71, 74–79
Adh1 gene 72
Afghanistan 5
Africa 8
Agrobacterium 20, 74, 76, 78, 79, 94, 96, 97, 210, 251–254
cv. AK Harrow 50, 51, 52
Albinism 116
Alfalfa 122, 210, 221, 225, 232, 234, 235
Allantoic acid 110, 225, 226, 228
Allantoin 110, 146, 225, 226, 228
Allantoinase 228
Allantoxanamide 110
Allergenic properties 137
Allopolyploidization 13
Allopurinol 110
cv. Altona 132
Aluminium tolerance 21
Amide synthesis 225–226
Amphiploids 13
Amplification polymorphisms 205
cv. Amsoy 132

Amylase 23, 132
Aneuploids 99
Aneutetraploids 13
Animal feed 132, 133
Anthocyanin pathway 78, 80–82
Antibiotic resistance 114, 120, 252, 253, 255
Antisense mutagenesis 192
Arabidopsis 38, 73, 76, 171, 180–182, 190, 210, 226
Argentina 15
Asparagine synthetase 225
Asparaginyl endopeptidase 139, 143
Aspartate aminotransferase 225
Atrazine 113–114, 120
atp6 60
Australia 2, 10
Azaguanine 108

Bacillus thuringiensis 98
Bacterial pustule 18
Barley 122
cv. Bedford 116
Beet armyworm 19
Bigeyed bug 19
Biotin carboxylase 131
Bowman-Birk inhibitor 23, 134, 146–147
Bradyrhizobium 194, 208, 209, 221

cv. Bragg 197
Brassica 59, 121, 143, 181
Brazil nut 142
bronze locus 72
Brown stem rot 18, 112
Brown rot 11, 113, 115

cv. Calland 116
cv. Camp 22
CAMV 35S promoter 76, 78–79, 255
Canada 3
cv. Canatto 22
Canavalia ensiformis 137
Canola 167, 177, 180–183
Carbohydrate content 131–132
Carbon metabolism 229–231, 237–238
Carboxyl transferase 131
Carrot 74, 108, 113
cdc2 208, 209
cv. Century 16, 175, 176, 178, 179
Chalcone synthase 81, 82, 221
Chimerism 117
China 2, 5, 7, 8, 16, 22
Chlamydomonas 59
Chlorophyll deficiency 21, 115, 117
Chlorophyll mutations 3
Chloroplast genome 57–59
cv. Chestnut 132
Cholinephosphotransferase 171, 176
Chromosome walking 141, 190
cv. Clark 3, 13, 44
Cocoa butter 167, 171
Coconut oil 168
cv. Coles 179
cv. Columbia 119
β-Conglycinin 134–141
Corn earworm 19
cv. Corsoy 110–111
Cosuppression 82, 183
Cotransformation 256–257
Cotton 113, 254
Cowpea 60, 222, 228
coxII 60
Crossability rate 11, 13
Curled leaf trait 117

cv. Cutler 133
Cyclin B 208, 209
Cyst nematode 18, 45
Cytogenetic traits 22
Cytokinins 233
Cytoplasmic male sterility 61–63

cv. Dare 141
Databases 9, 11, 52
Datura innoxia 113, 120
Days to bloom 118
Days to maturity 118
Diacylglycerol acyltransferase 171
Diaphorase 117
Dihydroflavonol reductase 81
Dihydrofolate reductase 75
Dissociation elements 70, 71, 75–79
DNA fingerprinting 101, 204–207
DNA markers 101
DNA polymerase 58
Downy mildew 18
cv. DPS 3546 197
*Dra*I 195
Drought tolerance 98
Dwarfness 21, 115, 117

*Eco*RI 202
cv. Elgin 178
Embryogenesis 115, 120, 121
cv. Emerald 129
Enhancer/Suppressor-mutator element 72, 74, 76, 78, 79
enod40 209
Enoyl-ACP reductase 170
cv. Enrei 197
Erucate 168, 173
Escherichia coli 76, 170
cv. Essex 129
Ethyl methanesulphonate 108, 109
Europe 8
exo gene 8

fad 171, 182
fan 175, 180

fap 178–180
fas 177, 180
Fatty acid content 45, 116, 129, 130, 165–184
Fertility restorer gene 62–63
Fertility variants 3
Flavonol glycosides 21
Flavour problems 133, 144–145, 149, 166
3′-Flavonoid hydroxylase 82
Flax 74, 173
Flower colour 21, 78, 80–81, 116
Flowering time 21
Friable callus 107
Fructose 131
cv. Funman 116
Fusarium solani 113

Galactopinitol 131
Galactose 131
Gene expression 73–74
Gene tagging 77–79
Gene transfer 97–98
Genetic mapping *see* Mapping
Genome duplication 40–41
Genome organization 37–40
Glucose 131
Glucuronidase 75, 77, 234, 252, 256
Glutamate synthase 225
Glutamine synthetase 192, 224, 225
Glyceollin 112–113, 221
Glycerol-3-P-acyltransferase 171
Glycine albicans 2
G. *arenaria* 2
G. *argyrea* 2
G. *canescens* 2
G. *clandestina* 2
G. *curvata* 2, 11, 12
G. *cyrtoloba* 2, 11, 12
G. *falcata* 2
G. *hirticaulis* 2
G. *lactovirens* 2
G. *latifolia* 2
G. *latrobeana* 2
G. *microphylla* 2
G. *pindanica* 2

G. *soja* 2, 3–7, 10, 11, 19, 20, 22–23, 95, 98, 194–197, 205, 206
G. *tabacina* 2, 13
G. *tomentella* 2, 12, 13, 98
Glycinin 134–141
Glyphosate 97, 252
GmENOD13 223
Growth habit 116, 117
gus 210

Happlopappus 108
Hard seededness 45
cv. Hark 50, 51
cv. Harosoy 3, 44, 50, 51
cv. Hawkeye 50, 51, 111
hcf-106 78
Herbicide resistance 21, 97, 113–114, 120, 252
cv. Hill 16
*Hind*III 23
*Hpa*II 118
Human food 17–18, 149
Hycanthone 108
5′, 3′-Hydroxylase 81
Hygromycin phosphotransferase 74, 79, 114

IMC129 182, 183
Impermeable seed coat trait 20
Indonesia 5
Inducible genes 232
Industrial oils 168
Inflorescence growth 21
Inosine monophosphate 110
Insect resistance 18, 19, 98
Intersubgeneric hybridization 98
Introductions 14, 15, 17, 22
In vitro selection
    antibiotic resistance 114
    biochemical traits 108–112
    disease resistance 112–113
    evaluation of variants 100–101
    herbicide resistance 113–114
    somaclonal variation 115–119, 120–122

Iron deficiency chlorosis 21, 111, 119
Isozymes 20, 22–23, 101, 117

Japan 2, 4, 6, 7, 8, 16, 22
Jojoba wax 168

Kanamycin 114, 120, 252, 253, 255
β-ketoacyl-ACP reductase 170
β-ketoacyl-ACP synthase 170, 179
knotted-1 gene 78
Korea 2, 4, 5, 7, 8, 22
Kunitz trypsin inhibitor 16, 23, 134, 146–147

Latitude 8, 10
Leaf
    form 3, 21, 115
    shape 116
    variegation 116
    weight 19
Leaflet number 115
Lectin 79–80, 134, 148
Lectin-null mutants 119
Leghaemoglobin 20, 192, 224, 232
Light intensity 129
cv. Lincoln 23
Linoleate 130, 165, 167, 170, 172, 173–176, 181
Linoleate desaturase 172, 182
Linolenate 165, 166, 167, 168, 170, 172, 173–176
Linolenate desaturase 172
Linolenic acid 21
Linseed oil 168
Lipoxygenase 16, 23, 134, 144–145
Lodging 14, 115, 118
Lotus corniculatus 234
L. japonicus 210
lps gene 233–234
Lupin 225
Lysophosphatidic acid acyltransferase 171
Lysophosphatidylcholine acyltransferase 172, 176

Maize 59, 73, 75, 76, 78, 80, 81, 82, 120, 142, 143, 236, 254
Malate dehydrogenase 117, 229–230
Maltose utilization trait 108–109
cv. Mandarin 23, 108
cv. Mandarin Ottawa 50, 51
Mannotriose 131
Mapping
    classical markers 41–44
    chloroplast genome 58
    database 52
    evaluation of variants 101
    future prospects 52–53
    genome duplication 40–41
    genome organization 37–40
    genotype analyses 45–52
    molecular genetic linkage 23
    molecular markers 41–44
    nodulation mutant genes 194–200
    quantitative trait loci 44–45, 94–95
    supernodulation region 199–200
Maturation time 21
Maturity groups 4, 7, 8, 9, 14
Medicago sativa 58
M. truncatula 193, 210
Methanol 109
Methionine 142, 143
Methotrexate 112
Methyl methanesulphonate 108
Methylation 74, 76, 118–119
Methylobacterium mesophilicum 109–110, 120
Mexican bean beetle 18
cv. Minsoy 118, 197
Miso 18
Mitochondrial DNA polymorphism 22–23
Mitochondrial genome 59–61
Molecular traits 22
Moth bean 226, 228
MSENOD2 233
*Msp*I 118
cv. Mukden 50, 51
Multiple shoots mutant 115
Mung bean 60

Mutable alleles 80–82
Mutagens 108
Mutation breeding 96
Mutator elements 72, 74, 78
Myoinositol 131

Nar phenotype 221
cv. Nattawa 22
Natto 18, 22
cv. Nattosan 22
ndv gene 233–234
Nematodes 18, 45
Neomycin phosphotransferase 74
Nitrogen assimilation 225–229, 238
nod139 mutant 209
Nod factors 220–221, 223, 233–234
Nodulation
    analysis of related genes
        208–209
    RFLP mapping of mutant genes
        194–200
    molecular genetics 192–194
Nodules
    carbon metabolism 229–231,
        237–238
    gene expression 222–223,
        232–236
    host defence responses 221–222
    initiation 220–221
    nitrogen assimilation 225–229,
        238
    osmoregulation 231–232, 239
    peribacteroid membrane 223–224
Nodulin genes 222–223, 232, 235, 236
Nodulins 192, 222, 224
cv. Noir I 118, 197
nts gene 192–193, 195, 199–200,
    206–207, 209
Nutrient efficiency 45

Odour problems 132, 144–145, 166
Oenothera 59
Oil
    biosynthesis 168–172
    breeding for quality 173–180

content 10, 45, 115, 118, 129–130,
    149
functionality 167
industrial 168
mutants 180–181
nutritional quality 167–168
oxidative stability 166
transgenic approaches 181–184
Oleate 165, 170, 172, 173–175, 181
Oleate desaturase 172, 180, 181, 182
Oleoyl-ACP thioesterase 181
opaque-2 gene 78
Open reading frames 61, 77
Organogenesis 115, 120, 121
Osmoregulation 231–232, 236, 239

PA mutants 178–179
pallida gene 81
Palmitate 165, 167, 170, 173–175,
    178–180, 181
cv. Pando 133
Papua New Guinea 2
Parsley 73
Particle bombardment 254–256
Pea 60, 73, 140, 142, 193, 222, 225
cv. Pearl 22
cv. Peking 96, 97, 197, 253
cv. Pella 175
Peribacteroid membrane 222,
    223–224
Pest resistance 16, 18–19
Petiole growth 21
Petunia 15, 73, 75, 78
Ph6 78
Phaseolus coccineus 61
P. polyanthus 61
P. vulgaris 40, 59, 60, 61–62, 137, 138,
    142, 193, 225
Phenylalanine ammonia lyase 81, 221
Phenylphosphoradiamidate 110
Phialophora gregata 112, 119
Philippines 2
Phomopsis seed decay 18
Phosphatidic acid phosphatase 171
Phosphoenolpyruvate
    carboxylase 225

Phosphoglucomutase 22
Phosphotransferase 74, 75, 78, 79
Photoautotrophic cultures 110–111
Photoperiod sensitivity 20
Photosynthase 229
Physiological traits 19–20
Phytic acid 23
Phytoalexins 112–113, 221, 222
Phytophthora megasperma 112–113
Phytophthora rot 11, 18
cv. Pickett 18
Pinitol 131
Plant height 14, 118
Pod
    abortion 13
    colour 21, 116
    number 129
    variegation 116
Polyploidization 40–41
Polyunsaturation 172, 181
Potato 73, 76
Poultry feed 132
Powdery mildew 11
PPFM bacteria see *Methylobacterium
    mesophilicum*
Proline synthesis 232
cv. Protana 133
Protein content 10, 45, 115, 118, 130,
    133–148
Protoplast electroporation 257–258
cv. Provar 133
PsENOD5 222
PsENOD12 222, 223, 233
Pseudomonas syringae 190
*Pst*I 195
Pubescence type 19, 21
Purine nucleosidase 236
Purine synthesis 226–227
pUTG-132a probe 195–199, 204
pvs 61–62

Qualitative trait tagging 95, 101–102

rab7 gene 221, 224
Raffinose 131, 132

Rapeseed 168, 173, 181
Recombinant inbred lines 206
Reproductive traits 45
Resistance
    antibiotics 114, 120, 252, 253, 255
    atrazine 120
    bacterial pustule 18
    brown spot 11, 113, 115–116
    brown stem rot 18, 112
    cyst nematode 18, 45
    downy mildew 18
    Fusarium solani 113
    glyphosate 252
    herbicides 97, 113–114, 120, 252
    hygromycin 114
    insects 18, 19, 98
    kanamycin 114, 120, 252, 253,
        255
    methotrexate 112
    Mexican bean beetle 18
    peanut mottle virus 18
    Phomopsis seed decay 18
    Phytophthora 11, 18, 113
    powdery mildew 11
    soybean mosaic virus 18
    soybean rust 11, 18
    viruses 11, 18, 98
Restriction enzymes 118
Restriction fragment length
    polymorphisms 23, 194–199
Retrotransposons 72–73
RH-42 gene 223, 233
RH-44 gene 233
Rhizobium 21, 233–237
Riboflavin 193
Rice 74, 210, 254
cv. Richland 50, 51
rol gene 210
Root fluorescence 21
Russia 2, 4, 7, 8, 22
Rust 11, 18

Salt tolerance 11, 98
Satellites 207–208
Seed
    carbohydrate 131–132

coat colour 81–82
development 128–131
fill 19, 128
oil 129–130, 149, 165–184
pigment 3
protease inhibitors 146–147
protein 130, 133–148
size 129
weight 20, 45, 118, 129
Selenastrum minutum 230
Septoria glycines 11, 113, 115, 120
Shoot morphogenesis 250–251
sle gene 38
Snapdragon 73, 78, 81, 82
Sodium arsenate 109
Somaclonal variation 99, 115–119,
120–122
Somatic embryogenesis 250–251
cv. Sooty 119
South Pacific 2, 8
Soybean looper 19
Stachyose 131, 132
Standability 14
Starch 131
Stearate 165, 167, 170, 173–174,
176–178, 181
Stearoyl-ACP desaturase 177, 181
Stem growth 21
Sterility 21, 61–63, 115, 117, 122
Storage proteins 23, 98, 134–143
Streptomycin phosphotransferase 75,
78
Sucrose 131
Sucrose synthase 224, 229, 236
Sudden death syndrome 113
Sulphydryl proteases 139
Sunflower 142, 173, 181
Supernodulation 192–200, 206–207,
209
Sweden 10

Taiwan 2, 4, 7
Telomeres 207–208
Temperature 130, 175, 177
Tetraploids 13, 116
Tgm1 element 80

Thick leaves 116
Thioguanine 108
Tissue culture 250–251
Tobacco 59, 71, 73, 74, 76, 80, 108, 112,
113, 143, 181, 234
Tofu 18
Tomato 74, 75, 190
Transcripts 76, 77
Transformation 94, 96–99,
249–258
Transgenic plants 79, 94, 96, 97, 181,
183, 210, 252–254, 257
Transposable elements 69–83
Transposase 71–73, 78
Triacylglycerol assembly 171
Tung oil 168
Twin seeds 115

Ultraviolet radiation 21, 108
Urease 134, 145–146
Urease-null mutants 109
Ureide metabolism 110, 225–229,
231
Uric acid 110
Uricase 110, 192, 224, 227–228, 236
USA 3, 15, 20, 91–92, 165
USDA Soybean Germplasm Collec-
tion 1, 3–10

Value-added traits 17
Variant evaluation 99–100
Variant selection 99–102
Velvetbean caterpillar 19
Vicia faba 61, 62, 143
Vietnam 5
Vigna aconitifolia 225
Virus
cauliflower mosaic 75, 78, 79, 255
peanut mottle 18
resistance 11, 18, 98
soybean mosaic 18
wheat dwarf 76
yellow mosaic 11, 18

waxy locus 70, 71
cv. Wayne 116
cv. Weber 175
Wheat 254
Wild species 2, 10, 11, 13, 98
cv. Williams 3, 16
Wrinkled leaf trait 115, 117, 121
Wrinkled pea trait 73–74

Xanthine dehydrogenase 110, 227, 228
*Xho*I 208

Yeast artificial chromosomes 190, 200–204
Yellow-edged cotyledons 117
Yield 14–17, 19, 92–94